Lecture Notes in Artificial Intelligence   8722

Subseries of Lecture Notes in Computer Science

LNAI Series Editors

Randy Goebel
  University of Alberta, Edmonton, Canada
Yuzuru Tanaka
  Hokkaido University, Sapporo, Japan
Wolfgang Wahlster
  DFKI and Saarland University, Saarbrücken, Germany

LNAI Founding Series Editor

Joerg Siekmann
  DFKI and Saarland University, Saarbrücken, Germany

Gennady Agre   Pascal Hitzler
Adila A. Krisnadhi   Sergei O. Kuznetsov (Eds.)

Artificial Intelligence: Methodology, Systems, and Applications

16th International Conference, AIMSA 2014
Varna, Bulgaria, September 11-13, 2014
Proceedings

Springer

Volume Editors

Gennady Agre
Bulgarian Academy of Sciences
Institute of Information and Communication Technologies
Sofia, Bulgaria
E-mail: agre@iinf.bas.bg

Pascal Hitzler
Wright State University
Dayton, OH, USA
E-mail: pascal.hitzler@wright.edu

Adila A. Krisnadhi
Wright State University, Dayton, OH, USA
and
University of Indonesia, Depok, Indonesia
E-mail: krisnadhi@gmail.com

Sergei O. Kuznetsov
National Research University
Higher School of Economics
Moscow, Russia
E-mail: skuznetsov@hse.ru

ISSN 0302-9743   e-ISSN 1611-3349
ISBN 978-3-319-10553-6   e-ISBN 978-3-319-10554-3
DOI 10.1007/978-3-319-10554-3
Springer Cham Heidelberg New York Dordrecht London

Library of Congress Control Number: 2014946289

LNCS Sublibrary: SL 7 – Artificial Intelligence

Typesetting: Camera-ready by author, data conversion by Scientific Publishing Services, Chennai, India

Printed on acid-free paper

Springer is part of Springer Science+Business Media (www.springer.com)

Preface

AIMSA 2014 was the 16th in a biennial series of AI conferences that have been held in Bulgaria since 1984. The series began as a forum for scientists from Eastern Europe to exchange ideas with researchers from other parts of the world, at a time when such meetings were difficult to arrange and attend. The conference has thrived for 30 years, and now functions as a place where AI researchers from all over the world can meet and present their research.

AIMSA continues to attract submissions from all over the world, with submissions from 27 countries. The range of topics is almost equally broad, from traditional areas such as computer vision and natural language processing to emerging areas such as mining the behavior of Web-based communities. It is good to know that the discipline is still broadening the range of areas that it includes at the same time as cementing the work that has already been done in its various established subfields.

The Program Committee selected just over 30% of the submissions as long papers, and further accepted 15 short papers for presentation at the conference. We are extremely grateful to the Program Committee and the additional reviewers, who reviewed the submissions thoroughly, fairly and very quickly.

Special thanks go to our invited speakers, Bernhard Ganter (TU Dresden), Boris G. Mirkin (Higher School of Economics, Moscow) and Diego Calvanese (Free University of Bozen-Bolzano). The invited talks were grouped around ontology design and application, whether using clustering and biclustering approaches (B.G. Mirkin), formal concept analysis, a branch of applied lattice theory (B. Ganter), or being concerned with ontology-based data access (D. Calvanese).

Finally, special thanks go to the AComIn project (Advanced Computing for Innovation, FP7 Capacity grant 316087) for the generous support for AIMSA 2014, as well as Bulgarian Artificial Intelligence Association (BAIA), and Institute of Information and Communication Technologies at Bulgarian Academy of Sciences (IICT-BAS) as sponsoring institutions of AIMSA 2014.

June 2014

Gennady Agre
Pascal Hitzler
Adila A. Krisnadhi
Sergei Kuznetsov

Organization

Program Committee

Gennady Agre	Institute of Information Technologies, Bulgarian Academy of Sciences, Bulgaria
Galia Angelova	Institute for Parallel Processing, Bulgarian Academy of Sciences, Bulgaria
Grigoris Antoniou	University of Huddersfield, UK
Sören Auer	Universität Leipzig, Germany
Sebastian Bader	MMIS, Computer Science, Rostock University, Germany
Roman Bartak	Charles University in Prague, Czech Republic
Christoph Beierle	University of Hagen, Germany
Meghyn Bienvenu	CNRS, Université Paris-Sud, France
Diego Calvanese	KRDB Research Centre, Free University of Bozen-Bolzano, Italy
Virginio Cantoni	Università di Pavia, Italy
Stefano A. Cerri	LIRMM: University of Montpellier and CNRS, France
Michelle Cheatham	Wright State University, USA
Davide Ciucci	University of Milan, Italy
Chris Cornelis	Ghent University, Belgium
Madalina Croitoru	LIRMM, University Montpellier II, France
Isabel Cruz	University of Illinois at Chicago, USA
Claudia D'Amato	Università di Bari, Italy
Artur D'Avila Garcez	City University London, UK
Darina Dicheva	Winston-Salem State University, USA
Ying Ding	Indiana University, USA
Danail Dochev	Institute of Information Technologies, Bulgarian Academy of Sciences, Bulgaria
Stefan Edelkamp	University of Bremen, Germany
Esra Erdem	Sabanci University, Turkey
Floriana Esposito	Università di Bari, Italy
William Michael Fitzgerald	EMC Information Systems International, Ireland
Miguel A. Gutiérrez-Naranjo	University of Sevilla, Spain
Barbara Hammer	Institute of Computer Science, Clausthal University of Technology, Germany
Pascal Hitzler	Wright State University, USA
Dmitry Ignatov	Higher School of Economics, Moscow, Russia

Grigory Kabatyansky	Institute for Information Transmission Problems, Russian Academy of Sciences, Russia
Mehdy Kaytoue	LIRIS - CNRS, and INSA Lyon, France
Gabriele Kern-Isberner	Technical University of Dortmund, Germany
Kristian Kersting	Technical University of Dortmund, Fraunhofer IAIS, Germany
Vladimir Khoroshevsky	Computing Center of Russian Academy of Science, Moscow, Russia
Matthias Knorr	CENTRIA, Universidade Nova de Lisboa, Portugal
Petia Koprinkova-Hristova	Bulgarian Academy of Sciences, Bulgaria
Irena Koprinska	The University of Sydney, Australia
Petar Kormushev	IIT Genoa, Italy
Adila A. Krisnadhi	Wright State University, USA, and Faculty of Computer Science, Universitas Indonesia, Indonesia
Kai-Uwe Kuehnberger	University of Osnabrück, Institute of Cognitive Science, Germany
Oliver Kutz	University of Bremen, SFB/TR 8 Spatial Cognition, Germany
Sergei O. Kuznetsov	National Research University Higher School of Economics, Moscow, Russia
Luis Lamb	Federal University of Rio Grande do Sul, Brazil
Evelina Lamma	ENDIF, University of Ferrara, Italy
Joohyung Lee	Arizona State University, USA
Yue Ma	Technical University of Dresden, Germany
Frederick Maier	Florida Institute for Human and Machine Cognition, USA
Riichiro Mizoguchi	Japan Advanced Institute of Science and Technology, Japan
Reinhard Muskens	Tilburg Center for Logic and Philosophy of Science, The Netherlands
Kazumi Nakamatsu	University of Hyogo, Japan
Amedeo Napoli	LORIA, CNRS-Inria Nancy, and University of Lorraine, France
Sergei Obiedkov	Higher School of Economics, Moscow, Russia
Manuel Ojeda-Aciego	Department of Applied Mathematics, University of Malaga, Spain
Guilin Qi	Southeast University, China
Allan Ramsay	School of Computer Science, University of Manchester, UK
Chedy Raïssi	LORIA, CNRS-Inria Nancy, France

Ioannis Refanidis	Department of Applied Informatics, University of Macedonia, Greece
Ute Schmid	University of Bamberg, Germany
Luciano Serafini	Fondazione Bruno Kessler, Italy
Dominik Slezak	University of Warsaw, Poland
Umberto Straccia	ISTI-CNR, Italy
Hannes Strass	Leipzig University, Germany
Doina Tatar	University Babes-Bolyai, Romania
Annette Ten Teije	Vrije Universiteit Amsterdam, The Netherlands
Dan Tufis	Institutul de Cercetari pentru Inteligenta Artificiala, Academia Romana, Romania
Petko Valtchev	UQAM, Université de Montréal, Canada
Tulay Yildirim	Yildiz Technical University, Turkey

Additional Reviewers

Batsakis, Sotiris	Kriegel, Francesco
Beek, Wouter	Meriçli, Tekin
Bellodi, Elena	Minervini, Pasquale
Borgo, Stefano	Mutharaju, Raghava
Cordero, Pablo	Nakov, Preslav
Fanizzi, Nicola	Osenova, Petya
Gavanelli, Marco	Papantoniou, Agissilaos
Gluhchev, Georgi	Redavid, Domenico
Hu, Yingjie	Rizzo, Giuseppe
Huan, Gao	Schwarzentruber, François
Kashnitsky, Yury	van Delden, André

Sponsoring Institutions

Bulgarian Artificial Intelligence Association (BAIA)
Institute of Information and Communication Technologies
at Bulgarian Academy of Sciences (IICT-BAS)

Keynote Presentation Abstracts

Scalable End-User Access to Big Data

Diego Calvanese

Free University of Bozen-Bolzano, Italy

Keynote Abstract

Ontologies allow one to describe complex domains at a high level of abstraction, providing end-users with an integrated coherent view over data sources that maintain the information of interest. In addition, ontologies provide mechanisms for performing automated inference over data taking into account domain knowledge, thus supporting a variety of data management tasks. Ontology-based Data Access (OBDA) is a recent paradigm concerned with providing access to data sources through a mediating ontology, which has gained increased attention both from the knowledge representation and from the database communities. OBDA poses significant challenges in the context of accessing large volumes of data with a complex structure and high dinamicity. It thus requires not only carefully tailored languages for expressing the ontology and the mapping to the data, but also suitably optimized algorithms for efficiently processing queries over the ontology by accessing the underlying data sources. In this talk we present the foundations of OBDA relying on lightweight ontology languages, and discuss novel theoretical and practical results for OBDA that are currently under development in the context of the FP7 IP project Optique. These results make it possible to scale the approach so as to cope with the challenges that arise in real world scenarios, e.g., those of two large European companies that provide use-cases for the Optique project.

About the Speaker

Diego Calvanese is a professor at the KRDB Research Centre for Knowledge and Data, Free University of Bozen-Bolzano, Italy. His research interests include formalisms for knowledge representation and reasoning, ontology languages, description logics, conceptual data modelling, data integration, graph data management, data-aware process verification, and service modelling and synthesis. He is actively involved in several national and international research projects in the above areas, and he is the author of more than 250 refereed publications, including ones in the most prestigious international journals and conferences in Databases and Artificial Intelligence. He is one of the editors of the Description Logic Handbook. In 2012–2013 he has been a visiting researcher at the Technical University of Vienna as Pauli Fellow of the "Wolfgang Pauli Institute". He will be the program chair of PODS 2015.

Formal Concepts for Learning and Education

Bernhard Ganter

Technical University of Dresden, Germany

Keynote Abstract

Formal Concept Analysis has an elaborate and deep mathematical foundation, which does not rely on numerical data. It is, so to speak, fierce qualitative mathematics, that builds on the algebraic theory of lattices and ordered sets. Since its emergence in the 1980s, not only the mathematical theory is now mature, but also a variety of algorithms and of practical applications in different areas. Conceptual hierarchies play a role e.g., in classification, in reasoning about ontologies, in knowledge acquisition and the theory of learning. Formal Concept Analysis provides not only a solid mathematical theory and effective algorithms; it also offers expressive graphics, which can support the communication of complex issues.

In our lecture we give an introduction to the basic ideas and recent developments of Formal Concept Analysis, a mathematical theory of concepts and concept hierarchies and then demonstrate the potential benefits and applications of this method with examples. We will also review some recent application methods that are currently being worked out. In particular we will present results on a "methodology of learning assignments" and on "conceptual exploration".

About the Speaker

 Bernhard Ganter is a pioneer of formal concept analysis. He received his PhD in 1974 from the University of Darmstadt, Germany, and became Professor in 1978. Currently, he is a Professor of Mathematics and the Dean of Science at the Technical University of Dresden, Germany. His research interests are in discrete mathematics, universal algebra, lattice theory, and formal concept analysis. He is a co-author of the first textbook and editor of several volumes on formal concept analysis.

Ontology as a Tool for Automated Interpretation

Boris G. Mirkin

National Research University, Higher School of Economics,
Moscow, Russia

Keynote Abstract

In the beginning, I am going to outline, in-brief, the current period of developments in the artificial intelligence research. This is of synthesis, in contrast to the sequence of previous periods (romanticism, deduction, and induction). Three more or less matured ontologies, and their use, will be reviewed: ACM CCS, SNOMED CT and GO. The popular strategy of interpretation of sets of finer granularity via the so-called overrepresented concepts will be mentioned. A method for generalization and interpretation of fuzzy/crisp query sets by parsimoniously lifting them to higher ranks of the hierarchy will be presented. Its current and potential applications will be discussed.

About the Speaker

 Boris Mirkin holds a PhD in Computer Science and DSc in Systems Analysis degrees from Russian Universities. In 1991–2010, he travelled through long-term visiting appointments in France, Germany, USA, and a teaching appointment as Professor of Computer Science, Birkbeck University of London, UK (2000–2010). He develops methods for clustering and interpretation of complex data within the "data recovery" perspective. Currently these approaches are being extended to automation of text analysis problems including the development and use of hierarchical ontologies. His latest publications: textbook "Core concepts in data analysis" (Springer 2011) and monograph "Clustering: A data recovery approach" (Chapman and Hall/CRC Press, 2012).

Table of Contents

Long Papers

Short Papers

Learning Probabilistic Semantic Network of Object-Oriented Action and Activity

Masayasu Atsumi

Department of Information Systems Science, Faculty of Engineering,
Soka University, 1-236 Tangi, Hachioji, Tokyo 192-8577, Japan
masayasu.atsumi@gmail.com

Abstract. This paper proposes a method of learning probabilistic semantic networks which represent visual features and semantic features of object-oriented actions and their contextual activities. In this method, visual motion feature classes of actions and activities are learned by an unsupervised Incremental Probabilistic Latent Component Analysis (I-PLCA) and these classes and their semantic tags in the form of case triplets are integrated into probabilistic semantic networks to visually recognize and verbally infer actions in the context of activities. Through experiments using video clips captured with the Kinect sensor, it is shown that the method can learn, recognize and infer object-oriented actions in the context of activities.

Keywords: action recognition, activity recognition, probabilistic learning, semantic network, probabilistic inference.

1 Introduction

It is necessary for a human support robot to understand what a person is doing in everyday living environment. In human motion in everyday life, there is a lot of motion that interacts with objects, which is referred to as an "object-oriented motion" in this research. The meaning of an object-oriented motion is determined not only a motion itself but also an object with which the motion interacts and this view corresponds to an affordance in which motion is dependent on object perception. In addition, each motion is performed in a context which is defined by a sequence of motions and a certain motion frequently occurs in some context and rarely occurs in other contexts. For example, a motion using a fork is frequently observed in a context of eating meals. In this research, each object-oriented motion is referred to as an action and a sequence of actions is referred to as an activity and it is assumed that an activity gives a context of an action and promotes action recognition. This assumption is consistent with findings that context improves category recognition of ambiguous objects in a scene [1] and requires an extension to action recognition of several methods [2,3] which incorporate context into object categorization. Though object-oriented actions and activities can be clustered into motion classes according to visual motion features and their semantic features can be labeled by using motion

G. Agre et al. (Eds.): AIMSA 2014, LNAI 8722, pp. 1–12, 2014.

labels and their target labels, motion classes and their labels do not have one-to-one correspondence. Therefore, in this research, motion classes are labeled with target synsets and motion synsets of case triplets in a form of "⟨target synset, case, motion synset⟩" and motion classes and synsets are probabilistically linked in probabilistic semantic networks of actions and activities, where a synset is a synonymous set of the WordNet [4] which represents a primitive semantic feature. In addition, contextual relationship between actions and activities is acquired as co-occurrence between them.

This paper proposes a method of learning probabilistic semantic networks which represent visual features and semantic features of object-oriented actions and their contextual activities. It also provides a probabilistic recognition and inference method of actions and activities based on probabilistic semantic networks. The main characteristics of the proposed method are the following: (1) visual motion feature classes of actions and activities are learned by an unsupervised "Incremental Probabilistic Latent Component Analysis (I-PLCA)" [5], (2) visual feature classes of motion and synsets of case triplets are integrated into probabilistic semantic networks to visually recognize and verbally infer actions and activities, and (3) actions are inferred in the context of activities through acquired co-occurrence between them.

As for related work, Kitani et al. [6] proposed a categorization method of primitive actions in video by leveraging related objects and relevant background features as context. Yao et al. [7] proposed a mutual context model to jointly recognize objects and human poses in still images of human-object interaction. Also in [8], they proposed a model to identify different object functionalities by estimating human poses and detecting objects in still images of different types of human-object interaction. One difference between our method and these existing methods is that our method not only uses an action and its target object as mutual context but also uses activities as context of actions. Another difference is that our method probabilistically infers actions and activities by using different semantic features linked with different visual motion features in probabilistic semantic networks.

2 Proposed Method

2.1 Overview

Human motion is captured as a temporal sequence of three-dimensional joint coordinates of a human skeleton, which can be captured with the Microsoft Kinect sensor. Since this research focuses on object-oriented motions of hands, a temporal sequence of three-dimensional coordinates of both hand joints relative to a shoulder center joint are computed from human skeleton data.

A motion feature of both hands is computed from a temporal sequence of relative three-dimensional coordinates of both hand joints by the following procedure. First, relative three-dimensional coordinates of both hand joints are spatially-quantized at a certain interval and a temporal sequence of quantized coordinates and their displacement are obtained as a motion representation. Next, actions are manually segmented from a temporal sequence of quantized coordinates and their displacement and they are semantically annotated with case triplets. An activity is also segmented as a sequence of actions and is semantically annotated with a case triplet. Then, for a temporal sequence of quantized coordinates and their displacement of an action, a motion feature of the action is computed as a motion histogram by firstly dividing the space around a shoulder center into modestly-sized regions, secondly calculating a displacement histogram of each region and lastly combining them into one histogram. A motion histogram of an activity is also calculated in the same way.

A probabilistic semantic network of actions is learned from a set of their motion histograms with case triplets. First of all, a set of action classes is generated from a set of motion histograms by the probabilistic clustering method I-PLCA. Here, an action class represents a motion feature of the action. Then, a probabilistic semantic network is generated as a network whose nodes consist of action classes and synsets of case triplets and are linked by joint probabilities of these classes and synsets. A probabilistic semantic network of activities is also learned from a set of their motion histograms with case triplets as a network whose nodes consist of activity classes and synsets of case triplets and are linked by joint probabilities of these classes and synsets. Here, an activity class is generated by the I-PLCA and represents a motion feature of the activity. For probabilistic semantic networks of actions and activities, co-occurrence between an action and an activity is obtained by calculating pointwise mutual information of their case triplets. A pair of probabilistic semantic networks of actions and activities with co-occurrence between them is referred to as the ACTNET in this paper. Fig. 1 shows an example of an ACTNET.

For a sequence of motion histograms of actions, actions and an activity are sequentially guessed by recognizing action and activity classes and inferring synsets of case triplets of them. First of all, an action class is recognized with the degree of confidence for a motion histogram of each action and at the same time an activity class is recognized with the degree of confidence for a sum of motion

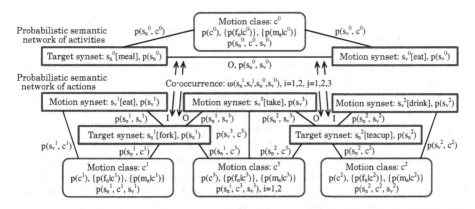

Fig. 1. An example of an ACTNET (Symbols in the figure are explained in the text.)

histograms of an action sequence. Then, synsets of case triplets of the actions and the activity are inferred from these classes and co-occurrence between them on probabilistic semantic networks.

2.2   Motion Feature of Action and Activity

Let $p^l = (p_x^l, p_y^l, p_z^l)$ and $p^r = (p_x^r, p_y^r, p_z^r)$ be relative quantized three-dimensional coordinates of left and right hands and let $d^l = (d_x^l, d_y^l, d_z^l)$ and $d^r = (d_x^r, d_y^r, d_z^r)$ be their displacement respectively. Here, l represents a left hand, r represents a right hand and the displacement is given by the difference of quantized coordinates between two successive frames. Let $\langle s_n[w_n], r, s_v[w_v] \rangle$ be a case triplet which is used to annotate a temporal sequence of quantized coordinates and their displacement of an action or an activity, where w_n is a noun which represents a target of motion and s_n is its synset, w_v is a verb which represents motion and s_v is its synset, and r is a case notation. Here, a synset is given by a synonymous set of the WordNet [4] and a case is currently one of the objective case (O), the instrumental case (I) and the locative case ($L[at|\ inside|\ around|\ above|\ below|\ beyond|\ from|\ to]$). For a temporal sequence of quantized coordinates and their displacement of an action, a case triplet of an activity which includes the action is also given in addition to a case triplet of the action. Then, for a motion $m = \{((p^l, d^l), (p^r, d^r))_t\}$, a motion histogram is constructed to represent a motion feature of both hands around a shoulder center as follows. Let B be a set of modestly-sized regions which is obtained by dividing the space around a shoulder center and let $|B|$ be the number of regions. For each region $b \in B$, a motion sub-histogram is computed for a set of coordinate and displacement data $\{(p^l, d^l)\}$ and $\{(p^r, d^r)\}$ whose coordinate p^l or p^r is located in the region b. A motion sub-histogram has 27 bins each bin of which corresponds to whether displacement is positive, zero or negative along x-, y-, and z-axes and is counted up according to values of displacement d^l and d^r. A motion histogram $h(m)$ is constructed as a $|B|$-plex histogram by combining these sub-histograms into one histogram so that the size of $h(m)$ is $27 \times |B|$.

2.3   Learning Probabilistic Semantic Network

An ACTNET is learned through generating probabilistic semantic networks of actions and activities followed by calculating co-occurrence between actions and activities. A probabilistic semantic network is generated from a set of motion histograms of actions or activities with their case triplets. Let $h(m_a) = [h_{m_a}(f_1), \ldots, h_{m_a}(f_{|F|})]$ be a motion histogram of a motion m with a case triplet a and let $H = \{h(m_a)\}$ be a set of motion histograms with case triplets. Here, $f_i \in F$ is a bin of a histogram and the size of a histogram is $|F| = 27 \times |B|$. A probabilistic semantic network is obtained through generation of motion classes by clustering a set of motion histograms H using I-PLCA [5] and derivation of a probabilistic network whose nodes consist of motion classes and synsets of case triplets and they are linked by joint probabilities of these classes and synsets.

The problem to be solved for generating motion classes is estimating probabilities $p(m_a, f_n) = \sum_c p(c) p(m_a|c) p(f_n|c)$, namely, a class probability distribution $\{p(c)|c \in C\}$, instance probability distributions $\{p(m_a|c)|m_a \in M \times A, c \in C\}$, probability distributions of class features $\{p(f_n|c)|f_n \in F, c \in C\}$, and the number of classes $|C|$ that maximize the following log-likelihood

$$L = \sum_{m_a} \sum_{f_n} h_{m_a}(f_n) \log p(m_a, f_n) \tag{1}$$

for a set of motion histogram $H = \{h(m_a)\}$, where C is a set of motion classes, M is a set of motions, A is a set of case triplets and m_a is a motion m with a case triplet a, that is, an instance of an action or an activity. These probability distributions and the number of classes are estimated by the tempered EM algorithm with subsequent class division. The process starts with one or a few classes, pauses at every certain number of EM iterations less than an upper limit and calculates the following dispersion index

$$\delta_c = \sum_{m_a} \left(\left(\sum_{f_n} |p(f_n|c) - \frac{h_{m_a}(f_n)}{\sum_{f'_n} h_{m_a}(f'_n)}| \right) \times p(m_a|c) \right) \tag{2}$$

for $\forall c \in C$. Then a class whose dispersion index takes the maximum value among all classes is divided into two classes. Let c^* be a source class to be divided and let c_1 and c_2 be target classes after division. Then, for a motion $m_a^* = \arg\max_{m_a} \{p(m_a|c^*)\}$ which has the maximum instance probability and its motion histogram $h(m_a^*) = [h_{m_a^*}(f_1), ..., h_{m_a^*}(f_{|F|})]$, one class c_1 is set by specifying a probability distribution of a class feature, an instance probability distribution and a class probability as

$$p(f_n|c_1) = \frac{h_{m_a^*}(f_n) + \kappa}{\sum_{f_{n'}} (h_{m_a^*}(f_{n'}) + \kappa)}, \quad \forall f_n \in F \tag{3}$$

$$p(m_a|c_1) = \begin{cases} p(m_a^*|c^*) & \cdots \ m_a = m_a^* \\ \frac{1 - p(m_a^*|c^*)}{|M| - 1} & \cdots \ \forall m_a(m_a \neq m_a^*) \end{cases} \tag{4}$$

$$p(c_1) = \frac{p(c^*)}{2} \tag{5}$$

respectively where κ is a positive correction coefficient. Another class c_2 is set by specifying a probability distribution of a class feature $\{p(f_n|c_2)|f_n \in F\}$ at random, an instance probability distribution $\{p(m_a|c_2)\}$ as 0 for m_a^* and an equal probability $\frac{1}{|M|-1}$ for $\forall m_a(m_a \neq m_a^*)$, and a class probability as $p(c_2) = \frac{p(c^*)}{2}$. This class division process is continued until dispersion indexes or class probabilities of all the classes become less than given thresholds. The temperature coefficient of the tempered EM is set to 1.0 until the number of classes is fixed and after that it is gradually decreased according to a given schedule until the EM algorithm converges and all the probability distributions are determined.

A probabilistic semantic network is derived from a class probability distribution $\{p(c)|c \in C\}$ and instance probability distributions $\{p(m_a|c)|m_a \in M \times A, c \in C\}$ associated with motion classes. The network nodes consist of motion class nodes and synset nodes. A synset node is derived from a target synset s_n or a motion synset s_v of a case triplet $\langle s_n[w_n], r, s_v[w_v]\rangle$ which is given by an instance m_a of an action or an activity where $a = \langle s_n[w_n], r, s_v[w_v]\rangle$. A motion class node is derived from a motion class $c \in C$ and has a class probability $p(c)$, a probability distribution of a class feature $\{p(f_n|c)|f_n \in F\}$, an instance probability distribution $\{p(m_a|c)|m_a \in M \times A\}$ and a set of joint probabilities each of which is a joint probability with a target synset s_n and a motion synset s_v of a case triplet that annotates the class c and is given by the expression (6)

$$p(s_n, c, s_v) = p(c) \quad \times \sum_{a=\langle s_n[*],*,s_v[*]\rangle} p(m_a|c) \tag{6}$$

where $*$ represents any word or any case. The network link between two nodes of a motion class c and a target synset s_n has a joint probability $p(s_n, c)$. The network link between two nodes of a motion class c and a motion synset s_v has a joint probability $p(s_v, c)$. The network link between two nodes of a target synset s_n and a motion synset s_v has a joint probability $p(s_n, s_v)$. The network nodes of a target synset s_n and a motion synset s_v have probabilities $p(s_n)$ and $p(s_v)$ respectively. These probabilities are computed by the expressions (7).

$$p(s_n, c) = \sum_{s_v} p(s_n, c, s_v), \ p(s_v, c) = \sum_{s_n} p(s_n, c, s_v)$$
$$p(s_n, s_v) = \sum_c p(s_n, c, s_v) \tag{7}$$
$$p(s_n) = \sum_c p(s_n, c), \ p(s_v) = \sum_c p(s_v, c)$$

In addition, a noun w_n of a target synset s_n and a verb w_v of a motion synset s_v are set to the target synset node and the motion synset node respectively.

Co-occurrence between actions and activities is computed between a pair of a target synset s_n and a motion synset s_v of an action and a pair of a target synset s_n^0 and a motion synset s_v^0 of an activity when the action has a case triplet $\langle s_n[w_n], r, s_v[w_n]\rangle$ and is included in the activity with a case triplet $\langle s_n^0[w_n^0], r^0, s_v^0[w_n^0]\rangle$. Let $p(s_n, s_v)$ be a joint probability of a target synset s_n and a motion synset s_v of an action and let $p(s_n^0, s_v^0)$ be a joint probability of a target synset s_n^0 and a motion synset s_v^0 of an activity. Then, co-occurrence between them is defined by the expression (8)

$$\omega(s_n, s_v, s_n^0, s_v^0) = \log \frac{p(s_n, s_v, s_n^0, s_v^0)}{p(s_n, s_v)p(s_n^0, s_v^0)} \tag{8}$$

where a joint probability $p(s_n, s_v, s_n^0, s_v^0)$ is calculated from action instances according to the expression (9)

$$p(s_n, s_v, s_n^0, s_v^0) = \sum_c \left(p(c) \quad \times \sum_{a=\langle s_n[*],*,s_v[*]\rangle@\langle s_n^0[*],*,s_v^0[*]\rangle} p(m_a|c)\right) \tag{9}$$

where $a = \langle s_n[*], *, s_v[*]\rangle@\langle s_n^0[*], *, s_v^0[*]\rangle$ means that an action m_a has a case triplet which matches a pattern $\langle s_n[*], *, s_v[*]\rangle$ and its contextual activity has a case triplet which matches a pattern $\langle s_n^0[*], *, s_v^0[*]\rangle$.

2.4   Recognition and Inference of Action and Activity

An ACTNET is used to recognize and infer a sequence of actions and an activity for a given sequence of motion histograms of actions. First of all, for each motion histogram, a motion class of an action is recognized with the degree of confidence and at the same time a motion class of an activity is recognized with the degree of confidence for a sequential sum of motion histograms of an action history. Then, synsets of case triplets of the actions and the contextual activity are inferred from these classes and co-occurrence between actions and activities.

Motion classes of an action and an activity are respectively recognized for a motion histogram and a sequential sum of motion histograms of an action history. Let $h(m) = [h_m(f_1), \ldots, h_m(f_{|F|})]$ be a motion histogram or a sum of motion histograms and let $\hat{h}(m) = [\hat{h}_m(f_1), \ldots, \hat{h}_m(f_{|F|})]$ be a distribution of it. Then, a motion class is obtained through calculating similarity between probability distributions of class features $\{p(f_n|c)|f_n \in F, c \in C\}$ and the distribution $\hat{h}(m)$ by the expression (10) and selecting the most similar class of all the classes C.

$$\beta(c, m) = 1 - \frac{\sum_{f_n} |p(f_n|c) - \hat{h}_m(f_n)|}{2} \tag{10}$$

This similarity provides the degree of confidence of a motion class.

For a selected class, a target synset and a motion synset are inferred with their degrees of confidence. Let c be a motion class of an action or an activity and let β be its degree of confidence. Then, a target synset s_n, a motion synset s_v and a pair of them are respectively inferred with their degrees of confidence $p(s_n|c) \times \beta$, $p(s_v|c) \times \beta$ and $p(s_n, s_v|c) \times \beta$ by following links from a node of the motion class c to adjacent synset nodes of s_n and s_v and retrieving probabilities maintained in those nodes and links of an ACTNET. When additional information about either a target synset or a motion synset is given, a motion synset or a target synset is respectively inferred with the degree of confidence $p(s_v|c, s_n) \times \beta$ or $p(s_n|c, s_v) \times \beta$ on an ACTNET. By introducing co-occurrence in the above inference, an action and an activity are interdependently inferred with the degree of confidence. Let c and β be a motion class of an action and its degree of confidence and let c^0 and β^0 be a motion class of an activity and its degree of confidence. Then, a pair of a target synset s_n and a motion synset s_v of the action and a pair of a target synset s_n^0 and a motion synset s_v^0 of the activity are inferred with the degree of confidence $\beta(s_n, s_v, s_n^0, s_v^0|c, c^0)$ that is calculated by the expression (11) which incorporates co-occurrence between them

$$\beta(s_n, s_v, s_n^0, s_v^0|c, c^0) = p(s_n, s_v|c) \times p(s_n^0, s_v^0|c^0) \times \frac{(\beta + \beta^0)}{2} + \lambda \times \omega(s_n, s_v, s_n^0, s_v^0) \tag{11}$$

where λ is a co-occurrence coefficient. When a pair of synsets (s_n^0, s_v^0) of the activity is fixed to (s_n^*, s_v^*) by additional information, a pair of synsets (s_n, s_v) of the action is inferred with the degree of confidence $\beta(s_n, s_v, s_n^*, s_v^*|c, c^0)$.

Table 1. Case triplets for activities and actions in a small data set

Activity	Action
<07573696-n[meal], O, 01166351-v[eat]>	<03383948-n[fork],O,01216670-v[take]> <03383948-n[fork],I,01166351-v[eat]> <03383948-n[fork],O,01494310-v[put]> <04398044-n[teapot],O,01216670-v[take]> <04398044-n[teapot],I,02070296-v[pour]> <04398044-n[teapot],O,01494310-v[put]> <04397452-n[teacup],O,01216670-v[take]> <04397452-n[teacup],I,01170052-v[drink]> <04397452-n[teacup],O,01494310-v[put]>
<03561345-n[illustration], O, 01684663-v[paint]>	<06415419-n[notebook],O,02311387-v[take-out]> <06415419-n[notebook],O,01346003-v[open]> <06415419-n[notebook],O,01291941-v[close]> <06415419-n[notebook],O,01308381-v[put-back]> <03908204-n[pencil],O,01216670-v[take]> <03908204-n[pencil],I,01684663-v[paint]> <03908204-n[pencil],O,01494310-v[put]>

3   Experiments

3.1   Experimental Framework

Experiments were conducted to evaluate learning, recognition and inference of object-oriented actions and activities on ACTNETs by using video clips captured with the Kinect Sensor. Relative three-dimensional coordinates of both hands are interpolated to about $30fps$ and are quantized at an interval of $1cm$. The space around a shoulder center is divided as follows for constructing a motion histogram. The front and the side in the vicinity of the body are both divided into 9 regions the size of which is $30cm$ on a side. The front and the side outside of these regions are respectively divided into 9 and 8 major regions and the back of the body is covered by just one region. As a result, the number of regions are 36 and the size of a motion histogram is $972(= 27 \times 36)$.

Two data sets of video clips were prepared for experiments. One is a small data set which is used to illustrate the basic capability of the proposed method in the experiment 1 and the other is a larger data set which is used to evaluate the performance of the proposed method in the experiment 2. A small data set contains 2 activities which are annotated by case triplets <07573696-n[meal],O,01166351-v[eat]> and <03561345-n[illustration],O,01684663-v[paint]>. The former activity "eating a meal" contains 3 objects and 9 object-oriented actions and the latter activity "painting an illustration" contains 2 objects and 7 object-oriented actions. The total number of actions is 16. Table 1 shows case triplets for activities and actions in this data set. Fig. 2 shows examples of motion quantization, that is, quantized coordinates and their displacement of both hands of some actions in this data set. A large data set contains 4 activities, 10 objects and

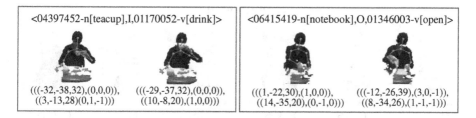

<04397452-n[teacup],I,01170052-v[drink]>	<06415419-n[notebook],O,01346003-v[open]>
(((-32,-38,32),(0,0,0)), ((3,-13,28)(0,1,-1))) (((-29,-37,32),(0,0,0)), ((10,-8,20),(1,0,0)))	(((1,-22,30),(1,0,0)), ((14,-35,20),(0,-1,0))) (((-12,-26,39),(3,0,-1)), ((8,-34,26),(1,-1,-1)))

Fig. 2. Examples of quantized motion of both hands in actions

Table 2. Activities and action sequences in a large data set

Activity	Action sequence
<clothes,O,wear>	<necktie,O,take>,<necktie,O,tie>,<jacket,O,take> <jacket,O,wear>,<scarf,O,take>,<scarf,O,wind>
<meal,O,eat>	<fork,O,take>,<fork,I,eat>,<fork,O,put> <teapot,O,pick-up>,<teapot,I,pour>,<teapot,O,put> <teacup,O,pick-up>,<teacup,I,drink>,<teacup,O,put>
<desk,O,clean-up>	<notebook-computer,O,close>,<mouse,O,put-back> <book,O,close>,<book,O,put-back> <mop,O,take>,<mop,I,wipe-up>
<report,O,write>	<notebook-computer,O,open>,<book,O,take> <book,O,open>,<book,O,turn>,<teacup,O,pick-up> <teacup,I,drink>,<teacup,O,put> <mouse,O,operate>,<notebook-computer,I,input>

27 object-oriented actions in total. Table 2 shows these activities and object-oriented action sequences in the abbreviated form without synsets. Fig. 3 shows snapshots of object-oriented action sequences in two activities in this data set.

The parameters were set as follows. In the I-PLCA, a threshold of the dispersion index for class division was 0.1, a threshold of the class probability for class division was 0.1 for a small data set and 0.05 for a large data set respectively, and a correction coefficient κ in the expression (3) was 1.0. In the tempered EM, a temperature coefficient was decreased by multiplying it by 0.95 at every 20 iterations until it became 0.8. A co-occurrence coefficient λ between an action and an activity in the expression (11) was set to 0.2.

3.2 Experimental Results

The left two rows of Table 3 shows the composition of an ACTNET which was learned in the experiment 1 using a small data set. The number of classes was automatically determined by the class division in the I-PLCA. Fig. 1 shows a part of this ACTNET and the detail of an activity network of this ACTNET is shown in Fig. 4. The left row of Table 4 shows results of recognition and inference for action sequences used for learning. The classification accuracy of activities

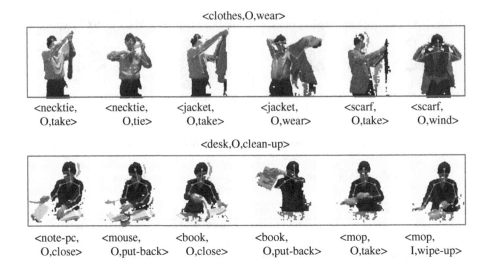

Fig. 3. Examples of action sequences in activities

Table 3. The composition of ACTNETs

Experiments	Experiment 1		Experiment 2	
	Activity	Action	Activity	Action
The number of classes	2	16	12	53.5
The number of target synsets	2	5	4	10
The number of motion synsets	2	10	4	16
The number of pairs of target and motion synsets	2	16	4	27

was 100%. The classification accuracy of actions was 81.3% when activities were used as context, whereas it was 75.0% when activities were not used as context. When additional information about object labels was given, the classification accuracy of actions was increased up to 93.8%.

In the experiment 2 using a large data set, 4 video clips were prepared for each of 4 activities and recognition and inference were evaluated for action sequences through 4-fold cross validation. The right two rows of Table 3 shows the composition of ACTNETs and the right row of Table 4 shows results of recognition and inference of actions and activities. The classification accuracy of activities was 93.8%. The classification accuracy of actions was 62.5% when activities were used as context, whereas it was 53.3% when activities were not used as context. When additional information about object labels was given, the classification accuracy of actions without and with contextual activities was respectively increased up to 75.8% and 83.4%. The percentage values in parentheses in Table 4 are the classification accuracy of the top two guesses of actions and activities.

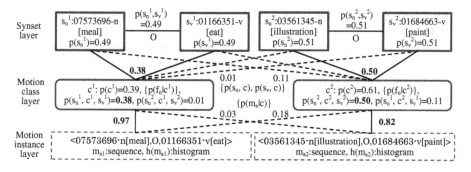

Fig. 4. An activity network of an ACTNET

Table 4. Results of recognition and inference

Experiments	Experiment 1	Experiment 2
Classification accuracy of activities	100%	93.8% (100%)
Classification accuracy of actions without contextual activities	75.0% (93.8%)	53.3% (59.2%)
Classification accuracy of actions with contextual activities	81.3% (100%)	62.5% (76.7%)
Classification accuracy of actions without contextual activities (when object labels are given)	93.8% (100%)	75.8% (85.8%)
Classification accuracy of actions with contextual activities (when object labels are given)	93.8% (100%)	83.4% (96.7%)

4   Discussion and Concluding Remarks

In data sets of experiments, there are the same actions for different objects and also different actions with similar motion, and these make it difficult to recognize object-oriented actions. In the experiment 2, classification accuracy is mainly decreased due to false recognition of 5 actions <notebook-computer,O, close>, <mouse,O,put-back>, <book,O,close>, <mop,O,take> and <teacup,O, pick-up>, which suffer the above difficulty and also have insufficient motion features since these actions are performed in a short time on a desk and skeleton extraction of the Kinect sensor is not accurate against the crowded background on the desk. In addition, in the data set of experiment 2, there are different activities which contains the same actions and just one false recognition of activities was occurred on one of these activities. As referential evaluation, though experimental setting and data sets are different, accuracy of similar action recognition by existing methods [7,8] is 48% ~ 59% and our method achieved better results.

In conclusion, we have proposed a learning method of a probabilistic semantic network ACTNET which integrates visual features and semantic features of object-oriented actions and their contextual activities and also provided a method using the ACTNET which visually recognize and verbally infer actions in the context of activities. Through two experiments, it has illustrated that the ACTNET can learn integrated probabilistic structure of visual and semantic features of object-oriented actions and activities and it has shown that activities and actions can be well recognized and inferred using the ACTNET, especially action understanding is improved using the context of activities and also additional information about objects.

Acknowledgment. This work was supported in part by Grant-in-Aid for Scientific Research (C) No.23500188 from Japan Society for Promotion of Science.

References

1. Bar, M.: Visual objects in context. Nature Reviews Neuroscience 5, 617–629 (2004)
2. Rabinovich, A., Vedaldi, C., Galleguillos, C., Wiewiora, E., Belongie, S.: Objects in context. In: Proc. of IEEE Int. Conf. on Computer Vision (2007)
3. Atsumi, M.: Object categorization in context based on probabilistic learning of classification tree with boosted features and co-occurrence structure. In: Bebis, G., et al. (eds.) ISVC 2013, Part I. LNCS, vol. 8033, pp. 416–426. Springer, Heidelberg (2013)
4. Isahara, H., Bond, F., Uchimoto, K., Utiyama, M., Kanzaki, K.: Development of Japanese WordNet. In: Proc. of the 6th Int. Conf. on Language Resources and Evaluation, pp. 2420–2423 (2008)
5. Atsumi, M.: Learning visual categories based on probabilistic latent component models with semi-supervised labeling. GSTF Int. Journal on Computing 2(1), 88–93 (2012)
6. Kitani, K., Okabe, T., Sato, Y.: Discovering primitive action categories by leveraging relevant visual context. In: Proc. of the IEEE Int. WS on Visual Surveillance (2008)
7. Yao, B., Fei-Fei, L.: Recognizing human-object interactions in still images by modeling the mutual context of objects and human poses. IEEE Trans. on Pattern Analysis and Machine Intelligence 34(9), 1691–1703 (2012)
8. Yao, B., Ma, J., Fei-Fei, L.: Discovering object functionality. In: Proc. of Int. Conf. on Computer Vision (2013)

Semantic-Aware Expert Partitioning

Veselka Boeva[1], Lilyana Boneva[1], and Elena Tsiporkova[2]

[1] Computer Systems and Technologies Department,
Technical University of Sofia, 4000 Plovdiv, Bulgaria
vboeva@tu-plovdiv.bg, lil2@abv.bg
[2] ICT & Software Engineering Group, Sirris, 1030 Brussels, Belgium
elena.tsiporkova@sirris.be

Abstract. In this paper, we present a novel semantic-aware clustering approach for partitioning of experts represented by lists of keywords. A common set of all different keywords is initially formed by pooling all the keywords of all the expert profiles. The semantic distance between each pair of keywords is then calculated and the keywords are partitioned by using a clustering algorithm. Each expert is further represented by a vector of membership degrees of the expert to the different clusters of keywords. The Euclidean distance between each pair of vectors is finally calculated and the experts are clustered by applying a suitable partitioning algorithm.

Keywords: expert location, expert partitioning, knowledge management.

1 Introduction

Expertise retrieval is not something new in the area of information retrieval. Finding the right person in an organization with the appropriate skills and knowledge is often crucial to the success of projects being undertaken [31]. Expert finders are usually integrated into organizational information systems, such as knowledge management systems, recommender systems, and computer supported collaborative work systems, to support collaborations on complex tasks [16]. Initial approaches propose tools that rely on people to self-assess their skills against a predefined set of keywords, and often employ heuristics generated manually based on current working practice [13,36]. Later approaches try to find expertise in specific types of documents, such as e-mails [9,11] or source code [31]. Instead of focusing only on specific document types systems that index and mine published intranet documents as sources of expertise evidence are discussed in [17]. In the recent years, research on identifying experts from online data sources has been gradually gaining interest [4,19,23,37,40,43]. For instance, Tsiporkova and Tourwé propose a prototype of a software tool implementing an entity resolution method for topic-centered expert identification based on bottom-up mining of online sources [40]. The tool extracts information from online sources in order to build a repository of expert profiles to be used for technology scouting purposes.

G. Agre et al. (Eds.): AIMSA 2014, LNAI 8722, pp. 13–24, 2014.
© Springer International Publishing Switzerland 2014

Many scientists who work on the expertise retrieval problem distinguish two information retrieval tasks: expert finding and expert profiling, where expert finding is the task of finding experts given a topic describing the required expertise [10], and expert profiling is the task of returning a list of topics that a person is knowledgeable about [3]. For instance, in [5,10] expertise retrieval is approached as an association finding task between topics and people. In Balog's PhD thesis, the expert finding and profiling tasks are addressed by the application of probabilistic generative models, specifically statistical language models [5].

Document clustering is a widely studied problem with many applications such as document organization, browsing, summarization, classification [1,28]. Clustering analysis is a process that partitions a set of objects into groups, or clusters in such a way that objects from the same cluster are similar and objects from different clusters are dissimilar. A text document can be represented either in the form of binary data, when we use the presence or absence of a word in the document in order to create a binary vector. A more enhanced representation would include refined weighting methods based on the frequencies of the individual words in the document, e.g., TF-IDF weighting [35]. However, the sparse and high dimensional representation of the different documents necessitate the design of text-specific algorithms for document representation and processing. Many techniques have been proposed to optimize document representation for improving the accuracy of matching a document with a query in the information retrieval domain [2,35]. Most of these techniques can also be used to improve document representation for clustering. Moreover, researchers have applied topic models to cluster documents. For example, clustering performance of probabilistic latent semantic analysis (PLSA) and Latent Dirichlet Allocation (LDA) has been investigated in [28]. LDA and PLSA are used to model the corpus and each topic is treated as a cluster. Documents are clustered by examining topic proportion vector.

In this work, we are concerned with the problem of how to cluster experts into groups according to the degree of their expertise similarity. The cluster hypothesis for document retrieval states that similar documents tend to be relevant to the same request [21]. In the context of expertise retrieval this can be re-stated that similar people tend to be experts on the same topics. Traditional clustering approaches assume that data objects to be clustered are independent and of identical class, and are often modelled by a fixed-length vector of feature/attribute values. The similarities among objects are assessed based on the attribute values of involved objects. However, the calculation of expertise similarity is a complicated task, since the expert expertise profiles usually consist of domain-specific keywords that describe their area of competence without any information for the best correspondence between the different keywords of two compared profiles. Therefore Boeva et al. propose to measure the similarity between two expertise profiles as the strength of the relations between the semantic concepts associated with the keywords of the two compared profiles [7]. In addition, they introduce the concept of expert's expertise sphere and show how the subject in question can be compared with the expertise profile of an individual expert and her/his sphere of expertise. In this paper, the problem is approached in a different way. Namely, it proposes a semantic-aware clustering approach for partitioning of a group of experts represented by lists of keywords. Initially, a common set of all different keywords is formed by pooling the keywords of all the expert profiles. Then the semantic distance between each pair of keywords is calculated and the keywords are partitioned by applying a selected clustering algorithm. Further, each expert is represented by a vector of membership degrees of the expert to the different clusters of keywords. Finally, the Euclidean distance between each pair of vectors is calculated and the experts are clustered by using some partitioning algorithm.

The rest of the paper is organized as follows. Section 2 briefly discusses the partitioning algorithms and describes the proposed semantic-aware clustering approach for partitioning of experts. Section 3 presents the initial evaluation of the proposed approach, which is applied to perform partitioning of researchers taking part in a scientific conference, and interprets the obtained clustering results. Section 4 is devoted to conclusions and future work.

2   Methods

In this section, we present our clustering method by first reviewing the characteristics of three widely-used groups of partitioning algorithms and then by describing how experts represented by lists of keywords can be clustered.

2.1   Partitioning Algorithms

Three partitioning algorithms are commonly used for the purpose of dividing data objects into k disjoint clusters [29]: k-means clustering, k-medians clustering and k-medoids clustering. All three methods start by initializing a set of k cluster centers, where k is preliminarily determined. Then, each object of the dataset is assigned to the cluster whose center is the nearest, and the cluster centers are recomputed. This process is repeated until the objects inside every cluster become as close to the center as possible and no further object item reassignments take place. The expectation-maximization (EM) algorithm [12] is commonly used for that purpose, *i.e.* to find the optimal partitioning into k groups. The three partitioning methods in question differ in how the cluster center is defined. In *k-means* clustering, the cluster center is defined as the mean data vector averaged over all objects in the cluster. For *k-medians* clustering the median is calculated for each dimension in the data vector instead. Finally, in *k-medoids* clustering [24], which is a robust version of the k-means, the cluster center is defined as the object which has the smallest sum of distances to the other objects in the cluster, *i.e.* this is the most centrally located point in a given cluster.

2.2 Semantic-Aware Expert Partitioning Approach

We propose herein a semantic-aware clustering approach that is used to partition experts into groups according to degree of their expertise similarity. It consists of three distinctive steps: 1) Construction of expert profiles via the extraction and association with each expert of a set of relevant keywords representing his/hers topics of interest; 2) Topic clustering based on pairwise semantic distance between the different keyword; 3) Clustering of experts based on their degree of association with the different topic clusters.

Construction of Expert Profiles. An expert profile may be quite complex and can, for example, be associated with information that includes: e-mail address, affiliation, a list of publications, co-authors etc. In view of this, an expert profile can be defined as a list of keywords, extracted from the available information about the expert in question, describing her/his area of expertise. The data needed for constructing the expert profiles could be extracted from various Web sources, *e.g.*, LinkedIn, the DBLP library, Microsoft Academic Search, Google Scholar Citation etc.

There exist several open tools for extracting data from public online sources. For instance, Python LinkedIn is a tool which can be used in order to execute the data extraction from LinkedIn. This is a package which provides a pure Python interface for the LinkedIn Connection, Profile, Search, Status, Messaging and Invitation APIs [32]. The DBLP database offers an easy access to the researchers' expertise since it describes each publication entry in an XML format and thus allowing easy parsing and information gathering for constructing the expert profiles [8].

The Stanford part-of-speech tagger [39] can be used to annotate the different words in the text collected for each expert with their specific part of speech. Next to the part of speech recognition, the tagger also defines whether a noun is plural, whether a verb is conjugated, etc. Further the annotated text can be reduced to a set of keywords (tags) by removing all the words tagged as articles, prepositions, verbs, and adverbs. Practically, only the nouns and the adjectives are retained and the final keyword set can be formed according to the following simple chunking algorithm:

- *adjective-noun(s) keywords:* a sequence of an adjective followed by a noun is considered as one compound keyword *e.g.* "supervised learning";
- *multiple nouns keywords:* a sequence of adjacent nouns is considered as one compound keyword *e.g.* "mixture model";
- *single noun keywords:* each of the remaining nouns forms a keyword on its own.

Clustering of Topics (Keywords). Assume that n different expert profiles are created in total and each expert profile i ($i = 1, 2, \ldots, n$) is represented by a list of p_i keywords. Further suppose that a set of m ($m << \sum_{i=1}^{n} p_i$) different keywords is formed by gathering all the keywords of all n expert profiles.

Then we can calculate the semantic distance between each pair of keywords by using, *e.g.*, the WordNet [14,30]. WordNet is a large lexical database of English. Nouns, verbs, adjectives and adverbs are grouped into sets of cognitive synonyms (synsets), each expressing a distinct concept. Synsets are interlinked by means of conceptual-semantic and lexical relations. WordNet's structure makes it a useful tool for computational linguistics and natural language processing.

Initially, the WordNet networks for the four different parts of speech were not linked to one another and the noun network was the first to be richly developed. This imposes some constrains on the use of WordNet ontology. Namely, most of the researchers who use it limit themselves to the noun network. However, not all keywords representing the expert profiles are nouns. In addition, the algorithms that can measure similarity between adjectives do not yield results for nouns hence the need for combined measure.

Let m_i be an arbitrary similarity measure and v be an arbitrary keyword. Then $m_i(v, v)$ gives the maximum possible score of m_i. We define MN_i as a normalized measure of m_i. Initially, for any two keywords v and w and for each used similarity measure m_i ($i = 1, 2, \ldots, r$) we compute its normalized measure $MN_i(v, w) = m_i(v, w)/m_i(v, v)$. One can easily see that if m_i takes non-negative values, then MN_i takes values in $[0, 1]$. In order to compute our own normalized measure MN combined from r different similarity measures $m_1, m_2, \ldots, m_r$, we first normalize independently each m_i using the above method and then define: $\alpha_1, \alpha_2, \ldots, \alpha_r$, such that α_i denotes the weight of i-th measure. In addition, $\alpha_1 + \alpha_2 + \ldots + \alpha_r = 1$. Further the normalized measure MN for any two keywords v and w is calculated as follows:

$$MN(v, w) = \alpha_1 MN_1(v, w) + \alpha_2 MN_2(v, w) + \ldots + \alpha_r MN_r(v, w).$$

It is clear that MN takes values in $[0, 1]$.

Once we have the distances (or similarity scores) calculated, the keywords can be clustered by applying the k-means (or other partitioning) algorithm which is explained in Section 2.1. Initially, the number of cluster centers is identified. As discussed in [15,38], this can be performed by running the selected clustering algorithm on the dataset in question for a range of different numbers of clusters. Subsequently, the quality of the obtained clustering solutions needs to be assessed in some way in order to identify the clustering scheme which best fits the considered datasets. For example, the internal validation measure that is presented in Section 3.2 or different one can be used as validity index to identify the best clustering scheme. Suppose that k cluster centers are determined for the set of keywords.

Clustering of Experts. As discussed above, the m keywords are grouped by the selected clustering algorithm into k clusters, *i.e.* a set of clusters $C_1, C_2, \ldots, C_k$ is produced. Let us denote by b_{ij} the number of keywords from the expert profile of expert i that belong to cluster C_j. Now each expert i ($i = 1, 2, \ldots, n$) can be represented by a vector $e_i = (e_{i1}, e_{i2}, \ldots, e_{ik})$, where $e_{ij} = b_{ij}/p_i$ ($j = 1, 2, \ldots, k$) and p_i is the total number of keywords in the expert profile representation. In this way, each expert i is represented by a k-length vector of membership degrees of the expert to k different clusters of keywords. Then we can calculate, *e.g.*, the Euclidean distance between each pair of vectors and group the experts by applying the k-means or other clustering algorithm.

3   Initial Evaluation

3.1   Test Data

We need test data that is tied to our specific task, namely the expert clustering. For this task, we use the test collection from a scientific conference [20] devoted to information technology in bio- and medical informatics. For each topic, participants (53 in total) of the corresponding conference session are regarded as experts on that topic. This is an easy way of obtaining topics and relevance judgements. A total of 5 topics (sessions) are created by the conference science committee. A list of researchers for these topics are also supplied, *i.e.*, names that are listed in the conference program on the sessions (topics) information. These researchers are considered as relevant experts, thus, used as the ground truth to benchmark the results of the proposed clustering method.

The data needed for constructing the researcher expertise profiles are extracted from Microsoft Academic Search, *i.e.*, a researcher profile is defined by a list of keywords used in the profile page of the author in question to describe her/his scientific area. Note that some of the used keywords are multiple-word terms, e.g. "Molecular Biology", "Data Mining", "Software Engineering", "Information Retrieval" etc. However, not all the multiple-word terms are present in WordNet ontology. Therefore, these keywords have been divided into their constituting words. The latter does not have effect on the quality of final expert clustering, because even the constituting words have been allocated in different clusters of keywords they are both included into the corresponding expert profiles and further are taken into account by the experts' membership degrees to those clusters.

3.2   Cluster Validation Measures

One of the most important issues in cluster analysis is the validation of clustering results. Essentially, the cluster validation techniques are designed to find the partitioning that best fits the underlying data, and should therefore be regarded as a key tool in the interpretation of clustering results. Since none of the clustering algorithms performs uniformly best under all scenarios, it is not reliable to use a single cluster validation measure, but instead to use at least two that reflect different aspects of a partitioning. In this sense, we have used two different validation measures. We apply the *Silhouette Index* (SI) for assessing compactness and separation properties of the obtained clustering solutions [34]. SI is also used as a validity index to identify the clustering scheme which best fits the test data described in the foregoing section. Furthermore, we use the *F-measure* for evaluating the accuracy of the generated clustering solutions [25].

Silhouette Index. The *Silhouette index* reflects the compactness and separation of clusters [34]. Suppose $C = \{C_1, C_2, \ldots, C_k\}$ is a clustering solution of the considered data set, which contains the attribute vectors of m objects. Then the SI is defined as

$$s(C) = \frac{1}{m} \sum_{i=1}^{m} (b_i - a_i)/\max\{a_i, b_i\},$$

where a_i represents the average distance of object i to the other objects of the cluster to which the object is assigned, and b_i represents the minimum of the average distances of object i to object of the other clusters. The values of Silhouette Index vary from -1 to 1.

F-measure. The *F-measure* is the harmonic mean of the precision and recall values for each cluster. Let us consider two clustering solutions $C = \{C_1, C_2, \ldots, C_k\}$ and $C' = \{C'_1, C'_2, \ldots, C'_l\}$ of the same data set. The first solution C is a known partition of the considered data set while the second one C' is a partition generated by the applied clustering algorithm. The F-measure for a cluster C'_j is then given as

$$F(C'_j) = \frac{2 |C_i \cap C'_j|}{|C_i| + |C'_j|},$$

where C_i is the cluster that contains the maximum number of objects from C'_j. The overall F-measure for clustering solution C' is defined as the mean of cluster-wise F-measure values, *i.e.* $F(C') = \frac{1}{l} \sum_{j=1}^{l} F_j$. For a perfect clustering, when $l = k$, the maximum value of the F-measure is 1.

3.3   Implementation and Availability

A free distributed Java library has been used to measure the word similarity - WordNet Similarity for Java (WS4J) [41]. A Java program using WS4J API has been applied to calculate a word similarity matrix for the keywords describing the expert profiles. The semantic relatedness algorithms implemented by the library have been used in our experiments [6,18,22,26,27,33,42]. As the score ranges of the algorithms vary in different intervals we have performed a normalization on all scores in order to obtain a final score in one and the same range - [0,1] (see Section 2.2). The weights are evenly distributed among the algorithms that produce a score for a given word pair. Some algorithms work for noun pair and other can be used on other parts of the speech. When an algorithm is not applicable an error score of -1 is returned and the corresponding algorithm is excluded from the calculation of the normalized measure. The algorithms that produce scores for a given word pair are used for calculating its normalized score as a mean of the produced scores. We do not give preference to any algorithm, because of the automation and the lack of any preliminary knowledge about the words being compared.

R scripts have been used to implement all the other experiments and to generate the result plots.

3.4   Experimental Results

Initially, a set of 44 different keywords is formed by gathering all the keywords of
all 53 expert profiles. Then the semantic distance between each pair of keywords
is calculated by using WordNet.

Once we have the normalized similarity scores calculated using the method
presented in Section 2.2, the keywords are partitioned by applying k-means clustering algorithm. The partitioning algorithms as k-means contain the number of
clusters (k) as a parameter and their major drawback is the lack of prior knowledge for that number to construct. Therefore, we have run k-means clustering
algorithm for all values of k between 2 and 20 and plot the values of the selected
index (Silhouette Index) obtained by each k as the function of k (see Figure 1(a)).
We search for the values of k at which a significant local change in value of the
index occurs [15]. These values are 4, 6 and 10. Thus, we apply the k-means
on the set of keywords for three different values of k $(k = 4, 6, 10)$. In this way,
three different clustering solutions for the set of keywords are produced. The
partition generated for $k = 10$ can be seen in Table 1. In fact, $k = 10$ is more
proper number of clusters for the set of keywords than $k = 4$ and $k = 6$. This is
supported by the higher SI scores produced on the clustering solutions of the set
of experts when the keywords are partitioned in 10 clusters (see Figure 1(b)).

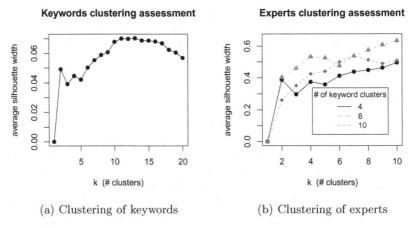

(a) Clustering of keywords   (b) Clustering of experts

Fig. 1. SI values generated by k-means clustering method on the set of keywords (a)
and on the set of experts (b) for different number of clusters

Further, each expert is represented by a vector of membership degrees of
the expert to the different clusters of keywords. Finally, the Euclidean distance
between each pair of vectors is calculated and the experts are clustered by using
the selected clustering algorithm. Thus k-means clustering algorithm has been
run on the set of experts for all values of k between 2 and 10 for each of the
three clustering solutions of the keywords. For each generated clustering solution
a Silhouette Index score is calculated and plotted in Figure 1(b). As can be
noticed, the optimal number of clusters for the set of experts are 4 and 7.

Table 1. Clustering of the set of keywords for $k = 10$

Clusters	Keywords
1	Algorithm, Engineering
2	Artificial Intelligence, Computer Science, Electrical Engineering, Computing
3	Mathematics, Electronics, Physiology, Neuroscience, Biochemistry, Chemistry, Biology, Molecular Biology
4	Database, Information, Software, Graphics, Botany
5	Medicine, Pharmacology, Ophthalmology, Toxicology, Distribute, Pattern
6	Data Mining, Retrieval, Energy
7	Learning, Theory, Pattern
8	World Wide Web, Machine
9	Security, Recognition, Privacy, Parallel
10	Zoology

Table 2. F-measure scores generated by k-means clustering method on the set of experts for $k = 4, 7$ for three different partitions of keywords ($k = 4, 6, 10$)

keywords clustering / experts clustering	$k = 4$	$k = 6$	$k = 10$
$k = 4$	0.439	0.439	0.432
$k = 7$	0.373	0.421	0.428

Next the F-measure is used to assess the accuracy of the clustering solutions generated on the set of experts for $k = 4, 7$ for three different partitions of keywords ($k = 4, 6, 10$), see Table 2. Each produced clustering solution is benchmarked to the (known) partition of the researchers explained in Section 3.1. The obtained scores are between 0.373 and 0.439. Notice that higher values are produced by the expert partitions generated for $k = 4$. However, there are no superior results with respect to the different clustering solutions of the keywords.

Finally, let us consider the clustering solution generated on the set of experts for $k = 4$ when the keywords are partitioned in 6 clusters. The experts have been grouped into four main clusters. **Cluster 1** contains 27 researchers most of who have expertise in Bioinformatics & Computational Biology, Artificial Intelligence, Data Mining and Machine Learning. Note that all the scientists with expertise in Bioinformatics & Computational Biology are allocated in this cluster. In addition, a clear sub cluster is formed by four experts all with only competence in Biochemistry. In fact, these are grouped in a separate cluster for

$k = 7$. Cluster 1 is the most heterogeneous cluster. This might be due to the fact that it contains many experts (20 such researchers) who have competence in more than two scientific areas. **Cluster 2** contains 9 experts who have competence in Engineering, Artificial Intelligence and Computer Science. This cluster is divided in two separate clusters for $k = 7$. **Cluster 3** contains 12 experts whose main expertise is in Databases and Software Engineering. This is very homogeneous cluster consisting of experts all having the keyword "Database" in her/his expertise profile. **Cluster 4** contains only 5 experts: three with expertise in Medicine, one in Ophthalmology and one in Toxicology, Pharmacology and Molecular Biology. Evidently, the considered clustering solution is a good partition of the researchers with respect to their scientific expertise.

4 Conclusion and Future Work

This paper elaborates on a novel semantic-aware approach for clustering of experts represented by lists of keywords. The proposed approach has initially been evaluated by applying the algorithm to partition of researchers taking part in a scientific conference. The produced clustering solutions have been validated by two different cluster validation measures. The obtained results demonstrate that the proposed approach is a robust clustering technique that is able to produce good quality clustering solutions.

For future work, the aim is to pursue further enhancement and validation of the proposed clustering approach applying alternative partitioning methods e.g. hierarchical clustering on richer expert profiles extracted from online sources e.g. LinkedIn, Google Scholar, the DBLP library, Microsoft Academic Search, etc. In addition, our future intention is to evaluate the scalability of the proposed approach. Presently, the method consists of two different clustering phases, which can be rather computationally expensive when the number of targeted experts grows. Another impact on scalability is also the degree of heterogeneity among the experts in terms of expertise. The higher this degree, the more topic clusters will be generated and therefore the vectors representing the experts will have higher dimension. It can also occur in this situation that many topic clusters are of little relevance to all of the experts. One possible way to tackle this problem is adapt the method to deal with sparse data.

References

1. Aggarwal, C., Zhai, C.: A survey of text clustering algorithms. In: Mining Text Data, pp. 77–128 (2012)
2. Baeza-Yates, R.A., Ribeiro-Neto, B.A.: Modern Information Retrieval - the concepts and technology behind search, 2nd edn. Pearson Education Ltd., Harlow (2011)
3. Balog, K., et al.: Broad expertise retrieval in sparse data environments. In: 30th Annual Int. ACM SIGIR Conference on Research and Development in Information Retrieval. ACM Press, New York (2007)

4. Balog, K., de Rijke, M.: Finding similar experts. In: 30th Annual Int. ACM SIGIR Conference on Research and Development in Information Retrieval, pp. 821–822. ACM Press, New York (2007)
5. Balog, K.: People search in the enterprise. PhD thesis, Amsterdam University (2008)
6. Banerjee, S., Pedersen, T.: An Adapted Lesk Algorithm for Word Sense Disambiguation Using WordNet. In: Gelbukh, A. (ed.) CICLing 2002. LNCS, vol. 2276, pp. 136–145. Springer, Heidelberg (2002)
7. Boeva, V., Krusheva, M., Tsiporkova, E.: Measuring Expertise Similarity in Expert Networks. In: 6th IEEE Int. Conf. on Intelligent Systems, pp. 53–57. IEEE (2012)
8. Buelens, S., Putman, M.: Identifying experts through a framework for knowledge extraction from public online sources. Master thesis, Gent University, Belgium (2011)
9. Campbell, C.S., Maglio, P.P., Cozzi, A., Dom, B.: Expertise identification using Bibliography 189 email communications. In: 12th International Conference on Information and Knowledge Management. ACM Press (2003)
10. Craswell, N., et al.: Overview of the TREC-2005 Enterprise Track. In: 14th Text Retrieval Conference (2006)
11. D'Amore, R.: Expertise community detection. In: 27th Annual Int. ACM SIGIR Conf. on Research and Development in Information Retrieval. ACM Press (2004)
12. Dempster, A.P., Laird, N.M., Rubin, D.B.: Maximum Likelihood from Incomplete Data via the EM Algorithm. J. of the Royal Statistical Society B 39(1), 1–38 (1977)
13. ECSCW99 Workshop. Beyond knowledge management: Managing expertise, http://www.informatik.uni-bonn.de/~prosec/ECSCW-XMWS/
14. Fellbaum, C.: WordNet: An Electronic Lexical Database. MIT Press, Cambridge (2001)
15. Halkidi, M., Batistakis, Y., Vazirgiannis, M.: On clustering validation techniques. Journal of Intelligent Information Systems 17(2) (2001)
16. Hattori, F., et al.: Socialware: Multiagent systems for supporting network communities. Communications of the ACM 42(3), 55–61 (1999)
17. Hawking, D.: Challenges in enterprise search. In: 15th Australasian Database Conference. Australian Computer Society, Inc. (2004)
18. Hirst, G., St-Onge, D.: Lexical Chains as Representations of Context for Detection and Correction of Malapropisms. In: WordNet: An Electronic Lexical Database, pp. 305–332. MIT Press (1998)
19. Hristoskova, A., Tsiporkova, E., Tourwé, T., Buelens, S., Putman, M., De Turck, F.: A Graph-based Disambiguation Approach for Construction of an Expert Repository from Public Online Sources. In: 5th IEEE Int. Conf. on Agents and Art. Int. (2013)
20. Böhm, C., Khuri, S., Lhotská, L., Pisanti, N. (eds.): ITBAM 2011. LNCS, vol. 6865. Springer, Heidelberg (2011)
21. Jardine, N., van Rijsbergen, C.J.: The use of hierarchic clustering in information retrieval. Information Storage and Retrieval 7, 217–240 (1971)
22. Jiang, J., Conrath, D.: Semantic Similarity Based on Corpus Statistics and Lexical Taxonomy. In: International Conference Research on Computational Linguistics, pp. 19–33 (1997)
23. Jung, H., Lee, M., Kang, I., Lee, S., Sung, W.: Finding topic-centric identified experts based on full text analysis. In: 2nd International ExpertFinder Workshop at the 6th International Semantic Web Conference, ISWC (2007)
24. Kaufman, L., Rousseeuw, P.J.: Finding Groups in Data: An Introduction to Cluster Analysis. Wiley, New York (1990)

25. Larsen, B., Aone, C.: Fast and Effective Text Mining Using Linear Time Document Clustering. In: KDD 1999, pp. 16–29 (1999)
26. Leacock, C., Chodorow, M.: Combining Local Context and WordNet Similarity for Word Sense Identification, pp. 265–283. MIT Press, Cambridge (1998)
27. Lin, D.: An Information-Theoretic Definition of Similarity. In: 15th International Conference on Machine Learning, ICML, pp. 296–304 (1998)
28. Lu, Y., Mei, Q., Zhai, C.: Investigating task performance of probabilistic topic models: an empirical study of plsa and lda. Information Retrieval 14(2), 178–203 (2011)
29. MacQueen, J.B.: Some methods for classification and analysis of multivariate observations. In: 5th Berkeley Symp. Math. Stat. Prob., vol. 1, pp. 281–297 (1967)
30. Miller, G.A.: WordNet: A Lexical Database for English. Communications of the ACM 38(11), 39–41 (1995)
31. Mockus, A., Herbsleb, J.D.: Expertise browser: a quantitative approach to identifying expertise. In: 24th Int. Conf. on Software Engineering. ACM Press (2002)
32. Python LinkedIn - a python wrapper around the LinkedIn API, http://code.google.com/p/python-linkedin/
33. Resnik, P.: Using Information Content to Evaluate Semantic Similarity in a Taxonomy. In: 14th International Joint Conference on Artificial Intelligence, vol. 1, pp. 448–453 (1995)
34. Rousseeuw, P.: Silhouettes: a graphical aid to the interpretation and validation of cluster analysis. Journal of Computational Applied Mathematics 20, 53–65 (1987)
35. Salton, G., Buckley, C.: Term Weighting Approaches in Automatic Text Retrieval. Information Processing and Management 24(5), 513–523 (1988)
36. Seid, D., Kobsa, A.: Demoir: A hybrid architecture for expertise modeling and recommender systems (2000)
37. Stankovic, M., Jovanovic, J., Laublet, P.: Linked data metrics for flexible expert search on the Open Web. In: Antoniou, G., Grobelnik, M., Simperl, E., Parsia, B., Plexousakis, D., De Leenheer, P., Pan, J. (eds.) ESWC 2011, Part I. LNCS, vol. 6643, pp. 108–123. Springer, Heidelberg (2011)
38. Theodoridis, S., Koutroubas, K.: Pattern recognition. Academic Press (1999)
39. Toutanova, K.: Enriching the knowledge sources used in a maximum entropy partofspeech tagger. In: Joint SIGDAT Conference on Empirical Methods in Natural Language Processing and Very Large Corpora, EMNLP/VLC-2000 (2000)
40. Tsiporkova, E., Tourwé, T.: Tool support for technology scouting using online sources. In: De Troyer, O., Bauzer Medeiros, C., Billen, R., Hallot, P., Simitsis, A., Van Mingroot, H. (eds.) ER Workshops 2011. LNCS, vol. 6999, pp. 371–376. Springer, Heidelberg (2011)
41. WordNet Similarity for Java (WS4J), https://code.google.com/p/ws4j/
42. Wu, Z., Palmer, M.: Verbs semantics and lexical selection. In: 32nd Annual Meeting on Association for Computational Linguistics, pp. 133–138 (1994)
43. Zhang, J., Tang, J., Li, J.: Expert finding in a social network. In: Kotagiri, R., Radha Krishna, P., Mohania, M., Nantajeewarawat, E. (eds.) DASFAA 2007. LNCS, vol. 4443, pp. 1066–1069. Springer, Heidelberg (2007)

User-Level Opinion Propagation Analysis
in Discussion Forum Threads

Dumitru-Clementin Cercel[1] and Ştefan Trăuşan-Matu[1,2]

[1] Faculty of Automatic Control and Computers, University POLITEHNICA of Bucharest,
Bucharest, Romania
[2] Romanian Academy Research Institute for Artificial Intelligence, Bucharest, Romania
{dumitru.cercel,stefan.trausan}@cs.pub.ro

Abstract. Online discussions such as forums are very popular and enable participants to read other users' previous interventions and also to express their own opinions on various subjects of interest. In online discussion forums, there is often a mixture of positive and negative opinions because users may have similar or conflicting opinions on the same subject. Therefore, it is challenging to track the flow of opinions over time in online discussion forums. Past research in the field of opinion propagation has dealt mainly with online social networks. In this paper, by contrast, we address the opinion propagation in discussion forum threads. We proposed a user-level opinion propagation analysis method in the discussion forum threads. This method establishes for a given time step whether the discussion will result in complete agreement between participants or in disparate and even contrary opinions.

Keywords: Opinion Propagation, Discussion Forum, Forum Thread, Discrete Opinion Space, Term-User Opinion Matrix, User-User Similarity Matrix.

1 Introduction

The concept of *opinion* as a core component of the opinion propagation problem is defined in [1] by means of four keywords (*"Topic"*, *"Holder"*, *"Claim"*, *"Sentiment"*): *"the Holder believes a Claim about the Topic, and in many cases associates a Sentiment, such as good or bad, with the belief"*. Several efforts have been made to research the opinion propagation problem in online social networks. In this paper, we perform an analysis of the opinion propagation in other forms of social media than the online social networks, i.e. in discussion forum threads. To the best of our knowledge, this subject has not been addressed in previous studies yet.

There are differences between our solution and other existing studies [2]. Previous studies have focused on the propagation of opinions in the networks (e.g. the Watts-Strogatz network model [3] or the Barabási-Albert network model [4]) that have properties (e.g. small-world and scale-free properties) similar to those of the online social networks [5, 6]. Moreover, in these studies, opinions are randomly assigned to individuals, and the interactions over time between neighboring individuals are also randomly established.

G. Agre et al. (Eds.): AIMSA 2014, LNAI 8722, pp. 25–36, 2014.
© Springer International Publishing Switzerland 2014

We propose a user-level method for the opinion propagation problem in discussion forum threads, taking into account the opinions written by users. For this purpose, we extract the users' opinions from their posts in the discussion. More concretely, our solution is based on the opinions written by users on diverse themes expressed by specific noun terms that are semantically related to the subject of the discussion. Thus, our solution consists in determining whether the participants in the discussion reach complete agreement or they continue to have disparate or even contrary opinions, and therefore impossible or very hard to adapt from one to the other. We evaluate our method by tracking the opinion propagation in the real-world discussion forum threads.

The paper is structured as follows. In Section 2, we discuss shortly the related work, and in Section 3, we formally define the user-level opinion propagation problem in discussion forums. In Section 4, we detail the system proposed to solve the problem we study. The experimental results are presented in Section 5. Finally, Section 6 is dedicated to the conclusions.

2   Related Work

There are several models of the opinion propagation process in online social networks. In the standard Voter model [2] defined on a regular lattice, each individual randomly selects one of his neighbors at the time step t, and adopts, at the time step $t+1$, the opinion of the neighbor selected at the time step t. The original Sznajd model [7] was described for a one-dimensional lattice and subsequently extended to complex networks [8]. In the case of complex networks, at each time step t, if two connected individuals hold the same opinions, then at the time step $t+1$ they convey the opinion to all their neighbors. In the Deffuant model [9], at each time step, two individuals randomly selected from the network change their own opinions only if the difference between opinions is less than a given threshold parameter ε. In the Hegselmann-Krause model [10], at each time step t, a randomly selected individual changes his opinion, which becomes equal to the arithmetic average of the opinions of all his neighbors. For more details about the opinion propagation problem in online social networks, see [11].

3   Problem Formulation

Before defining the user-level opinion propagation problem in discussion forums, we introduce some necessary definitions and notations. Moreover, this section provides a comprehensive discussion of the online forums in order to understand the target data of our research. Formally, an online forum is composed of a set of discussion threads defined as follows:

Definition 1 (Discussion Thread). *Given a discussion forum website, we describe a discussion forum thread (simplified "discussion thread") at the time step t_τ, $\tau \in \mathbb{N}^*$, by a tuple $(T_{DT}, S_{DT}, U_{DT}(t_\tau), P_{DT}(t_\tau), R_{DT}(t_\tau))$, where $t_1, t_2, ..., t_\tau, ...$ are discrete time steps corresponding to events that occur in a discussion thread, i.e. adding new posts.*

A discussion thread is specifically characterized by a subject S_{DT}, and generally by a topic T_{DT}. Thus, each discussion thread is about a particular subject S_{DT}, and many such discussion threads are grouped under the same topic T_{DT}. The posts $p \in P_{DT}(t_\tau)$ represent interventions of the users $u \in U_{DT}(t_\tau)$ on the subject S_{DT} in the respective discussion thread until the time step t_τ. Each post $p \in P_{DT}(t_\tau)$ is a reply to another previously written post.

At the beginning of the discussion thread, which corresponds to the time step t_1, a user sends a message, and then, in the following time steps t_τ, $\tau \in \mathbb{N}$, $\tau \geq 2$, other users intervene in the discussion. In online social networks, such as Facebook, every user only interacts with users with whom he has explicitly formed friendships. In contrast, in a discussion thread, a user $u \in U_{DT}(t_\tau)$ interacts, at the time step t_τ, with any other users who participate in the discussion until the time step $t_{\tau-1}$, even if he doesn't know them, which means that the user $u \in U_{DT}(t_\tau)$ can reply to the previous posts they have written. In other words, all messages sent by all users are visible to every other participant in the discussion thread.

In a discussion thread, at the time step t_τ, two types of relations can be defined: $R_{UP}(t_\tau)$ and $R_{PP}(t_\tau)$, $R_{DT}(t_\tau) = R_{UP}(t_\tau) \cup R_{PP}(t_\tau)$. A binary partial relation $R_{UP}(t_\tau) \subset U_{DT}(t_\tau) \times P_{DT}(t_\tau)$, called *belong-to relation*, is defined as the relation between the sets $U_{DT}(t_\tau)$ and $P_{DT}(t_\tau)$ so that, if $(u, p) \in R_{UP}(t_\tau)$, the post $p \in P_{DT}(t_\tau)$ was written by the user $u \in U_{DT}(t_\tau)$. A binary partial relation $R_{PP}(t_\tau) \subset P_{DT}(t_\tau) \times P_{DT}(t_\tau)$, called *reply-to relation*, is established between the posts from the set $P_{DT}(t_\tau)$ so that, if $(p_1, p_2) \in R_{PP}(t_\tau)$, the post $p_1 \in P_{DT}(t_\tau)$ replies to another post $p_2 \in P_{DT}(t_\tau)$.

Taking into account all the above explanations, for two instances $(T_{DT}, S_{DT}, U_{DT}(t_\tau), P_{DT}(t_\tau), R_{DT}(t_\tau))$ and $(T_{DT}, S_{DT}, U_{DT}(t_{\tau'}), P_{DT}(t_{\tau'}), R_{DT}(t_{\tau'}))$ of the same discussion thread at the time steps t_τ and $t_{\tau'}$, $t_{\tau'} > t_\tau$, we can write the following relations: $U_{DT}(t_\tau) \subseteq U_{DT}(t_{\tau'})$, $P_{DT}(t_\tau) \subset P_{DT}(t_{\tau'})$, $R_{UP}(t_\tau) \subset R_{UP}(t_{\tau'})$, and $R_{PP}(t_\tau) \subset R_{PP}(t_{\tau'})$.

The relationships established between users in an online social network enable us to represent the community of users by a social graph $G_s = (V_s, E_s)$, where V_s is the set of users, and E_s is the set of relationships that connect them. In contrast to an online social network, a discussion thread is represented by an oriented graph structure, using the reply-to relation R_{PP}.

A discussion thread comprises a series of posts exchanged between users about the same subject S_{DT}. The subject S_{DT} enables users who participate in the discussion to express their opinions on diverse noun terms. Moreover, these noun terms are semantically related to one of the noun terms that appear in the subject S_{DT}. Consequently, we can write the following definition:

Definition 2 (Noun Term Vocabulary). *In a discussion thread $(T_{DT}, S_{DT}, U_{DT}(t_\tau), P_{DT}(t_\tau), R_{DT}(t_\tau))$, all distinct noun terms on which the users $U_{DT}(t_\tau)$ expressed their opinions and that are semantically related to one of the noun terms by which the subject S_{DT} is written define a noun term vocabulary. We use $V^d_{DT}(t_\tau) = \{s_1, s_2, ..., s_d\}$ to denote a d-dimensional vocabulary of noun terms shared by all users in the discussion thread until the time step t_τ, $\tau \in \mathbb{N}^*$.*

The size and the structure of the noun term vocabulary may vary over time. New posts that are added to the discussion thread may contain opinions on new noun terms.

In what follows, we make the assumption that the users' opinions are modeled by numerical values.

Definition 3 (Discrete Opinion Space). *Given at the time step t_τ, $\tau \in \mathbb{N}^*$, the discussion thread $(T_{DT}, S_{DT}, U_{DT}(t_\tau), P_{DT}(t_\tau), R_{DT}(t_\tau))$ from a discussion forum website and the corresponding d-dimensional vocabulary of noun terms $V^d_{DT}(t_\tau)$, then the opinions of each user $u \in U_{DT}(t_\tau)$ on the noun terms from the vocabulary $V^d_{DT}(t_\tau)$ can be represented by a vector in a d-dimensional discrete opinion space $OS^d_{DT} = \{-1, 0, +1\}^d$.*

In a d-dimensional opinion space, each axis corresponds to a noun term, and the values on each axis correspond to the users' possible opinions on the respective noun term. In the particular case in which the opinion space is discrete, opinions are values on the discretized axes.

Let $o^d_{DT}(t_\tau) = [o_1 \; o_2 \; ... \; o_d]^T$ denote a vector at the time step t_τ, $\tau \in \mathbb{N}^*$, in the d-dimensional discrete opinion space OS^d_{DT}, and that corresponds to a user $u \in U_{DT}(t_\tau)$. Each entry o_k in the vector o^d_{DT} represents the opinion held by the user $u \in U_{DT}(t_\tau)$ on each noun term s_k from the vocabulary $V^d_{DT}(t_\tau)$ until the time step t_τ, $\tau \in \mathbb{N}^*$.

Opinion entries o_k may take one of the following values: -1, 0, or +1, where the value -1 denotes a negative opinion, the value 0 denotes a neutral opinion, and the value +1 denotes a positive opinion. If the user $u \in U_{DT}(t_\tau)$ does not express his opinion on the noun term $s_k \in V^d_{DT}(t_\tau)$ until the time step t_τ, $\tau \in \mathbb{N}^*$, then we consider the value 0 for the entry $o_k \in o^d_{DT}(t_\tau)$. If until the time step t_τ, $\tau \in \mathbb{N}$, the user $u \in U_{DT}(t_\tau)$ give more opinions on a noun term, then only his last opinion is taken into consideration.

Inspired by the work of Stavrianou [12], we redefine the representation of discussion threads by using a data structure in the form of a graph that takes into account the opinions expressed by users in their posts as follows:

Definition 4 (Post-Reply Opinion Graph). *Given at the time step t_τ, $\tau \in \mathbb{N}^*$, a discussion thread $(T_{DT}, S_{DT}, U_{DT}(t_\tau), P_{DT}(t_\tau), R_{DT}(t_\tau))$ from a discussion forum website, then the discussion thread is associated with an oriented graph $G_{DT}(t_\tau)(V_{DT}(t_\tau), E_{DT}(t_\tau))$, called post-reply opinion graph, where:*

- *$V_{DT}(t_\tau) = \{v_j \mid v_j = (v_j^p, v_j^u, v_j^{tm}, v_j^{op}), v_j^p \in P_{DT}(t_\tau), v_j^u \in U_{DT}(t_\tau), v_j^{tm} \in \mathbb{N}, v_j^{op} = [o_1 \; o_2 \; ... \; o_m]^T, m \le d, v_j^{op} \subset OS^m_{DT}, j \in [1, n]\}$ is the set of vertices in the graph $G_{DT}(t_\tau)$, and n denotes the number of nodes in the post-reply opinion graph $G_{DT}(t_\tau)$;*
- *$E_{DT}(t_\tau) = \{e_1, e_2, ..., _{n-1}\}$ is the set of edges so that, if $e = (v_i, v_j) \in E_{DT}(t_\tau)$, then the post v_i is a reply to the post v_j.*

In the post-reply opinion graph $G_{DT}(t_\tau)$, each vertex $v_j \in V_{DT}(t_\tau)$ corresponds to a post v_j^p written by a user v_j^u at the timestamp v_j^{tm}. The post v_j^p expresses opinions in the m-dimensional opinion vector v_j^{op}, where m is less than or equal to the dimension d of the noun term vocabulary $V^d_{DT}(t_\tau)$. Let us consider two more definitions as follows:

Definition 5 (Term-User Opinion Matrix). *Given at the time step t_τ, $\tau \in \mathbb{N}^*$, the discussion thread $(T_{DT}, S_{DT}, U_{DT}(t_\tau), P_{DT}(t_\tau), R_{DT}(t_\tau))$ from a discussion forum website, the corresponding post-reply opinion graph $G_{DT}(t_\tau)(V_{DT}(t_\tau), E_{DT}(t_\tau))$, and the d-dimensional vocabulary of noun terms $V^d{}_{DT}(t_\tau)$, then we build a $d \times n$ term-user matrix $A_{T-U}(t_\tau) = [A_1(t_\tau) A_2(t_\tau) \dots A_n(t_\tau)]$, where n denotes the number of users in the set $U_{DT}(t_\tau)$. Each column $A_u(t_\tau) = [a_{1,u}(t_\tau) \; a_{2,u}(t_\tau) \dots a_{d,u}(t_\tau)]^T$ corresponds to a user $u \in U_{DT}(t_\tau)$ and denotes the d-dimensional opinion vector of the user $u \in U_{DT}(t_\tau)$ in the opinion space $OS^d{}_{DT}$.*

Definition 6 (User-User Similarity Matrix). *Given at the time step t_τ, $\tau \in \mathbb{N}^*$, the discussion thread $(T_{DT}, S_{DT}, U_{DT}(t_\tau), P_{DT}(t_\tau), R_{DT}(t_\tau))$ from a discussion forum website, the corresponding post-reply opinion graph $G_{DT}(t_\tau)(V_{DT}(t_\tau), E_{DT}(t_\tau))$, the d-dimensional vocabulary of noun terms $V^d{}_{DT}(t_\tau)$, the $d \times n$ term-user opinion matrix $A_{T-U}(t_\tau)$, and a similarity measure between two vectors, then we build a $n \times n$ user-user matrix $B_{U-U}(t_\tau)$. The entry of the row k^{th} and of the column h^{th} of the matrix $B_{U-U}(t_\tau)$ is denoted by $b_{k,h}(t_\tau)$ and represents the similarity between the users k and $u \in U_{DT}(t_\tau)$ from the perspective of the opinions expressed by these users.*

More concretely, the entry $b_{k,h}(t_\tau)$ measures the similarity between the column vectors $A_k(t_\tau) = [a_{1,k}(t_\tau) \; a_{2,k}(t_\tau) \dots a_{d,k}(t_\tau)]^T$ and $A_h(t_\tau) = [a_{1,h}(t_\tau) \; a_{2,h}(t_\tau) \dots a_{d,h}(t_\tau)]^T$ of the matrix $A_{T-U}(t_\tau)$. We calculate the similarity between two vectors by using the cosine of the angle formed between them as follows:

$$sim_{\cos}(A_k(t_i), A_h(t_i)) = \frac{\sum_{i=1}^{n} a_{ik} \times a_{ih}}{\sqrt{\sum_{i=1}^{n}(a_{ik})^2} \times \sqrt{\sum_{i=1}^{n}(a_{ik})^2}} \qquad (1)$$

The cosine measure returns a score between -1 and +1. In this case, the value -1 means complete disagreement between users (the users have opinions with opposite sentiments on the same target term), while value +1 means complete agreement between users (the users have opinions with similar sentiments on the same target term).

Finally, we can formally define the problem of user-level opinion propagation in a discussion thread, by extending the definition taken from [13]:

Definition 7 (The Problem of User-Level Opinion Propagation in Discussion Forum Threads). *Given, at the time step t_τ, $\tau \in \mathbb{N}^*$, the discussion thread $(T_{DT}, S_{DT}, U_{DT}(t_\tau), P_{DT}(t_\tau), R_{DT}(t_\tau))$ from a discussion forum website, the corresponding post-reply opinion graph $G_{DT}(t_\tau)(V_{DT}(t_\tau), E_{DT}(t_\tau))$, the d-dimensional vocabulary of noun terms $V^d{}_{DT}(t_\tau)$, the $d \times n$ term-user opinion matrix $A_{T-U}(t_\tau)$, the $n \times n$ user-user matrix $B_{U-U}(t_\tau) = \{b_{k,h}(t_\tau)\}$, and a subset of individuals $U'{}_{DT}(t_\tau) \subset U_{DT}(t_\tau)$ that have similar opinion vectors at any time step t_i, $t_i \geq t_\tau$, i.e. $b_{u2u1} \leq \varepsilon_1$, $\forall u_2, u_1 \in U'{}_{DT}(t_\tau)$, and that initiated the opinion propagation at the time step t_τ, then a user $u \in U_{DT}(t_j) \setminus U'{}_{DT}(t_\tau)$ is considered to be influenced by the opinion propagation at the time step t_j only if this condition is met:*

$$b_{u3,u}(t_\tau) \leq b_{u1,u}(t_j) + \varepsilon_2 \qquad (2)$$

where $u_3 \in U_{DT}(t_j) \setminus U'_{DT}(t_r)$, $\forall u_l \in U'_{DT}(t_r)$, and the parameters $\varepsilon_1, \varepsilon_2$ can be learned from a corpus or heuristically set.

The inequality (2) indicates that, if the user $u \in U_{DT}(t_j) \setminus U'_{DT}(t_r)$ is influenced at the time step t_j, $t_j > t_r$, due to an opinion propagation initiated by a particular subset of users $U'_{DT}(t_r)$ at the time step t_r, then the opinion of the user u should be similar to the opinion of the subset of users $U'_{DT}(t_r)$, irrespective of the opinions of other users $U_{DT}(t_j) \setminus U'_{DT}(t_r)$ in the discussion thread.

4   Proposed System

The system that implements our method consists of several modules, grouped as shown in Fig. 1. First, using the Stanford CoreNLP tool (http://nlp.stanford. edu/software/corenlp.shtml), specific natural language processing techniques are applied to each post in the discussion thread [14]: tokenization, part of speech tagging, syntactic parsing, and coreference resolution.

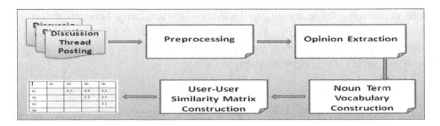

Fig. 1. General architecture of the user-level opinion propagation problem in a discussion forum thread

In what follows, we will consider that opinion words can be used as parts of speech such as adjectives, adverbs, and verbs. In order to identify opinion words from posts, the relevant Stanford dependency relations have been utilized [15]: "dobj", "nsubj", "amod", "acomp", "advmod", "xcomp", and "neg". From an opinion mining perspective, these dependency relations show the relations between noun terms and the words that contain opinions about the noun terms.

In order to construct the noun term vocabulary for the discussion thread at the time step t_r, we use Algorithm 1. The first step in Algorithm 1 extracts pairs (noun_term, opinion_word) from the discussion thread, by using the dependency relations considered for opinion mining (A1 : 1). Then, we apply a method to identify the nouns terms in the subject of the discussion thread (A1 : 2-9). Finally, we remove the noun terms that are not relevant to the discussion thread (A1 : 10-15) from the noun terms previously identified by the algorithm (A1 : 1). For this purpose, we use the Wu-Palmer measure [16] to calculate the similarity between two nouns. The noun terms for which the Wu-Palmer measure returns a score higher than zero will form the noun term vocabulary at the time step t_r.

Algorithm 1 (A1): Noun-Term Vocabulary Construction

Input: $P_{DT}(t_\tau) = \{p_\tau\}_{\tau \in N}$ – the set of posts in the discussion thread at the time step t_τ;
Input: $S_{DT} = \{w_k\}_{k \in N}$ – the subject of the discussion thread;
Output: $V^d_{DT}(t_\tau)$ – the noun term vocabulary of the discussion thread at the time step t_τ;

```
 1:  Ω        ←        MiningDependencyRelationsfromDiscussionThread
(P_DT(t_τ))// Ω = {(h_m, d_n)}_{m,n∈N}
 2:  for each word w_k in S_DT do
 3:      if (checkNoun(w_k)  and  length(w_k)  > 3) then
 4:          w_k ← lemmatization(w_k)
 5:          w_k ← lowercase(w_k)
 6:      else
 7:          S_DT ← S_DT \ w_k
 8:      end if
 9:  end for
10:  for each pair (h_m, d_n) in Ω do
```

11: $\quad sim \leftarrow \sum_{w_k \in S_{DTl}} (sim_{wu}(h_m, w_k))$

```
12:  if sim != 0 and h_m ∉ V^d_DT(t_τ) then
13:      V^d_DT(t_τ) ← V^d_DT(t_τ) ∪ (h_m)
14:  end if
15:  end for
```

To identify the orientation of the opinion words, we selected four opinion lexicons, which will be used in our experiments: SentiWordNet [17], Micro-WNOp [18], MPQA subjectivity lexicon [19], and Bing Liu's opinion lexicon [20]. The reason for choosing to test by using several opinion lexicons is that there is no opinion lexicon to return the exact sentiments of the opinion words, taking into account both the context of usage of the opinion words and the domain in which they are used. In the user-level opinion propagation problem that we study, the individuals' opinions are modeled by numerical values. After extracting the opinion words from posts, it becomes necessary to map each opinion word onto a numerical value: -1, 0, or +1, where a value of -1 corresponds to an opinion word of negative polarity, a value of 0 corresponds to an opinion word of neutral polarity, and a value of +1 corresponds to an opinion word of positive polarity. The opinion words from the SentiWordNet and Micro-WNOp opinion lexicons are assigned sentiment scores in a certain domain. In other opinion lexicons (MPQA subjectivity lexicon and Bing Liu's opinion lexicon), the opinion words are not assigned sentiment scores. We use specific algorithms for each opinion lexicon in order to map the opinion words onto numerical values. Finally, we construct the opinion vector for each participant in the discussion and also the user-user similarity matrix B_{U-U}.

5   Experiments

We evaluate our method on real-world discussion threads from the Internet Argument Corpus (IAC) created by Walker et al. [21]. This corpus contains discussion threads from the discussion forum website http://www.4forums.com. The IAC is a dataset freely available, in which each discussion thread is saved in the JSON format. The discussion threads from the IAC have structure characteristics as presented in section 3. Moreover, the data selected in this corpus for each discussion thread enable us to create the corresponding post-reply opinion graph. Each discussion thread comprises a set of posts, and each post in a discussion thread contains several components as follows: the author of the post, the post text, the parent post, and the timestamp. Moreover, each discussion thread is created under a topic.

Without losing the generality of our problem stated in Definition 7, in what follows we study the case in which $t_\tau = t_1$ and the set $U'_{DT}(t_1) = \{u_1\}$. More precisely, we consider that the opinion propagation process is initiated, at the time step t_1, by the first user (called u_1) that sends a post to the respective discussion thread. The target individuals of the opinion propagation process are all users that intervene in the discussion afterward, at the time step $t_j > t_{\tau i}$, and read the first user's posts. If, at the time step $t_j > t_\tau$, the equation $1 - \varepsilon \leq b_{u,u1}(t_j) \leq 1$ is true for all users u in the set $U_{DT}(t_j) \setminus \{u_1\}$, then the opinion propagation process is characterized by complete agreement between the users that participated in the discussion thread.

We selected two discussion threads from the topic "Existence of God" of the Internet Argument Corpus. We consider that the structure of the two discussion threads corresponds to three discrete time steps. Table 1 shows both the discrete time steps t_j and the number of users $U_{DT}(t_j)$ who sent posts to each discussion thread until the time step t_j.

Table 1. Statistics on the experimental corpus

Discussion Thread Identifier	Discrete Time Steps (t_j)	Number of Users ($U_{DT}(t_j)$)
Discussion Thread 1	t_{49}	7
	t_{100}	14
	t_{149}	15
Discussion Thread 2	t_{60}	18
	t_{121}	19
	t_{192}	20

Figure 2 shows the similarity between the opinion vector corresponding to the user u_1 and the opinion vectors corresponding to other users that intervene in the discussion afterward, for both discussion threads and for each of the four opinion lexicons. The other users are represented on the X-axis. More precisely, these similarity values correspond to the first line in the user-user similarity matrix B_{U-U}. In Fig. 2 (a-c), we can observe that the opinion vector similarities between the users u_1 and u_5 decrease over time. In Fig. 2 (d-f), we can observe that the opinion vector similarities between the users u_1 and u_4 decrease over time, and the opinion vector similarities between the users u_1 and u_{12} increase over time.

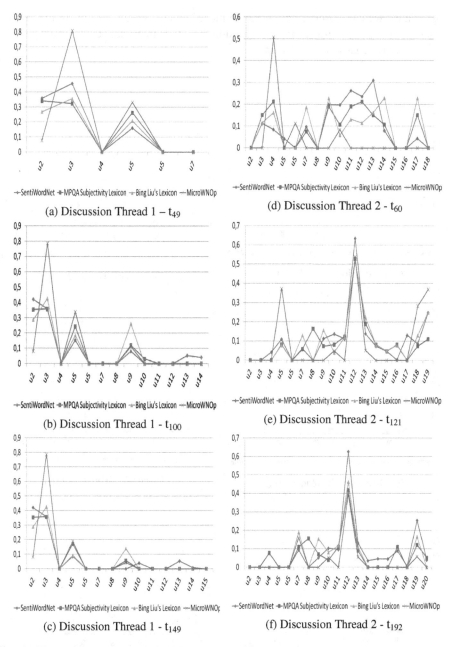

Fig. 2. The opinion vector similarities corresponding to the user who has initiated the discussion and to the users who join the discussion over time in the discussion thread 1(a-c) and in the discussion thread 2 (d-f)

In Fig. 3, we graphically show all the similarity values in the user-user similarity matrix $B_{U\text{-}U}$ for both discussion threads and for each of the four opinion lexicons. In

Fig. 3 (a-c), we can observe that there are more values higher than zero in the matrix B_{U-U} over time. In contrast, in the discussion thread 2, there are fewer values higher than zero in the matrix B_{U-U} over time.

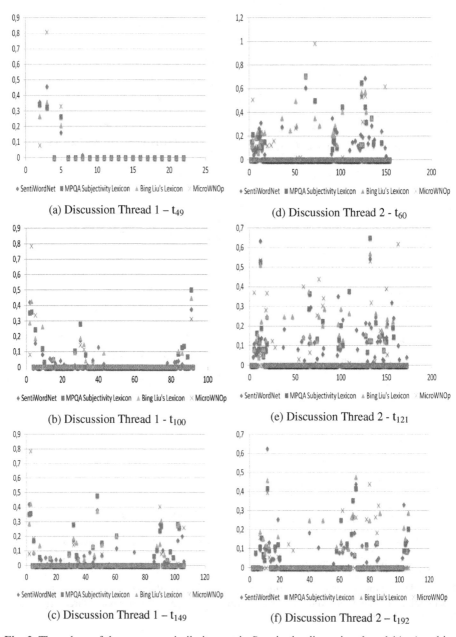

Fig. 3. The values of the user-user similarity matrix B_{U-U} in the discussion thread 1(a-c) and in the discussion thread 2 (d-f) over time

6   Conclusions

Quantifying the propagation of opinions, either in online social networks or discussion forum threads, is a challenging research topic. In recent years, the opinion propagation in social media has become a widespread phenomenon. There are multiple applications of this phenomenon such as viral marketing and election campaigns, and thus the detection of the opinion propagation is a current research problem. In this paper, we present a method for studying the propagation of opinions in discussion forum threads. The analysis is performed at user level. In discussion forum threads, many changes may occur over time since new users intervene in the discussion and opinions about the pros and cons of the same noun term are formed. Our method differs from the network-based models of opinion propagation proposed for online social networks, in the way of reflecting the discussion orientation, i.e. the posts are triggered by previous post(s). We tested the method proposed by us for the opinion propagation problem on real-world forum threads.

References

1. Kim, S.-M., Hovy, E.: Determining the sentiment of opinions. In: Proceedings of the 20th International Conference on Computational Linguistics, p. 1367. Association for Computational Linguistics, Geneva (2004)
2. Fushimi, T., Saito, K., Kimura, M., Motoda, H., Ohara, K.: Finding Relation between PageRank and Voter Model. In: Kang, B.-H., Richards, D. (eds.) PKAW 2010. LNCS, vol. 6232, pp. 208–222. Springer, Heidelberg (2010)
3. Watts, D.J., Strogatz, S.H.: Collective dynamics of small-world networks. Nature 393(6684), 440–442 (1998)
4. Barabási, A.-L., Albert, R.: Emergence of Scaling in Random Networks. Science 286(5439), 509–512 (1999)
5. Mislove, A., et al.: Measurement and analysis of online social networks. In: Proceedings of the 7th ACM SIGCOMM Conference on Internet Measurement, pp. 29–42. ACM, San Diego (2007)
6. Java, A., Song, X., Finin, T., Tseng, B.: Why We Twitter: An Analysis of a Microblogging Community. In: Zhang, H., Spiliopoulou, M., Mobasher, B., Giles, C.L., McCallum, A., Nasraoui, O., Srivastava, J., Yen, J. (eds.) WebKDD 2007. LNCS, vol. 5439, pp. 118–138. Springer, Heidelberg (2009)
7. Sznajd, J., Sznajd-Weron, K.: Opinion evolution in closed community. International Journal of Modern Physics C 11(06), 1157–1165 (2000)
8. da F. Costa, L., Rodrigues, F.A.: Surviving opinions in Sznajd models on complex networks. International Journal of Modern Physics C 16(11), 1785–1792 (2005)
9. Deffuant, G., et al.: Mixing beliefs among interacting agents. Advances in Complex Systems 3(1-4), 87–98 (2000)
10. Hegselmann, R., Krause, U.: Opinion Dynamics and Bounded Confidence, Models, Analysis and Simulation. Journal of Artificial Societies and Social Simulation 5(3), 2 (2002)
11. Cercel, D.-C., Trausan-Matu, S.: Opinion Propagation in Online Social Networks: A Survey. In: Proceedings of the 4th International Conference on Web Intelligence, Mining and Semantics (WIMS 2014), pp. 1–10. ACM, Thessaloniki (2014)

12. Stavrianou, A.: Modeling and mining of web discussions, University of Lyon, France, PhD Thesis (2010)
13. Zafarani, R., Cole, W.D., Liu, H.: Sentiment propagation in social networks: A case study in liveJournal. In: Chai, S.-K., Salerno, J.J., Mabry, P.L. (eds.) SBP 2010. LNCS, vol. 6007, pp. 413–420. Springer, Heidelberg (2010)
14. Manning, C.D., Schutze, H.: Foundations of statistical natural language processing, p. 680. MIT Press (1999)
15. Somprasertsri, G., Lalitrojwong, P.: Mining Feature-Opinion in Online Customer Reviews for Opinion Summarization (2010)
16. Wu, Z., Palmer, M.: Verbs semantics and lexical selection. In: Proceedings of the 32nd Annual Meeting on Association for Computational Linguistics, pp. 133–138. Association for Computational Linguistics, Las Cruces (1994)
17. Baccianella, A.E.S., Sebastiani, F.: SentiWordNet 3.0: An Enhanced Lexical Resource for Sentiment Analysis and Opinion Mining. In: Proceedings of the Seventh Conference on International Language Resources and Evaluation (LREC 2010). European Language Resources Association (ELRA), Valletta (2010)
18. Cerini, S., et al.: Micro-WNOp: A gold standard for the evaluation of automatically compiled lexical resources for opinion mining. In: Sans, A. (ed.) Language Resources and Linguistic Theory: Typology, Second Language Acquisition, English Linguistics. Franco Angeli Editore (2007)
19. Wilson, T., Wiebe, J., Hoffmann, P.: Recognizing contextual polarity in phrase-level sentiment analysis. In: Proceedings of the Conference on Human Language Technology and Empirical Methods in Natural Language Processing, pp. 347–354. Association for Computational Linguistics, Vancouver (2005)
20. Hu, M., Liu, B.: Mining and summarizing customer reviews. In: Proceedings of the Tenth ACM SIGKDD International Conference on Knowledge Discovery and Data Mining, pp. 168–177. ACM, Seattle (2004)
21. Walker, M.A., et al.: A Corpus for Research on Deliberation and Debate. In: LREC. European Language Resources Association, ELRA (2012)

Social News Feed Recommender

Milen Chechev and Ivan Koychev

Sofia University "Kliment Ohridski", Bulgaria
{milen.chechev,koychev}@fmi.uni-sofia.bg

Abstract. This paper presents research on social news recommendation at the biggest social network Facebook. The recommendation strategies which are used are based on content and social trust as the trust is selected as more reliable for recommendation. In order the news to get old in time a decay factor for the score is proposed. Both offline and online evaluation are made as the feedbacks shows that users find the application interesting and useful.

Keywords: recommender systems, social networks, news recommender.

1   Introduction

Recommender systems and social networks are two of the fastest growing areas in the information society at the beginning of 21 century. Recommender systems shows their strengths in helping users at online stores as Amazon[1] and more and more web sites start to use them to provide better user experience. Social networks and News web sites heavily exploit recommender system strategies on order to understand user's needs and to suggest proper content.

Facebook is the biggest social network as at the beginning of 2014 it already have more than 1.3 billion users, number which is increased with 22% last year[5]. The increased popularity of the social network is caused of the simplicity of connecting with people and sharing different e-content with them. However user's friend's publications create a news stream which can contain large amount of news, larger than user can manage. This is the place where recommender system should help to identify interesting publications. It ranks the interesting news at the beginning of the list where they can be easily accessed by the user.

In [1] authors investigate if the users need additional tool for managing their news feed. They asked 114 Facebook users about their habits and needs at the social network. Users provided valuable feedback and identify the need of a custom news feed application, which provides options for information filtering/retrieval and recommendation. Users want more flexible ways to manage their news feed and better explanation of the automation scoring of their news feed.

This paper continues the research at [1] and present custom news feed application at Facebook. Different recommendation approaches are applied as the best one is selected for online experiments.

---

[1] http://amazon.com/

G. Agre et al. (Eds.): AIMSA 2014, LNAI 8722, pp. 37–46, 2014.

The paper has the following structure: section 2 present related work, section 3 present the recommendation approaches which are applied and describe a decay factor for the news score, section 4 present both offline and online evaluation on the system and section 5 gives conclusions and future work.

2   Related Work

Recommender systems can be classified to 6 types – content based, collaborative filtering, knowledge based, demographic, community based and hybrid recommendation systems [2]. Research shows that people prefer to receive recommendations from friends rather than from people which are close to their profile [3]. The community based recommenders are often described as trust based recommenders. Authors at [4] show good survey of the different trust approaches and domains for recommendation. The trust at the social network can be two types – explicit and implicit. Explicit trust is when the user explicitly provide list of trusted users. Implicit trust is calculated from the user activities. Examples of trust network with explicit trust are the review's network Epinions [18] and Golberk's FilmTrust [7]. Massa and Avesani [9] study the dissemination of the trust and distrust at the social networks. Their research shows that the recommendation strategies based on trust provide good results even when the information is insufficient for the others recommendation approaches.

The Social network Facebook can also be considered as network of trust, but the trust showed there is implicit and have to be aggregated from the user's activity and collaboration. There are already several recommendation options integrated at the social network. The news feed is offered at two views – "Most Recent news" or "Top News". First view shows the news sorted on their publication time. It's comfortable for users which often read their news stream otherwise Facebook suggest the view "Top News". It sort the news based on several factors like how often the user use the social network, what are the relations between the user and the user who has published the news, etc. [1].

Authors at [8] make analysis on recommendation at social networks and show that the main features that can be used are users shared preferences, trust, reputation and tie strength. They also describing the approach for combining this features but it's not clear how exactly the trust is aggregated. There is also some research at recommending news at Facebook without using the notion of trust. Authors at [9] show Facebook news recommender system which recommend news to the user with the use of content based or collaborative filtering. Authors at [13] present custom Facebook application which recommend news based on collaborative filtering integration and social features.

3   Social News Feed Recommender

Facebook users identify several problems at the current news feed reader as the lack of search options and options for manual refinement of the recommendations [1]. In order to face these problems we have created Facebook application which harvest news from user's friends and provide them in custom order. There are two main challenges at news recommendation – calculation of the rank of the news and decay the rank when the news gets older.

3.1   Content Based Recommendation

As part of the research content based recommendation is evaluated in order to calculate the similarity of the news to the user. Content based filtering works at Vector Space model[15] as the news are presented with vector of their keywords. The user profile is learned from the keywords of the news which the user liked, commented or published. All keywords are evaluated with the tf-idf metric which gives bigger score to rare words at all documents and frequent at a document. We are using Roccio algorithm [14] to learn user profile. Once the profile is learned we calculate similarity between user's vectors and object's vectors with cosine similarity (formula 1)

$$similarity(d_a, u) = \frac{d_a \cdot u}{|d_a||u|} = \frac{\sum_{i=1}^{m} d_{a,i} u_i}{\sqrt{\sum_{i=1}^{m} d_{a,i}^2} \sqrt{\sum_{i=1}^{m} u_i^2}} \qquad (1)$$

At the formula above u is the keyword vector of a user, d_a is a keyword vector for document a, m is the dimension of the vector space, $d_{a,i}$ is the tf-idf score for the keyword i at document a and u_i is the tf-idf score for the keyword i at user profile u.

3.2   Facebook Social Trust

Unlike Epinions at Facebook the average user have hundreds of connections, so we cannot use these connections as explicit. We calculate the trust according to user interactions with their friends. The interaction between users is made with the actions-like and comment of the news published by friends. The authors of [10] define the trust as symmetric variable, but in reality the trust asymmetrical, so we are defining it with formula 2.

$$trust_{ij} = \frac{\alpha.\, like_{ij} + \beta.\, comment_{ij} + \gamma.\, share_{ij}}{published_j} + \mu.\frac{activity_{ij}}{activity_i} \qquad (2)$$

At the formula we mark with $trust_{ij}$ the trust of user i at user j, $like_{ij}$ is the number of the objects published by j and liked by i, respectively $comment_{ij}$ and $share_{ij}$ are objects published by j and commented or shared by i, $published_j$ is the number of objects published by j, $activity_i$ is the number of objects which user i interact with, $activity_{ij}$ is the number of objects which are liked or commented by both user i and j. $\alpha, \beta, \gamma, \mu$ are parameters which are selected by gradient descent algorithm in order to optimize the trust to the liked objects at the train set. For calculation of trust of user i at object x is used formula 3 where we aggregate the trust of user i to the author of object x and the trust to all users which interact with x. This value is then normalized with the maximum value of trust which is calculated as sum of the trust of user i to all its friends (formula 4).

$$trust_{ix} = \frac{trust_{autor_x} + \sum_{j \in \{users\ interacted\ with\ x\}} trust_{ij}}{maxtrust_i} \tag{3}$$

$$maxtrust_i = \sum_{j \in friends_i} trust_{ij} \tag{4}$$

3.3   Hybridization

As part of the research hybrid method is created to combine content and trust based recommendations. The combination is made with formula 5:

$$score_{ix} = \lambda trust_{ix} + (1 - \lambda)similarity_{ix} \tag{5}$$

The expected effect from this combination is that the hybrid algorithm have to take the strengths from both the content and trust preferences of the user and to provide relevant objects which are close to what user already interact with and also to be able to provide new content which is completely new but is published by the user trust sources. The parameter λ depend on the concrete case as for the experiment the value is selected with gradient descent algorithm.

3.4   Time Factor

When the domain of the recommendation is news we always have to deal with the problem that items are getting old and their score have to be decreased in time. Google News[2], Reddit[3] and Hacker news[4] are great examples of news recommender sites. There isn't official information about the recommendation approach for Google News, but there are sources with the formulas which Hacker news and Reddit use [6,11,12]. Both of them use up-votes and down-votes to calculate the score of the news and after that the score is decreased in time as Reddit used formula 6 where U is the number of up-votes, D is the number of down votes and t_{post} is the published time from fixed timestamp.

$$log_{10} \max(1, |U - D|) + \frac{(U-D)t_{post}}{45000} \tag{6}$$

Hacker news use different formula as in it the penalty term P is introduced which help in decreasing the score of vote pools and other news which the system identify as not important. Hacker news used formula 7 as t_{now} is the time in which the score is calculated.

---

[2] http://news.google.com/
[3] http://reddit.com/
[4] http://news.ycombinator.com/

$$\frac{(U - D - 1)^{\alpha}}{(t_{now} - t_{post})^{\gamma}} P \tag{7}$$

In our case we are taking objects from Facebook news feed and we have to take into account 2 different times – published time and last interaction time. The post which is already old can be still interesting because of new interaction with it – new likes or comments. We want the score of the news to decrease in time and if there is no interaction to the news the score to decrease faster.

We are using formula (8) where $time_{updated}$ is the time of last friend's interaction with the post and score is the score which is calculated for the post by some of the recommendation strategies above. In the formula we treat differently the published date and the last interaction date, as the last interaction date cause bigger decay for the news. The idea is that if the post is old, but there is interaction with it the score is decay slowly, but when the interaction to the post halts the score start to decrease promptly.

$$score\left(\frac{1}{2(t_{now} - t_{post} + 1)} + \frac{1}{2(t_{now} - t_{updated} + 1)^{2}}\right) \tag{8}$$

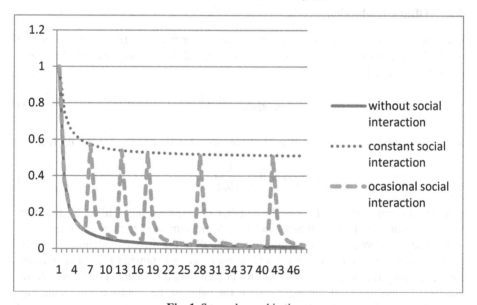

Fig. 1. Score changed in time

The time is calculated at hours so the score of the object is decreased hourly. The interaction with the object is crucial for his score as if the object receive continuous interaction it will receive good score even if the published date is old. Figure 1 show how the score is changed in time as X-axis is the time and Y-axis shows what part of the score will be used. Figure 1 shows that if there is no interaction with a news its score decrease with 90% for 6 hours, but if there was an constant interaction it score will be decreased only with 40% in addition if the interaction with the post is occasional the score is increased when the interaction is made but without continuous interaction few hours later its value is decreased close to the curve with no interaction.

4   Evaluation

The algorithms were evaluated with both offline and online experiments. The offline experiments are evaluated on dataset contains historical information about 25 Facebook users and their 6548 friends as for each of the users we have access through the Facebook API to the last 1000 objects liked by them and around 300 posts for each of user's friends. The data set is split to training and testing dataset as the training set contains oldest 90% of user's likes and the testing set contains the newest 10% of the likes.

The online evaluation is made with the same users as the real time recommendations are given and the interaction with them was measured. In addition users answer to several questions about the recommendations which they have received.

4.1   Offline Evaluation

For evaluation we use well known metrics from information retrieval– precision and recall. The precision is the fraction of news retrieved that are marked with like from the user (formula 9).

$$precision = \frac{|D_{sr}|}{|D_s|} \tag{9}$$

Recall is the fraction of news that are liked by user that are retrieved (formula 10).

$$recall = \frac{|D_{sr}|}{|D_r|} \tag{10}$$

In the formulas above $|D_{sr}|$ is the number of retrieved news liked by the user, $|D_r|$ is the number of all liked news and $|D_s|$ is the number of the retrieved news.

Evaluation of precision and recall is made for all of the recommendation strategies, as because we are calculating it on several users we are showing the average value. Figure 2 shows the precision and recall for content based filtering. The X-axis shows number of news retrieved and the Y-axis is the value of the precision and recall. Figure 2 show that when 100 objects are retrieved we have 0.1 for both precision and recall. When the number of the results is increased the precision dropped below 0.05 but the recall reach only 0.33. The results shows that the content of the news is not so predictive for the user's preferences at the social network.

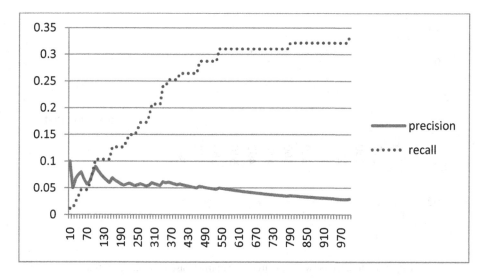

Fig. 2. Precision and recall for content based recommendation

Next figure 3 shows the precision and recall for the Trust based recommendation. The results are definitely better than the results from content based filtering as when recommending 30 objects the precision is 0.8, for 100 objects we have 0.74 recall and 0.65 precision as the recall is increased quickly to 0.86 for 140 objects.

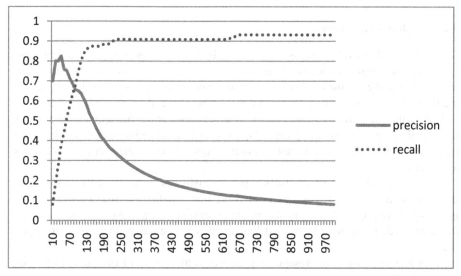

Fig. 3. Precision and recall for trust based recommendation

The hybrid method doesn't show much improvement than the trust based method so for better visualization figure 4 present the difference between the hybrid method and the trust based method.

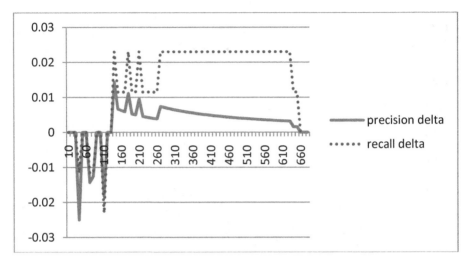

Fig. 4. Difference between the hybrid and trust based recommendations

Results shows that the precision of the hybrid method is lower than the precision of trust based method for the first 100 news, but after that there is a constant improvement. This results speak that its better if we make the recommendations for the first items based only on trust and to combine it with content based recommendations only to improve the news order at the lower part of the news list.

4.2   Online Evaluation

During the online evaluation users received 100 news based on their current news feed. The users can use the application to read the news to comment, liked and share news. Despite the offline evaluation is made without taking into account the publishing time of the news at the online evaluation we are using also and the time factor.

Before the evaluation all users received instructions about the application functionality and recommendation approach. Options for manual refinement of the trust metrics are also provided and presented to users. Figure 5 shows the user interface of the application.

The evaluation group of 25 users express satisfaction from the application as all they reviewed all presented news and used the functionality to write comments, liked news and share post. All of them also used the presented functionality for searching at the news feed. 13 of them have used functionality of manual refinement of user profile.

After test users tried the system they answered to a series of questions to provide explicit feedback for the system. They generally liked the application although 5 users recommend better design for it. The main features which users liked are the search functionality and the options for manual refinement of the profile although there were some suggestions about improving the usability of the component for manual refinement of the users trust.

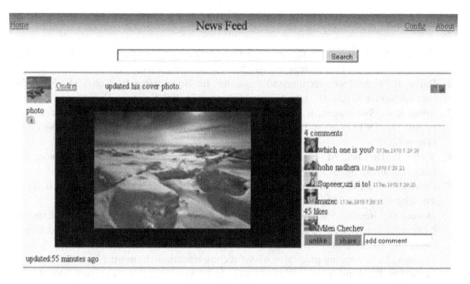

Fig. 5. News Feed Application

5   Conclusions and Future Work

Presented research showed a comparison of content based and trust based recommendation at the social network Facebook. The content based filtering provides quite poor results as the content of the news seems to be insignificant when describing user's interest. User's preferences are heavily influenced by their friend's publications as the content of the information is not as important as the users which interact with it. New formulas for calculating trust at social network Facebook and decreasing the score in time was introduced as results shows that both gives decent results and satisfy user needs. The online evaluation shows that users generally liked the prototype application. The application provides options for explicit feedback to the system through manual refinement of user profile which can is used for better recommendations.

Future research will be focused on more extensive online evaluation and experiments with different options for user interaction as providing options for immediate feedback to the system for good or bad recommendation, gathering implicit data about the time which user spend on the different posts, etc. This information is going be used for an automatic refinement of the results and user profile which would provide different experience to the user.

Acknowledgements. This work was supported by the European Social Fund through the Human Resource Development Operational Program under contract BG051PO001-3.3.06-0052 (2012/2014).

References

1. Chechev, M., Koychev, I.: Recommendations in Social Networks: an Extra Feature or an Essential Need. In: Proceedings of MIE 2013, Sofia, Bulgaria (2013)
2. Burke, R.: Hybrid web recommender systems. In: Brusilovsky, P., Kobsa, A., Nejdl, W. (eds.) Adaptive Web 2007. LNCS, vol. 4321, pp. 377–408. Springer, Heidelberg (2007)
3. Sinha, R.R., Swearingen, K.: Comparing Recommendations Made by Online Systems and Friends. In: DELOS Workshop: Personalisation and Recommender Systems in Digital Libraries (2001)
4. Victor, P., De Cock, M., Cornelis, C.: Trust and recommendations. In: Recommender Systems Handbook, pp. 645–675. Springer US (2011)
5. Facebook Statistics. (January 1, 2014), http://www.statisticbrain.com/facebook-statistics/ (visited May 16, 2014)
6. Dover, D.: Reddit, Stumbleupon, Del.icio.us and Hacker News Algorithms Exposed! http://moz.com/blog/reddit-stumbleupon-delicious-and-hacker-news-algorithms-exposed (visited May 16, 2014)
7. Golbeck, J.: Generating predictive movie recommendations from trust in social networks. In: Stølen, K., Winsborough, W.H., Martinelli, F., Massacci, F. (eds.) iTrust 2006. LNCS, vol. 3986, pp. 93–104. Springer, Heidelberg (2006)
8. Arazy, O., Kumar, N., Shapira, B.: Improving Social Recommender Systems. IT Professional 11(4), 38–44 (2009)
9. Agrawal, M., Karimzadehgan, M., Zhai, C.: An online news recommender system for social networks. Urbana 51, 61801 (2009)
10. Chen, W., Fong, S.: Social Network Collaborative Filtering Framework and Online Trust Factors: a Case Study on Facebook. International Journal of Web Applications 3 (November 2011)
11. Salihefendic, A.: How Hacker News ranking algorithm works, http://amix.dk/blog/post/19574 (visited May 16, 2014)
12. Salihefendic, A.: How Reddit ranking algorithms work, http://amix.dk/blog/post/19588 (visited May 16, 2014)
13. Noel, J., Sanner, S., Tran, K.N., Christen, P., Xie, L., Bonilla, E.V., . . .Della Penna, N.: New objective functions for social collaborative filtering. In: Proceedings of the 21st International Conference on World Wide Web, pp. 859–868. ACM (2012)
14. Rocchio, J.: Relevance Feedback Information Retrieval. In: The SMART Retrieval System - Experiments in Automated Document Processing. Prentice-Hall (1971)
15. Salton, G., Wong, A., Yang, C.: A vector space model for automatic indexing. Communications of the ACM (1975)

Boolean Matrix Factorisation for Collaborative Filtering: An FCA-Based Approach

Dmitry I. Ignatov[1], Elena Nenova[2],
Natalia Konstantinova[3], and Andrey V. Konstantinov[1]

[1] National Research University Higher School of Economics, Moscow, Russia
dignatov@hse.ru
[2] Imhonet, Moscow, Russia
[3] University of Wolverhampton, UK

Abstract. We propose a new approach for Collaborative filtering which is based on Boolean Matrix Factorisation (BMF) and Formal Concept Analysis. In a series of experiments on real data (MovieLens dataset) we compare the approach with an SVD-based one in terms of Mean Average Error (MAE). One of the experimental consequences is that it is enough to have a binary-scaled rating data to obtain almost the same quality in terms of MAE by BMF as for the SVD-based algorithm in case of non-scaled data.

Keywords: Boolean Matrix Factorisation, Formal Concept Analysis, Singular Value Decomposition, Recommender Algorithms.

1 Introduction

Recommender Systems have recently become one of the most popular subareas of Machine Learning and Data Mining. In fact, the recommender algorithms based on matrix factorisation techniques (MF) are now considered industry standard [1].

Among the most frequently used types of Matrix Factorisation we should definitely mention Singular Value Decomposition (SVD) [2] and its various modifications like Probabilistic Latent Semantic Analysis (PLSA) [3] and SVD++ [4]. However, the existing similar techniques, for example, non-negative matrix factorisation (NMF) [5] and Boolean matrix factorisation (BMF) [6], seem to be less studied in the context of Recommender Systems. An approach similar to the matrix factorization is biclustering which was also successfully applied in recommender system domain [7,8]. For example, Formal Concept Analysis [9] can be also used as a biclustering technique and there are already several examples of its applications in recommenders' algorithms [10,11].

The aim of this paper is to compare the recommendation quality of the aforementioned techniques on the real datasets and try to investigate methods' interrelationship. It is especially interesting to conduct experiments and compare recommendation quality in case of an input matrix with numeric values and in case of a Boolean matrix in terms of Precision and Recall as well as MAE.

G. Agre et al. (Eds.): AIMSA 2014, LNAI 8722, pp. 47–58, 2014.

Moreover, one of the useful properties of matrix factorisation is its ability to keep reliable recommendation quality even in case of dropping some insufficient factors. For BMF this issue is experimentally investigated in section 4.

The novelty of the paper is defined by the fact that it is the first time when BMF based on Formal Concept Analysis [9] is investigated in the context of Recommender Systems.

The practical significance of the paper is determined by the demand of recommender systems' industry, that is focused on gaining reliable quality in terms of Mean Average Error (MAE), Precision and Recall as well as competitive time performance of the investigated method.

The rest of the paper consists of five sections. Section 2 is an introductory review of the existing MF-based recommender approaches. In section 3 we describe our recommender algorithm which is based on Boolean matrix factorisation using closed sets of users and items (that is FCA). Section 4 contains methodology of our experiments and results of experimental comparison of two MF-based recommender algorithms by means of cross-validation in terms of MAE and F-measure. The last section concludes the paper.

2   Introductory Review of Some Matrix Factorisation Approaches

In this section we briefly describe two approaches to the decomposition of both real-valued and Boolean matrices.

2.1   Singular Value Decomposition (SVD)

Singular Value Decomposition (SVD) is a decomposition of a rectangular matrix $A \in \mathbb{R}^{m \times n} (m > n)$ into a product of three matrices

$$A = U \begin{pmatrix} \Sigma \\ 0 \end{pmatrix} V^T, \tag{1}$$

where $U \in \mathbb{R}^{m \times m}$ and $V \in \mathbb{R}^{n \times n}$ are orthogonal matrices, and $\Sigma \in \mathbb{R}^{n \times n}$ is a diagonal matrix such that $\Sigma = diag(\sigma_1, \ldots, \sigma_n)$ and $\sigma_1 \geq \sigma_2 \geq \ldots \geq \sigma_n \geq 0$. The columns of the matrix U and V are called singular vectors, and the numbers σ_i are singular values [2].

In the context of recommendation systems rows of U and V can be interpreted as vectors of user's and items's loyalty (attitude) to a certain topic (factor), and the corresponding singular values as importance of the topic among the others. The main disadvantage lies in the fact that the matrix may contain both positive and negative numbers; the last ones are difficult to interpret.

The advantage of SVD for recommendation systems is that this method allows to obtain a vector of user's loyalty to certain topics for a new user without SVD decomposition of the whole matrix.

The evaluation of computational complexity of SVD according to [12] is $O(mn^2)$ floating-point operations if $m \geq n$ or more precisely $2mn^2 + 2n^3$.

2.2 Boolean Matrix Factorisation (BMF) Based on Formal Concept Analysis (FCA)

Basic FCA Definitions Formal Concept Analysis (FCA) is a branch of applied mathematics and it studies (formal) concepts and their hierarchy [9]. The adjective "formal" indicates a strict mathematical definition of a pair of sets, called, the extent and intent. This formalisation is possible because of the use of the algebraic lattice theory.

DEFINITION 1. *Formal context* K is a triple (G, M, I), where G is a set of *objects*, M is a set of *attributes* , $I \subseteq G \times M$ is a binary relation.

The binary relation I is interpreted as follows: for $g \in G$, $m \in M$ we write gIm if the object g has the attribute m.

For a formal context $\mathbb{K} = (G, M, I)$ and any $A \subseteq G$ and $B \subseteq M$ a pair of mappings is defined:

$$A' = \{m \in M \mid gIm \ \text{ for all } g \in A\}, \quad B' = \{g \in G \mid gIm \ \text{ for all } m \in B\},$$

these mappings define Galois connection between partially ordered sets $(2^G, \subseteq)$ and $(2^M, \subseteq)$ on disjunctive union of G and M. The set A is called *closed set*, if $A'' = A$ [13].

DEFINITION 2. A *formal concept* of the formal context $\mathbb{K} = (G, M, I)$ is a pair (A, B), where $A \subseteq G$, $B \subseteq M$, $A' = B$ and $B' = A$. The set A is called the *extent*, and B is the *intent* of the formal concept (A, B).

It is evident that the extent and intent of any formal concept are closed sets. The set of all formal concepts of a context $\mathbb{K}$ is denoted by $\mathfrak{B}(G, M, I)$.

The state-of-the-art surveys on advances in FCA theory and its applications can be found in [14,15].

Description of FCA-Based BMF. Boolean matrix factorization (BMF) is a decomposition of the original matrix $I \in \{0, 1\}^{n \times m}$, where $I_{ij} \in \{0, 1\}$, into a Boolean matrix product $P \circ Q$ of binary matrices $P \in \{0, 1\}^{n \times k}$ and $Q \in \{0, 1\}^{k \times m}$ for the smallest possible number of k. We define boolean matrix product as follows:

$$(P \circ Q)_{ij} = \bigvee_{l=1}^{k} P_{il} \cdot Q_{lj},$$

where $\bigvee$ denotes disjunction, and $\cdot$ conjunction.

Matrix I can be considered a matrix of binary relations between set X of objects (users), and a set Y of attributes (items that users have evaluated). We assume that xIy iff the user x evaluated object y. The triple (X, Y, I) clearly forms a formal context.

Consider a set $\mathcal{F} \subseteq \mathcal{B}(X, Y, I)$, a subset of all formal concepts of context (X, Y, I), and introduce matrices $P_{\mathcal{F}}$ and $Q_{\mathcal{F}}$:

$$(P_{\mathcal{F}})_{il} = \begin{cases} 1, i \in A_l, \\ 0, i \notin A_l, \end{cases} \quad (Q_{\mathcal{F}})_{lj} = \begin{cases} 1, j \in B_l, \\ 0, j \notin B_l. \end{cases}$$

where (A_l, B_l) is a formal concept from F. We can consider decomposition of the matrix I into binary matrix product $P_{\mathcal{F}}$ and $Q_{\mathcal{F}}$ as described above. The following theorems are proved in [6]:

Theorem 1. (Universality of formal concepts as factors). For every I there is $\mathcal{F} \subseteq \mathcal{B}(X, Y, I)$, such that $I = P_{\mathcal{F}} \circ Q_{\mathcal{F}}$.

Theorem 2. (Optimality of formal concepts as factors). Let $I = P \circ Q$ for $n \times k$ and $k \times m$ binary matrices P and Q. Then there exists a $\mathcal{F} \subseteq \mathcal{B}(X, Y, I)$ of formal concepts of I such that $|\mathcal{F}| \leq k$ and for the $n \times |mathcalF|$ and $|mathcalF| \times m$ binary matrices $P_{\mathcal{F}}$ and $Q_{\mathcal{F}}$ we have $I = P_{\mathcal{F}} \circ Q_{\mathcal{F}}$.

There are several algorithms for finding $P_{\mathcal{F}}$ and $Q_{\mathcal{F}}$ by calculating formal concepts based on these theorems [6].

The algorithm we use (Algoritm 2 from [6]) avoids computation of all possible formal concepts and therefore works much faster [6]. Time estimation of the calculations in the worst case yields $O(k|G||M|^3)$, where k is the number of found factors, $|G|$ is the number of objects, $|M|$ is the number of attributes.

2.3   General Scheme of User-Based Recommendations

Once a matrix of rates is factorized we need to learn how to compute recommendations for users and to evaluate whether a particular method handles this task well.

For the factorized matrices already well-known algorithm based on the similarity of users can be applied, where for finding K nearest neighbours we use not the original matrix of ratings $A \in \mathbb{R}^{m \times n}$, but the matrix $U \in \mathbb{R}^{m \times f}$, where m is a number of users, and f is a number of factors. After the selection of K users, which are the most similar to a given user, based on the factors that are peculiar to them, it is possible, based on collaborative filtering formulas to calculate the projected rates for a given user.

After generation of recommendations the performance of the recommender system can be estimated by measures such as Mean Absolute Error (MAE), Precision and Recall.

3   A Recommender Algorithm Using FCA-Based BMF

3.1   kNN-Based Algorithm

Collaborative recommender systems try to predict the utility (in our case rates) of items for a particular user based on the items previously rated by other users.

Memory-based algorithms make rating predictions based on the entire collection of previously rated items by the users. That is, the value of the unknown rating $r_{c,s}$ for a user c and item s is usually computed as an aggregate of the ratings of some other (usually, the K most similar) users for the same item s:

$$r_{c,s} = aggr_{c' \in \widehat{C}} r_{c',s},$$

where $\widehat{C}$ denotes a set of K users that are the most similar to user c, who have rated item s. For example, the function $aggr$ may have the following form [16]

$$r_{c,s} = k \sum_{c' \in \widehat{C}} sim(c', c) \times r_{c',s},$$

where k serves as a normalizing factor and selected as $k = 1/\sum_{c' \in \widehat{C}} sim(c, c')$.

Similarity measure between users c and c', $sim(c, c')$, is essentially an inverse distance measure and is used as a weight, i.e., the more similar users c and c' are, the more weight rating $r_{c',s}$ will carry in the prediction of $r_{c,s}$.

Similarity between two users is based on their ratings of items that both users have rated. The two most popular approaches are *correlation* and *cosine-based*.

To apply this approach in case of FCA-based BMF recommender algorithm we simply consider the user-factor matrices obtained after factorisation of the initial data as an input.

3.2 Scaling

In order to move from a matrix of ratings R where $R_{ij} \in \{0, 1, 2, 3, 4, 5\}$ to a Boolean matrix, and use the results of Boolean matrix factorisation, scaling is required. It is well known that scaling is a matter of expert interpretation of the original data. In this paper, we use binary scaling with different thresholds and compare the results in terms of MAE.

1. $I_{ij} = 1$ if $R_{ij} > 0$, else $I_{ij} = 0$ (user i rates item j).
2. $I_{ij} = 1$ if $R_{ij} > 1$, else $I_{ij} = 0$.
3. $I_{ij} = 1$ if $R_{ij} > 2$, else $I_{ij} = 0$.
4. $I_{ij} = 1$ if $R_{ij} > 3$, else $I_{ij} = 0$.

4 Experiments

To test our hypotheses and study the behavior of recommendations based on the factorisation of a ratings matrix by different methods we used MovieLens data. We used a part of the data, containing 100,000 ratings, and considered only users who have given more than 20 ratings.

The user ratings are split into two sets, a training set consisting of 80 000 ratings, and a test set consisting of 20 000 ratings. The original data matrix has the size of 943×1682, where the number of rows is the number of users and the number of columns is the number of rated movies (each movie has at least one vote).

4.1 The Number of Factors That Cover $p\%$ of Evaluations in an Input Data for SVD and BMF

The main purpose of matrix factorisation is a reduction of matrices dimensionality. Therefore we examine how the number of factors varies depending on the method of factorization, and depending on p % of the data that is covered by factorization. For BMF the coverage of a matrix is calculated as the ratio of the number of ratings covered by Boolean factorization to the total number of ratings.

$$\frac{|covered\_ratings|}{|all\_ratings|} \cdot 100\% \approx p_{BMF}\%, \tag{2}$$

For SVD we use the following formula:

$$\frac{\sum_{i=1}^{K} \sigma_i^2}{\sum \sigma_i^2} \cdot 100\% \approx p_{SVD}\%, \tag{3}$$

where K is the number of factors selected.

Table 1. Number of factors for SVD and BMF at different coverage level

p%	100%	80%	60%
SVD	943	175	67
BMF	1302	402	223

4.2   MAE-Based Recommender Quality Comparison of SVD and BMF for Various Levels of Evaluations Coverage

The main purpose of matrix factorisation is the reduction of matrices dimensionality. As a result some part of the original data remains not covered, so it was interesting to explore how the quality of recommendations changes based on different factorisations, depending on the proportion of the data covered by factors.

Two methods of matrix factorisation were considered: BMF and SVD. The fraction of data covered by factors is defined in subsections 2 and 3.

Fig. 1 shows that MAE_{SVD60}, calculated for the model based on 60% of factors, is not very different from MAE_{SVD80}, calculated for the model built for 80% factors. At the same time, for the recommendations based on a Boolean factorization covering 60% and 80% of the data respectively, it is clear that increasing the number of factors improves MAE, as shown in Fig. 2.

Table 2 shows that MAE for recommendations built on a Boolean factorisation covering 80 % of the data, for the number of neighbours less than 50, is better than the MAE for recommendations built on SVD factorization. It is also easy to see that difference of MAE_{SVD80} and MAE_{BMF80} from MAE_{all} is no more than $1-7\%$.

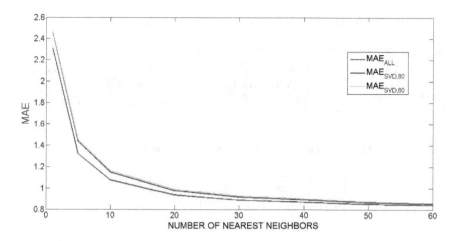

Fig. 1. MAE dependence on the percentage of the data covered by SVD-decomposition, and the number of nearest neighbours

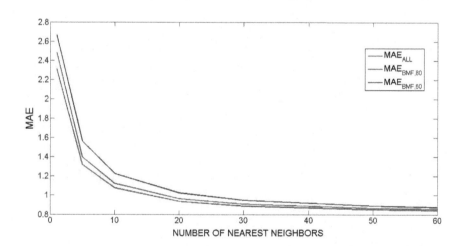

Fig. 2. MAE dependence on the percentage of the data covered by BMF-decomposition, and the number of nearest neighbours

4.3 Comparison of kNN-Based Approach and BMF by Precision and Recall

Besides comparison of the algorithms using MAE, other evaluation metrics can also be exploited, for example, $Recall = \frac{|objects\_in\_recommendation \cap objects\_in\_test|}{|objects\_in\_test|}$,

$Precision = \frac{|objects\_in\_recommendation \cap objects\_in\_test|}{|objects\_in\_recommendation|}$ and $F1 = \frac{2 \cdot Recall \cdot Precision}{Recall + Precision}$.

Usually the larger $F1$ (F-measure) is, the better is recommendation algorithm.

Figure 3 shows the dependence of the evaluation metric on the percentage of data covered by BMF-decomposition, and the number of nearest neighbours.

Table 2. MAE for SVD and BMF at 80% coverage level

Number of neighbours	1	5	10	20	30	50	60
MAE_{SVD80}	**2,4604**	1.4355	1.1479	0.9750	0.9148	0.8652	**0.8534**
MAE_{BMF80}	2.4813	**1.3960**	**1.1215**	**0.9624**	**0.9093**	**0.8650**	0.8552
MAE_{all}	2.3091	1.3185	1.0744	0.9350	0.8864	0.8509	0.8410

Table 3. The rating distribution of Movie Lens data

Rating	1	2	3	4	5
Part of all rates %	6.1	11.4	27.2	34.1	21.2

The number of objects to recommend was chosen to be 20. The figure shows that the recommendation based on the Boolean decomposition, is worse than recommendations generated on the full matrix of ratings.

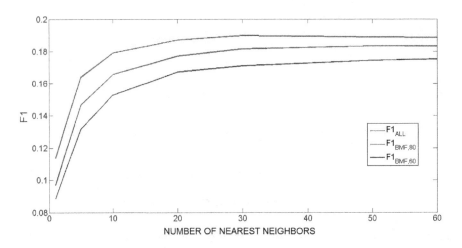

Fig. 3. *F1 dependence on the percentage of data covered by BMF-decomposition, and the number of nearest neighbours*

4.4 Influence of Scaling on the Recommendations Quality for BMF in Terms of MAE

Another aspect that was interesting to examine was the impact of scaling described in subsection 3.2 on the quality of recommendations. All four options from 3.2 of scaling were considered. The distribution of ratings in the data is shown in Table 3

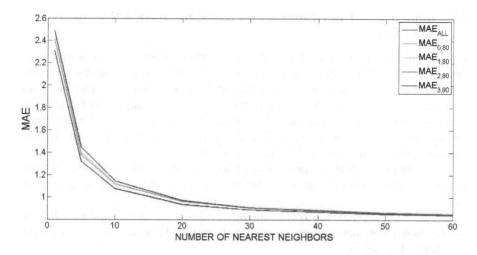

Fig. 4. MAE dependence on scaling and number of nearest neighbours for 80% coverage

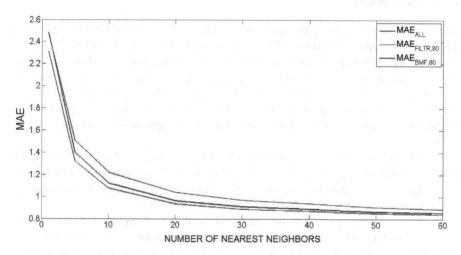

Fig. 5. MAE dependence on data filtration algorithm and the number of nearest neighbours

For each of the Boolean matrices we have calculated its Boolean factorisation, covering 80 % of the data. Then recommendations are calculated just like in subsection 4.2. It can be seen on Figure 4.4 that MAE_1 is almost the same as MAE_0, and $MAE_{2,3}$ is better than MAE_1, when the number of nearest neighbours is more than 30.

4.5   Influence of Data Filtering on MAE for BMF kNN-Based Approach

Besides the ability to search for K nearest neighbours not in the full matrix of ratings $A \in \mathbb{R}^{n \times m}$, but in the matrix $U \in \mathbb{R}^{m \times f}$, where m is a number of users, and f is a number of factors, Boolean matrix factorization can be used for data filtering. Since the algorithm as an output returns not only users-factors and factors-objects matrices, but also the ratings that were not used for factoring, we can try to search for users, similar to the target user, based on the matrix consisting only of ratings used for the factorisation.

Just as before to find the nearest neighbours cosine measure is used, and the predicted ratings are calculated as the weighted sum of the ratings of nearest users.

Figure 5 shows that the recommendations built on user-factor matrix, are better than recommendations, constructed on matrix of ratings filtered with boolean factorization.

5   Conclusion

In the paper we considered two methods of Matrix Factorisation which are suitable for Recommender Systems. They were compared on real datasets. We investigated BMF behaviour as part of recommender algorithm. We also conducted several experiments on recommender quality comparison with numeric matrices, user-factor and factor-item matrices in terms of F-measure and MAE. We showed that MAE of our BMF-based approach is not greater than MAE of SVD-based approach for the same coverage percentage of BMF and p-level of SVD.

We have also investigated how data filtering, namely scaling, influences on recommendations' quality. In terms of MAE, the BMF-based collaborative filtering algorithm demonstrates almost the same level of quality before scaling (full information) and after (considerable information loss).

Even though the reported results were obtained on a freely abvailable datasets and therefore they can be easily reproduced, in case of another datasets (different type of items or data size and its density) additional tests seems to be necessary.

As a future research direction we would like to investigate the proposed approaches in case of graded and triadic data [17,18] and reveal whether there are some benefits for the algorithm's quality in usage of least-squares data imputation techniques [19]. In the context of matrix factorisation we would also like to test our approach for the quality assessment of the recommender algorithms that we performed on some basic algorithms (see bimodal cross-validation in [20]).

Acknowledgments. We would like to thank Radim Belohlavek, Vilem Vychodil, Sergei Kuznetsov, Vladimir Bobrikov and anonymous reviewers for their comments, remarks and explicit and implicit help during the paper preparations. This work was supported by the Basic Research Program at the National Research University Higher School of Economics in 2012-2014 and performed in the Laboratory of Intelligent Systems and Structural Analysis. First author was also supported by Russian Foundation for Basic Research (grant #13-07-00504).

References

1. Koren, Y., Bell, R., Volinsky, C.: Matrix factorization techniques for recommender systems. Computer 42(8), 30–37 (2009)
2. Elden, L.: Matrix Methods in Data Mining and Pattern Recognition. Society for Industrial and Applied Mathematics (2007)
3. Hofmann, T.: Unsupervised learning by probabilistic latent semantic analysis. Machine Learning 42(1-2), 177–196 (2001)
4. Koren, Y.: Factorization meets the neighborhood: A multifaceted collaborative filtering model. In: Proceedings of the 14th ACM SIGKDD International Conference on Knowledge Discovery and Data Mining, KDD 2008, pp. 426–434. ACM, New York (2008)
5. Lin, C.J.: Projected gradient methods for nonnegative matrix factorization. Neural Comput. 19(10), 2756–2779 (2007)
6. Belohlavek, R., Vychodil, V.: Discovery of optimal factors in binary data via a novel method of matrix decomposition. Journal of Computer and System Sciences 76(1), 3–20 (2010), Special Issue on Intelligent Data Analysis
7. Symeonidis, P., Nanopoulos, A., Papadopoulos, A., Manolopoulos, Y.: Nearest-biclusters collaborative filtering based on constant and coherent values. Information Retrieval 11(1), 51–75 (2008)
8. Ignatov, D.I., Kuznetsov, S.O., Poelmans, J.: Concept-based biclustering for internet advertisement. In: 2012 IEEE 12th International Conference on Data Mining Workshops (ICDMW), pp. 123–130 (December 2012)
9. Ganter, B., Wille, R.: Formal Concept Analysis: Mathematical Foundations. Springer, Heidelberg (1999)
10. du Boucher-Ryan, P., Bridge, D.: Collaborative recommending using formal concept analysis. Knowledge-Based Systems 19(5), 309–315 (2006); AI 2005 SI
11. Ignatov, D.I., Kuznetsov, S.O.: Concept-based recommendations for internet advertisement. In: Belohlavek, R., Kuznetsov, S.O. (eds.) Proc. of the Sixth International Conference on Concept Lattices and Their Applications (CLA 2008), pp. 157–166, Palacky University, Olomouc (2008)
12. Trefethen, L.N., Bau, D.: Numerical Linear Algebra, 3rd edn. SIAM (1997)
13. Birkhoff, G.: Lattice Theory, 11th printing edn. Harvard University, Cambridge (2011)
14. Poelmans, J., Ignatov, D.I., Kuznetsov, S.O., Dedene, G.: Formal concept analysis in knowledge processing: A survey on applications. Expert Syst. Appl. 40(16), 6538–6560 (2013)
15. Poelmans, J., Kuznetsov, S.O., Ignatov, D.I., Dedene, G.: Formal concept analysis in knowledge processing: A survey on models and techniques. Expert Syst. Appl. 40(16), 6601–6623 (2013)
16. Adomavicius, G., Tuzhilin, A.: Toward the next generation of recommender systems: A survey of the state-of-the-art and possible extensions. IEEE Trans. on Knowl. and Data Eng. 17(6), 734–749 (2005)

17. Belohlavek, R., Osicka, P.: Triadic concept lattices of data with graded attributes. International Journal of General Systems 41(2), 93–108 (2012)
18. Belohlavek, R.: Optimal decompositions of matrices with entries from residuated lattices. Journal of Logic and Computation (2011)
19. Wasito, I., Mirkin, B.: Nearest neighbours in least-squares data imputation algorithms with different missing patterns. Comput. Stat. Data Anal. 50(4), 926–949 (2006)
20. Ignatov, D.I., Poelmans, J., Dedene, G., Viaene, S.: A new cross-validation technique to evaluate quality of recommender systems. In: Kundu, M.K., Mitra, S., Mazumdar, D., Pal, S.K. (eds.) PerMIn 2012. LNCS, vol. 7143, pp. 195–202. Springer, Heidelberg (2012)

Semi-supervised Image Segmentation

Gergana Angelova Lazarova

Sofia University "St. Kliment Ohridski", Bulgaria
gerganal@fmi.uni-sofia.bg

Abstract. In this paper, a semi-supervised multi-view teaching algorithm based on Bayesian learning is proposed for image segmentation. Beforehand, only a small amount of pixels should be classified by a teacher. The rest of the pixels are used as unlabeled examples. The algorithm uses two views and learns a separate classifier on each view. The first view contains the coordinates of the pixels and the second – the RGB values of the points in the image. Only the weaker classifier is improved by an addition of more examples to the pool of labelled examples. The performance of the algorithm for image segmentation is compared to a supervised classifier and shows very good results.

Keywords: semi-supervised learning, co-training, segmentation.

1   Introduction

Nowadays, we have access to massive amounts of abundant and diverse data. Nevertheless, not always, consists this data of only labelled examples. Recently, there has been significant interest in semi-supervised learning. In this area, we search for fast algorithms which can achieve at the same time good classification accuracy. Sometimes, labelling an example turns out to be a time-consuming process, requiring extra knowledge and skilled experts in the field.

The aim of this paper is to segment an image into pre-defined classes, having only few labelled pixels. A teacher gives us a basic idea of the classifier we want to learn, classifying few points of the image. Two sources of information are used: $view_1$'s features consist of the coordinates of the pixels ($View_1 = (X, Y)$) and $view_2$'s features are the RGB values ($View_2 = (R, G, B)$). $View_2$'s classifier is improved by an addition of more labelled examples based on $view_1$'s classifier. The final decision for each example's classification is taken based on both classifiers. This semi-supervised teaching algorithm used for segmentation is a modification of the standard co-training algorithm and represents a modern approach towards image segmentation.

Originally, co-training [1] was used to cluster faculty web pages into categories. The training set consisted of only a small amount of labelled examples. Two views were used - the first contained the words on the web pages. The latter was formed by the hyperlinks pointing to the web pages. Human recognition, using multiple-source data is another example of co-training application, combining both voice recognition and face recognition. Even recommender systems can benefit from the algorithm when there is scarce rating history of the users. The user profile can be divided into

G. Agre et al. (Eds.): AIMSA 2014, LNAI 8722, pp. 59–68, 2014.
© Springer International Publishing Switzerland 2014

separate views. For instance, such a model can have two views X = (X1, X2), where X1 = "articles a user has rated", X2 = "user browsing history".

Co-training has proven to be successful in many areas where there exists a natural split in more than one view. It outperforms a single standard learner when the following criteria are met [1]:

(1) Each view (set of features) is sufficient for classification
(2) The two views (feature sets of each instance) are conditionally independent given the class.
$$P(X1|Y, X2) = P(X1| Y), P(X2|Y, X1) = P(X2| Y)$$

Practically, it is very uncommon to see such uncorrelated views. Even the words in the web-page example above do not fully follow the rule. It is not often possible to find a natural split, division of the attributes, and it depends on the very data. Even when this condition does not fully hold, co-training proves to be effective.

Since 1998, when Blum and Mitchell [1] published the core algorithm, there has been considerable interest in the algorithm. It proved to be useful and has been applied in many areas – not only for web page classification, but also for object and scene recognition [9], and statistical parsing[10]. Jafar Tanha, Maarten van Someren and Hamideh Afsarmanesh [2] proposed an ensemble version of the algorithm. The ensemble is used to estimate the probability of incorrect labelling, deciding if adding a set of unlabeled data will reduce the error of a component classifier or not. The multi-view algorithm used for image segmentation in this paper differs substantially as it tries to increase only the list of labelled examples of the weaker classifier, based on the most confident examples of the other learner.

Semi-supervised clustering is another modern field of research combining both labelled and unlabeled examples. "*Cluster-then-label*" [4] first clusters the instances and later uses the labelled examples in each cluster for classification. The proposed innovative modification of co-training for image segmentation differs a lot as it does not perform a clustering algorithm but based on two views decides what the label of an instance is.

2   Multi-view Learning

The algorithm used for image segmentation is a two-view multi-class algorithm and learns two classifiers, each consisting of less attributes than the original set. It exploits information from two sources of data – dividing the attributes into two views (view$_1$, view$_2$). Let each instance X consist of two views X = (X1, X2). Both X1 and X2 represent feature sets.

Let D be the set of both labelled and unlabelled examples. D1 consists of only labelled ones and D_2 of the unlabelled examples. $D = D_1 \cup D_2$.

$$D_1 = \{(x_i, y_i)\}_{i=1}^{nl} , D_2 = \{x_j\}_{j=1}^{nu},$$

Usually, the number of instances without classifications u is much bigger than that of the labelled examples l ($u >> l$).

The two views can use different learners for the underlying classifiers. For the experimental results a Naïve Bayes Classifier and a supervised learner, based on multivariate normal distribution were chosen. The Naïve Bayes Classifier is preferred when fast execution speed is required because it does not perform complex computations.

2.1   Supervised Learning

a) Naïve Bayes Classifier (NB)

This classifier is a simple probabilistic classifier and relies on the preposition that the attributes are independent.

$$P(y_j)P(x_i \mid y_j) = P(y_j)\prod_{k=1}^{m} P(a_k \mid y_j)$$

In order to classify new examples it chooses the hypothesis that is most probable (maximum a posteriori). The corresponding classifier is the function f^* defined as:

$$f^* = \arg\max_{y_j} P(y_j)P(x_i \mid y_j)$$

b) Supervised learning, based on Multivariate Normal Distribution (MND-SL)[4].

The multivariate normal distribution is often used to describe any set of correlated real-valued random variables each of which clusters around a mean value. Let the parameter of the model be: $\theta = (\pi, \mu, \Sigma)$, π – prior class probability, μ – mean vector, Σ – covariance matrix. When the examples in the dataset are independent and identically distributed:

$$P(D \mid \theta) = \prod_{i=1}^{l} P(x_i, y_i \mid \theta) = \prod_{i=1}^{l} P(y_i \mid \theta)P(x_i \mid y_i, \theta).$$

In order to find the local maximum and the parameters of the model, we set the derivative to 0 and find θ:

$$\pi_j = \frac{l_j}{l}, \mu_j = \frac{1}{l_j}\sum_{i:yi=j} x_i, \Sigma_j = \frac{1}{l_j}\sum_{i:yi=j} (x_i - \mu_j)(x_i - \mu_j)^T$$

Where lj is the number of examples having classification y_j and l is the number of all labelled examples.

To classify a new example x_i, $P(y_i \mid x_i, \theta)$ should be calculated for each class y_i.

$$P(y_i \mid x_i, \theta) = \frac{P(y_i)P(x_i \mid \theta, y_i)}{P(x_i)} = \frac{P(y_i)N(x_i, \mu_y, \Sigma_y)}{P(x_i)}$$

$$N(x,\mu_y,\Sigma_y) = \frac{1}{(2\pi)^{\frac{D}{2}} |\Sigma y|^{\frac{1}{2}}} e^{\frac{-(x-\mu_y)^T \Sigma_y^{-1}(x-\mu_y)}{2}}$$

2.2   Co-training

The original algorithm, published by Blum and Mitchell [1], uses two classifiers (L1 and L2). Let U1 and U2 correspond to $view_1$ and $view_2$ of the examples in the training set but only those which have labels. The algorithm augments the set of labeled examples of each classifier, based on the other learner's predictions. Both classifiers are expected to agree on new example labels.

Original Algorithm:

1.   Learn L1 using U1, Learn L2 using U2

2.   Probabilistically label all unlabeled examples using L1. Add L1's most confident examples to U2

      Probabilistically label all unlabeled examples using L2. Add L2's most confident examples to U1

3.   Go to 1 until there are no more unlabeled examples or some other stop criterion is met – maximum number of iteration, etc.

2.3   Multi-view Teaching Algorithm (MTA)

The multi-view teaching algorithm is a modification of the standard Co-Training algorithm. The proposed modification improves only the weaker of the classifiers.
    Let L1 and L2 be two supervised classifiers. They can be based on *NB* or *MND-SL*. Let L2 be the weaker learner, with worse classification accuracy. Let U2 be the set of labelled examples of classifier L2. W contains all the unlabeled examples. At step 2 the algorithm learns L1 and finds the parameter θ_1. At step 3, L1's most confident examples of each class are added to L2's pool of labelled examples.

Algorithm:
1.   Initialization:
         - U1 = $view_1(D_1)$ – contains the attributes of view1 (only labelled examples);
         - U2 = $view_2(D_1)$ – contains the attributes of view2 (only labelled examples);
         - W = D_2 = all the unlabeled examples;
2.   Learn L1 based on $view_1$. Find θ_1
3.   Add more labelled examples to L2
         - For each class y_i and each example x_i in W calculate: $P(y_i|x_i,\theta_1)$ and find
           the most probable classification $y* = \text{argmax}_{yi} P(y_i|x_i,\theta_1)$
         - For each class y_i find the most confident examples and if they exceed some
           threshold add them to U2

- Learn L2(U2), find θ_2

4. Exit, Return θ_1 and θ_2 and classify new examples, based on $P(y_i \mid x_i, \theta_1)$ and

$P(y_i \mid x_i, \theta_2)$

The set of unlabeled instances (W) can be further divided into bins, and executed for every bin in parallel. This optimization technique will be extremely useful for large-size images with lots of unlabeled pixels.

Another point worth mentioning, is that the algorithm tries to find the most confident examples of L1 but not all of them are added to the pool of labelled examples. If the fitness of some examples do not exceed some threshold, they are not used at this point. If the first learner is not confident about some unlabeled example, this example is not further used in order not to worsen the second learner L2.

The algorithm is not an iterative procedure (the original co-training algorithm is), which makes it suitable for image/movie processing, tracking objects, etc.

2.4   Combining the Views

The results of the two classifiers are intuitively combined so that a final decision about the classification of new examples is made. Blum and Mitchell proposed a simple framework where the final class probability of a new example is the multiplication of the two views' probabilities:

$$P(y_j \mid x_i) = P(y_j \mid x_i, \theta_1) P(y_j \mid x_i, \theta_2)$$

The experimental results of the multi-view teaching algorithm were held based on this approach. One standard tuning procedure was added:

$$\arg\max_{y_j} P(y_j \mid x_i, \theta_1) P(y_j \mid x_i, \theta_2) =$$
$$\arg\max_{y_j} \log P(y_j \mid x_i, \theta_1) + \log P(y_j \mid x_i, \theta_2)$$

For future research a weighting system can also be integrated, based on the relevance of the two classifiers, especially when we have prior knowledge that one of the classifiers is stronger than the other. Ensemble methods can also be applied [2].

3   Image Segmentation

Image segmentation is the process of partitioning a digital image into regions or categories, which correspond to different classes. Every pixel in an image is allocated to one of the categories. Pixels with the same label (classification) share common visual characteristics.

For image segmentation various approaches exist – thresholding, the maximum entropy method, Otsu's method [13] (maximum variance), different unsupervised clustering methods (EM-algorithms [14], kMeans [15]). No prior information about the clusters is known.

Semi-supervised image segmentation concerns how to obtain the segmentation from a partially labelled image. It can be extremely useful when we already have information about the nature of the distribution of the points. A teacher should label few points of each class, giving the algorithm the idea of the clusters. The aim is to augment the training set with more labelled examples, reaching a better predictor. The described multi-view teaching algorithm proved to succeed in it.

For image segmentation naturally two views are defined:

1. $view_1 = (X, Y)$: The feature set consists of two attributes X and Y(correspond to the coordinates of the pixels)

2. $view_2 = (R, G, B)$: The feature set consists of three attributes Red, Green, Blue (correspond to the RGB values of the pixels)

The major limitation of image segmentation is that the higher the number of points, the bigger the algorithm complexity. Consequently, iterative procedures (unsupervised methods – the EM algorithm, kMeans) are not appropriate because of this complexity. Modern cameras support bigger and bigger resolution parameters and new simple and fast algorithms should be applied, especially if we are processing thousands of images or movies scenes.

The original co-training algorithm was also not suitable for this area of application not only because of the large number of pixels in the dataset, but also because it requires two strong classifiers, each sufficient for decent learning. The classifier, based on $view_2$ which consists of the RGB values, is especially susceptible to insufficient training examples. The reason is that the same color pixels can be more easily seen on random image parts. The better classifier is the one trained on $view_1$.

3.1   Experimental Framework

The original image was segmented by a teacher into pre-defined classes. In Fig. 1, 2 and 3. (a) is a sample image and (b) - the distribution of pixels between the classes. Image (b) was used for evaluation of the multi-view teaching algorithm and represents the desired classifications of image (a)'s pixels.

Fig. 1. (a) - original image, (b) – desired segmentation

Fig. 2. (a) - original image, (b) – desired segmentation

Fig. 3. (a) - original image, (b) – desired segmentation

Construction of the training and test sets:

- The training set consists of a fraction of labelled examples from the original dataset. Randomly a small amount of pixels are chosen, they are added to D_1. The rest of the instances have their classifications removed. These unlabeled examples are added to D_2. A final training set is constructed: $D = D_1 \cup D_2$.
- The test set contains all the examples in D_2. We will evaluate the algorithm based on these examples, using the original classifications with respect to image (b).

For classification error estimation a Monte cross-validation [11] was used, multiple random samples were averaged.

A k-folded cross-validation can also be applied. As we require limited and pre-defined size of the labelled examples, the parameter k depends on it. The examples should be divided into folds, so that one fold contains all the labelled examples. K is the number of all examples divided by the number of labelled ones. The smaller the number of labelled examples, the bigger the number of folds. Furthermore, to reduce variability, multiple rounds of cross-validation should be performed using different partitions, and the validation results should be averaged over the rounds.

3.2   Experimental Results

a) *Multi-view teaching algorithm based on Naïve Bayes Classifiers(MTA) vs Supervised Naïve Bayes Classifier (NB)*

The image from Fig. 1. consists of 50700 pixels. At each cross-validation step only a small amount of labelled pixels are used. Multiple tests were held depending on the number of labelled examples (4, 6, 10, 16, 20, 50 pixels). For example, the teacher labels four points in the image (two for each class) and the rest 50696 examples are used as unlabeled.

The performance of the multi-view teaching algorithm (MTA) based on Naïve Bayes Classifiers is compared to its supervised equivalent (NB), depending on the number of labelled examples. The averaged classification accuracy after the process of cross-validation can be seen in Table 1.

Table 1. Comparison of the two algorithms, based on the number of labelled pixels (Image 1)

Algorithm	4	6	10	16	20	50
NB	63.30%	76.23%	85.44%	89.57%	90.33%	92.37%
MTA	68.62%	81.30%	88.14%	90.74%	91.24%	92.51%

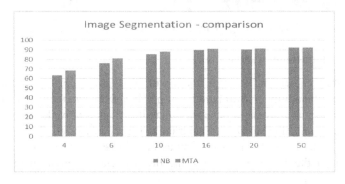

Fig. 4. MTA vs NB – classification accuracy comparison, based on the amount of labelled examples

It can be seen from Fig.4. that the multi-view teaching algorithm outperforms the Naïve Bayes classifier. It especially dominates when the number of labelled examples is small and the supervised classifier suffers from insufficient number of training examples. Consequently, it can be very useful in areas, in which labelling an example is a difficult or time-consuming process (bioinformatics, etc).

The performance of the algorithms for the three images (Fig.1, Fig.2, Fig.3) is compared in Table 2. 16 initial points are labelled (8 for each class). It was experimented with multiple thresholds and was chosen such that at least 10 examples could pass it and be further used as labelled examples.

Table 2. Comparison of the two algorithms, based on 16 initial labelled examples

	MTA	NB
Image 1	90.74%	89.57%
Image 2	80.76%	78.82%
Image 3	90.10%	89.12%

For the implementation of the multi-view teaching algorithm for image segmentation, opencv [12] was used. The cross-validation steps would be hard to perform, without the optimization of the library. Little research has been performed in this area so far, due to the tedious evaluation process, concerning the small amount of labelled examples and the large number of cross-validation steps. The performance of the algorithm is evaluated based on different samples of randomly chosen labelled examples. The teacher, who will be responsible for this process is expected to label representative examples which will lead to better classification accuracy.

Another point worth mentioning, is execution speed. It has already been mentioned that the original co-training algorithm is not appropriate for datasets of large amounts of examples. For example, an 18-megapixels image has 4896x3672 points, which makes 17,978,112 examples in the dataset. Another major execution speed advantage of the multi-view algorithm is the partition of attributes into views which can also impose a major speedup for some classification algorithms (the complexity depends on the number of attributes used for learning). Divide and conquer has been a positive strategy in many informatics fields and it is a cure for the curse of dimensionality.

b) *Multi-view teaching algorithm based on MND-SL(MTA-MD) vs Supervised MND - SL (MD)*

The performance of the multi-view teaching algorithm based on two *MND-SL* classifiers has also been evaluated for the images from Fig. 1, Fig. 2 and Fig. 3. 16 initial labelled examples were used. It was experimented with multiple thresholds and was chosen such that at least 10 examples could pass it and be further used as labeled examples. The performance of the algorithms is compared in Table 3. It can be seen from the table that the multi-view teaching algorithm outperforms its supervised equivalent. The reason for the scope of the improvements lies in the very multivariate distribution which is also susceptible to the small number of examples. A fine-tuning procedure was also used so that the covariance matrix becomes invertible: $\Sigma_y = \Sigma_y + eps$, where *eps* is small.

Table 3. Comparison of the two algorithms, 16 initial labelled examples

	MTA-MD	MD
Image 1	84.36%	79.22%
Image 2	79.14%	73.74%
Image 3	86.02%	80.18%

4 Conclusions

In this paper a state-of-the-art method (multi-view teaching algorithm) for semi-supervised image segmentation was proposed. It requires prior knowledge about the classes, an expert who labels few examples. Given the results from the previous section, it can be concluded that even with a small number of labelled instances, good classification accuracy can be achieved. The multi-view teaching algorithm, based on two Bayes classifiers turned out to be considerably useful for the segmentation process. Furthermore, the error of the algorithm is smaller than that of its supervised equivalent. The algorithm can be further applied in domains with big data, collections of numerous images, because of the advantages of the proposed framework.

Acknowledgments. This work was supported by the European Social Fund through the Human Resource Development Operational Programme under contract BG051PO001-3.3.06-0052 (2012/2014).

References

1. Blum, A., Mitchell, T.: Combining labeled and unlabeled data with co-training. In: Proceedings of the Eleventh Annual Conference on Computational Learning Theory, COLT 1998, pp. 92–100. ACM, New York (1998)
2. Tanha, J., van Someren, M., Afsarmanesh, H.: Ensemble based co-training. In: De Causmaecker, P., Maervoet, J., Messelis, T., Verbeeck, K., Vermeulen, T. (eds.) Proceedings of the 23rd Benelux Conference on Artificial Intelligence, Ghent, Belgium, November 3-4, vol. 23, pp. 223–231. BNAIC (2011)
3. Nigam, K., Ghani, R.: Analyzing the Effectiveness and Applicability of Co-training. In: Proceedings of the Ninth International Conference on Information and Knowledge Management, pp. 86–93. ACM, NY (2000), CiteSeerX: 10.1.1.37.4669
4. Zhu, X., Goldberg, A.B.: Introduction to Semi-Supervised Learning. Synthesis Lectures on Artificial Intelligence and Machine Learning. Morgan & Claypool Publishers (2009)
5. Chapelle, O., Scholkopf, B., Zien, A.: Semi-supervised Learning. MIT Press (2006)
6. Zhu, X., Ghahramani, Z., Lafferty, J.: Semi-supervised learning using Gaussian fields and harmonic functions. In: The 20th International Conference on Machine Learning, ICML (2003)
7. Cortez, P., Cerdeira, A., Almeida, F., Matos, T., Reis, J.: Modeling wine preferences by data mining from physicochemical properties. Decision Support Systems 47(4), 547–553 (2009)
8. Balcan, M.-F., Blum, A., Yang, K.: Co-training and expansion: Towards bridging theory and practice. In: Saul, L.K., Weiss, Y., Bottou, L. (eds.) Advances in Neural Information Processing Systems 17, Cambridge, MA (2005)
9. Han, X.-H., Chen, Y.-W., Ruan, X.: Multi-class Co-training Learning for Object and Scene Recognition. In: MVA 2011, pp. 67–70 (2011)
10. Sarkar, A.: Applying Co-Training Methods to Statistical Parsing. In: Proceedings of the 2nd Meeting of the North American Association for Computational Linguistics: NAACL 2001, Pittsburgh, PA, June 2-7, pp. 175–182 (2001)
11. Monte Carlo Crossvalidation, http://en.wikipedia.org/wiki/Cross-validation_%28statistics%29
12. OpenCv, http://opencv.org/
13. Otsu, N.: A threshold selection method from gray-level histograms. IEEE Transactions on Systems, Man, and Cybernetics 9(1), 919–926 (1979)
14. Carson, C., Belongie, S., Greenspan, H., Malik, J.: Blobworld: "Image segmentation using expectation-maximization and its application to image querying". IEEE Trans. Pattern Anal. and Machine Intell. 24(8), 1026–1038 (2002)
15. Dehariya, V.K., Shrivastava, S.K., Jain, R.C.: Clustering of Image Data Set Using K-Means and Fuzzy K-Means Algorithms. In: International Conference on CICN, pp. 386–391 (2010)
16. Grady, L., Funka-Lea, G.: Multi-label image segmentation for medical applications based on graph-theoretic electrical potentials. In: Sonka, M., Kakadiaris, I.A., Kybic, J. (eds.) CVAMIA-MMBIA 2004. LNCS, vol. 3117, pp. 230–245. Springer, Heidelberg (2004)

Analysis of Rumor Spreading in Communities Based on Modified SIR Model in Microblog

Jie Liu*, Kai Niu, Zhiqiang He, and Jiaru Lin

Key Lab of Universal Wireless Communications
Beijing University of Posts and Telecommunications, Beijing, China
liujieauto@163.com, {niukai,hezq,jrlin}@bupt.edu.cn

Abstract. Rumor spreading as a basic mechanism for information on online social network has a significant impact on people's life. In Web 2.0 media age, microblog has become a popular means for people to gain new information. Rumor as false information inevitably become a part of this new media. In this study, a modified rumor spreading model called SIRe is introduced, which compared to traditional rumor spreading model, have included the stifler's broadcasting effect and social intimacy degree between people. In order to verify the reasonableness of SIRe model, real rumor spreading data set and microblog network structure data set are obtained using Sina API. Then rumor predicting results using different models are compared. Finally, for the purpose of finding the characteristics of rumor spreading in community scale, a clustering method is used to discover the user communities. Analysis results have revealed that communities with higher closeness centrality tend to have higher max ratio of spreaders, and scattered immunization is better than centralized immunization, resulting lower max ratio of spreaders. Both the results and explanations are shown in this paper.

Keywords: rumor spreading, microblog, the SIR model, rumor predicting.

1 Introduction

Rumor, transmits among people, is refered to as an unverified account or explanation of events circulating among people and affects event or issue of public concern. Traditionally, rumor is disseminated through dialogue in real social activities, some classic rumor spreading models are introduced to simulate this process. But in the new media age, microblog has become a new means for people to receive and transmit message. In microblog, both the transmit method and interact method are changed. Information is transmitted through twist and retwist. when a message is published, all the user's friends have the chance to

---

* This work is supported by National Science and Technology Major Project of China (No.2012ZX03004005-002) and National Natural Science Foundation of China (61171100).

G. Agre et al. (Eds.): AIMSA 2014, LNAI 8722, pp. 69–79, 2014.

see it. The interact between two persons have become public to much more audience. MicroBlog is a social media, which means when a rumor spreads, the social characteristics play an important part. In this paper, a modified rumor model called SIRe is introduced to include the broadcasting effect and social intimacy effect of rumor spreading process in microblog.

A very important aspect in this field is the dynamics of information spreading in networks. There are some classic model about information spreading, such as DK model [1,2] and MK model. Many researchers have studied the dynamics of rumor spreading in detail [5,8]. Galam proposed a novel model called Galam's Model to simulate rumor propagation [3], and then Ellero et al. introduced a new scheme to improve Galam's Model [7]. However, these models are confined to word-of-mouth information exchanging and are not suitable for describing rumor spreading on large-scale social networks. Wang et al[4,6] introduced two new measuring indexes of rumor spreading effect the peak value of spreaders and the final proportion of stiflers. However, these researches ignore the community effect of social network. In community view, new rumor spreading characteristics will be discovered.

Considering the rumor spreading mechanism on online social network are still in infancy, in this paper, several contributions to the study of rumor dynamics on online social network are made. First of all, a new model called SIRe is introduced in Section 2, which compared with previous models, considering the broadcasting effect of stifler and social intimacy degree. Spreading equations and interpretations are established to describe and support the model. Then, in Section 3, using dynamic equations established above, parameters of different models are obtained by fitting actual rumor spreading data set. And predicting results are compared. In Section 4, SIRe model is used in the network which is constructed by communities to study rumor spreading characteristics of communities. Finally, conclusions are drawn in Section 5.

2   SIRe Rumor Spread Model for Online Social Network

The rumor model began in the 1960s when Daley and Kendall proposed the basic DK model. Then more appropriate MK model is proposed. In MK model, the population is subdivided into three groups: those who are unaware of the rumor(ignorants), those who spread the rumor(spreaders) and those who are aware of the rumor but choose not to spread it(stiflers). However, there are different features when rumor spreads on online social network like Sina microblog.

When a stifler finds a tweet which is a rumor including false information twisted or retwisted by an spreader, and this stifler is against the tweet, it is easy to retwist the message by preceding with an comment and addressing the original author with @. This comment may contain a picture with negative expression or rejection information, which shows clear negative attitude toward the rumor. More important, different from MK or other traditional rumor model, this negative message will not only be seen by the original spreader, but also by all the friends of the stiflers. The population of microblog users are subdivided into three groups, ignorant, spreaders, and stiflers(represented by I, S, and R ,respectively). Some of the unique characteristics of microbolg is summarized below.

1) The possibility of an ignorant who receives a rumor from a spreader to become a spreader is closely related to the social intimacy between the two persons. If the social intimacy is high, which means the two persons have more trust in each other, the rumor is more likely to transmit. One important way to measure the intimacy is the common friend number between the two persons. More common friend means more common social activity they may involve and more common value and knowledge they may share. So the rumor have more chance to be shared as trustful message.

2) Traditionally, a stifler can only have the chance to influence the spreader who have the intend to show him the rumor. But the message which flows in Sina microblog is mainly in the form of retwist, which means when a person publishes a message, it will be seen by all his friends. So When a stifler receives a rumor from a spreader, his reject of the rumor which mainly in the form of retwist will be seen by all the stifler's friends, resulting the transform of spreaders and ignorants to become stifler at certain possibility.

3) A stifler will not publish his rejection of the rumor until he receives a rumor. So the rumor message become a trigger for stifler to publish clarifying message.

Considering the process and unique characteristics of microblog, the new model called SIRe is introduced to better describe the process of this new media form. The spreading mechanism of SIRe model is shown in Fig 1.

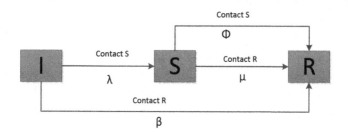

Fig. 1. Rumor spreading mechanism of SIRe

In SIRe model, at each time step, each person in the network adopts one of the three states: ignorant, spreader and stifler. If someone in the state of ignorant who receives a rumor, the ignorant will accept it with the acceptant probability and transform into another state, spreader. When a stifler receives a rumor, because stifler is described as rationalist who has the knowledge that this rumor contains false information or doubts the rumor before he verifies it. The stifler will retwist the message but preceding with negative information which shows clear attitude indicating the rumor is credible. Lacking the enthusiasm spreader has, a stifler will not twist a message until he receives one. ignorant and spreader of the stifler's friends will receive the message from stifler, and at certain possibility to become stifler. When a spreader receive a rumor from another spreader, the first one will become a stifler at certain possibility because he considers the rumor has lost spreading value. In the light of the SIRe rumor spreading process elaborated above, the evolution of the state density satisfies the following set of coupled differential equations:

$$\begin{cases} \frac{dI(t)}{dt} = -\lambda k I(t)S(t) - \beta k I(t)R(t) \\ \frac{dS(t)}{dt} = \lambda k I(t)S(t) - \mu k S(t)R(t) - \phi k S(t)S(t) \\ \frac{dR(t)}{dt} = S\beta k I(t)R(t) + \mu k S(t)R(t) + \phi k S(t)S(t) \end{cases} \tag{1}$$

I(t), S(t), R(t) are the proportion of three groups respectively and k represents the average degree of the network. It is obviously to find

$$I(t) + S(t) + R(t) = 1 \tag{2}$$

In equation (1), the acceptant probability λ is a variable which is important to the spreading process. In SIRe model, it is defined as

$$\begin{cases} \lambda = 1 - (1-\alpha) * e^{(-0.1\omega S_{(u_i,u_j)})} \\ S_{u_i,u_j} = |B(u_i) \bigcap B(u_j)| \end{cases} \tag{3}$$

S_{u_i,u_j} is the social intimacy degree between user i and user j. $B(u)$ is the set of bilateral friends of user u. ω indicates the influence of social relationships. α is to reflect the attraction of this rumor. Fig. 2 shows λ as a function of m, given different *omega* and set $\alpha=0.5$.

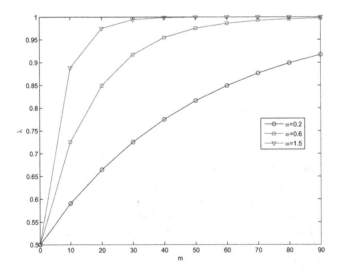

Fig. 2. Acceptant probability λ

3   Model Comparison

In this section, we intend to explore the validity of the SIRe model by comparing the prediction accuracy of different model. According to the model definition, if a model is more accuracy in predicting the rumor spreading process, it is considered to be more in line with the characteristics of microblog.

Two kinds of data set are obtained using Sina microblog API for meeting our goals. Data set 1 is consist of 10 trending topics which turned out to be false information between February 10th to March 10th 2014. It is used to train the model parameters and verify predictions of rumor spreading process. Data set 2 is collected to acquire the network structure of microblog. It is represent by an undirected graph $G(V, E, C)$. Every node $v_i \in V$ represents one user of microblog. If user i has user j as a friend, then a path $e_{ij} \in E$ exists from v_i to v_j. The nodes $v_i \in V$ are divided into different communities which represent by $c_i \in C$. The community property will be uncovered by cluster method in Section 4. Data set 2 is consist of 32686 users and used to stimulate different models to test the predict result.

In order to predict the rumor spreading process, appropriate value of parameters in MK and SIRe are needed. To accomplish this, we perform a multi-parameter least square fit by using a software called OpenLU[9]. Because the acquire of parameters is based on Equations 1 and Equation 2, the network structure is not included in this procedure. So when SIRe is applied, the acceptant probability λ can only be obtained as a constant. The time for training is the first w time points and the subsequent $m - w$ is used to test and verify the model predicting result. Because the predicting ability will be worse with the increase of m, $m - w$ will only be the half of w.

For a tweet tw, we predict its retwisting amount at every time point starting from $w+1$ to m time points, and define the overall predication error with respect to a tweet as:

$$Error(tw) = \frac{\sum_{t=w+1}^{m}(R_t - P_t)^2}{\sum_{t=w+1}^{m}R_t^2} \qquad (4)$$

where R_t and P_t are the real and predicted values at time point t, w is the time for training, and m is the end time points for predicting in the experimental data.

The mean average prediction error upon totally p tweets which in this paper is 10 is defined as:

$$E_{MRE} = \frac{\sum_{i=1}^{p} Error(tw_i)}{p} \qquad (5)$$

To illustrate the method, one simulation result used to fit actual data is plotted in Fig. 3 below.

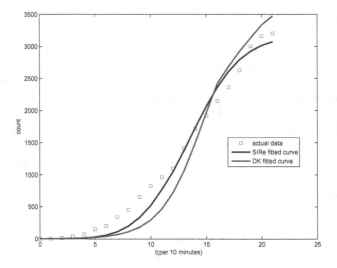

Fig. 3. The model fitting result to obtain parameters for predicting

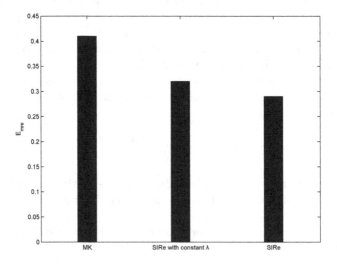

Fig. 4. Mean average errors of different model prediction method($w=22,n-w=11$)

As shown in Fig.3, SIRe model fits better than MK model to actual data. This multi-parameter least square fit is to obtain the appropriate parameters of different model. The acceptant probability λ of SIRe obtained using this method is a constant. So three models which are MK, SIRe with constant λ and SIRe are used to predict the rumor spreading. Then the mean average prediction errors are calculated and shown in Fig.4.

Fig.4 illustrates that SIRe model have better predicting ability. The rumor spreading process stated in SIRe is more in line with microbolg reality, that is the message of stifler can be seen by more people than only one spreader stated in MK model. Also, the acceptant probability considering the social intimacy degree have better predicting accuracy than only a constant. That shows when rumor spreads in micorblog network, social influence should be considered. Better prediction ability proves that SIRe model can better describe rumor spreading in microblog than traditional MK model.

4   Simulation and Analysis of Rumor Spreading in Communities

Sina microblog is a social platform which enables users to improve communication. While the whole user network is considered to be an entirely interactive platform, communities are formed in it. The users in these communities tend to communicate more and have more trust inside than outside. So when a rumor transmits, the characters of community have huge influence to rumor spreading result. In this paper, a clustering method is applied to find user communities in microblog. The algorithm is divided into two phases which are repeated iteratively and this method is used on data set 2. Each of the user in data set 2 is assigned a community after clustering.

In order to find the characteristics of rumor spreading in communities, the SIRe model is used to transmit rumor on the network which is composed of user communities. In this paper, the simulation is based on a platform coded by Java and it is specified to accomplish this task. Each of the node in the network is an agent, which is in the state of spreader, ignorant or stifler. The state transformation is according to the rules of SIRe.

4.1   Communities with Higher Mean Closeness Centrality Have Higher Max Ratio of Spreaders

For the sake of describing the closeness of a community, a parameter called closeness centrality is computed. The parameter is to characterize the difficulty for a node to reach other node through the network. Larger closeness centrality means the closeness of the network is higher, which indicates a node is closer to another node in the network, so information is transmitted faster in a network since less links is needed to relay. The closeness centrality of a node n_i is defined as:

$$C(n_i) = \frac{n-1}{\sum_{j=1}^{n} d_{n_i n_j}} \tag{6}$$

Where $d_{n_i n_j}$ is the shortest path distance from node i to node j, n is the total node number of the network, while $n-1$ represent the maximum possible number of nodes adjacent to node i. The mean closeness centrality of a community is the mean value of closeness centrality in this community.

Based on the mean closeness centrality, communities found in data set 2 is divided into three categories and are shown in table below.

Table 1. communities assigned to different categories according to mean closeness centrality

	category 1	category 2	category 3
mean closeness centrality	0.3~0.5	0.2~0.3	0.1~0.2
community number	5	18	21

when rumor spreads in the network, it is found that communities with higher mean closeness centrality are more likely to access the rumor and have higher max ratio of spreaders. Simulation results are shown in Fig. 5. In fact, the mean closeness centrality is usual pulled up by a few verified users who have very high closeness centrality. To evaluate the effect of verified users, those users in category 1 are removed and stimulation results are shown in Fig. 6. For the purpose of reducing the effect of random, the initial spreader is selected in 3 different categories, and each simulation is run for 10 times and the mean value is computed.

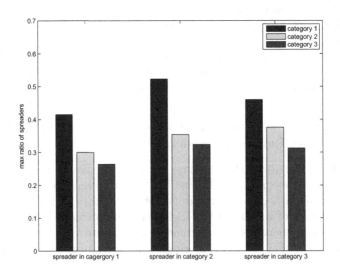

Fig. 5. In SIRe model max ratio of spreaders for different category communities

As shown in Fig. 5, in these simulations, communities with higher mean closeness centrality are more vulnerable to rumor. This parameter is to measure the closeness of communities, which means rumor in category 1 need less links to relay to reach another user than in category 2 and 3. So spreaders need less time steps to reach the whole community member in category 1, and both the number spreader have effect on and the social intimacy degree are high. At each time step, more ignorant will accept the rumor and in less time steps max ratio of spreaders is achieved in a higher mean closeness centrality community.

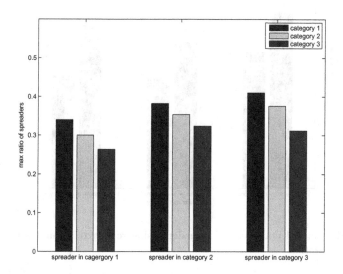

Fig. 6. In SIRe model max ratio of spreaders for different category communities(extreme high closeness centrality users removed in category 1)

As shown in Fig. 6, verified users who have very high closeness centrality are the central node for rumor spreading. Once removed, the max ratio of spreaders in category 1 greatly declines. This simulation result shows us to control rumor spreading in microblog, more attention should be paid to very high closeness centrality people.

4.2 Scattered Immunization Has Better Immunization Effect Than Centralized Immunization

Immunization is an important way to reduce the damage of rumor. Traditionally, immunization has been studied concerning the node degree or the closeness to the infection node. In this paper, a new method of immunization called scattered immunization is proposed considering the community property.

The main idea of scattered immunization is when several immune nodes are assigned in the whole network, each of the immune nodes are assigned in different communities. To make the effect of this immunization strategy more notable, centralized immunization is conducted in which all of the immune nodes are in the same community. Each of the results is repeat 10 times and the mean value is computed. The result is shown in Fig. 7.

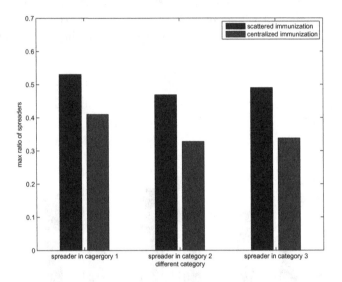

Fig. 7. In SIRe model max ratio of spreaders using scattered immunization and centralized immunization

Scattered immunization can reduce the spreading of rumor a lot compared to centralized immunization. When scattered immunization is used, since stifler are separated in different communities, these communities can react to the rumor more effective. That is to say, if a rumor is circulating in the whole network, these communities are acting as rumor spreading amplifier. If a rumor enters a community and a stifler is inside, the rumor will trigger the stifler's reaction at an early stage, creating more stiflers, thus can discourage the spreading of rumor. Although centralized immunization will reduce the spreading effect of the community which initial stiflers are all in, other communities will be more vulnerable to the rumor. Combining the effect, scatted immunization compared to centralized immunization, can greatly reduce the max ratio of spreaders.

5 Conclusion

In this paper, a new model more suitable for microblog is introduced. The stifler's broadcasting effect and social intimacy degree effect are considered. Model comparing which measures the ability to predict rumor spreading process shows that SIRe model is more accurate in describing network like microblog. Then SIRe model is conducted in network of communities for simulation. We found that, communities with higher mean closeness centrality will have higher max infection ration. And scattered immunization have better immune effect than centralized immunization, resulting lower max infection ratio. This result can more in-depth show us the way to clarify a rumor.

This research helps to reveal the true rumor disseminating mechanism in microblog. Rumor spreading effect on communities is simulated and show characteristics involving max ratio of spreaders and immunization. It provides a new access to revealing the rumor spreading.

References

1. Daley, D.J., Kendall, D.G.: Epidemics and rumours. Nature 204(4963), 1118 (1964)
2. Daley, D.J., Kendall, D.G.: Stochastic rumours. IMA Journal of Applied Mathematics 1(1), 42–55 (1965)
3. Ellero, A., Fasano, G., Sorato, A.: A modified galam's model for word-of-mouth information exchange. Physica A: Statistical Mechanics and its Applications 388(18), 3901–3910 (2009)
4. http://www.forcal.net
5. Moreno, Y., Nekovee, M., Pacheco, A.F.: Dynamics of rumor spreading in complex networks. Physical Review E 69(6), 066130 (2004)
6. Pan, Z.F., Wang, X.F., Li, X.: Simulation investigation on rumor spreading on scale-free network with tunable clustering. Journal of System Simulation 8, 2346–2348 (2006)
7. Wang, X., Li, X., Chen, G.: The theory and application of complex network. Tech. rep., Tsinghua University (2006) (in Chinese)
8. Zanette, D.H.: Dynamics of rumor propagation on small-world networks. Physical Review E 65, 041908 (2002)

Modeling a System for Decision Support in Snow Avalanche Warning Using Balanced Random Forest and Weighted Random Forest

Sibylle Möhle[1,2], Michael Bründl[2], and Christoph Beierle[1]

[1] Dept. of Computer Science, University of Hagen, 58084 Hagen, Germany
smoehle@acm.org, beierle@fernuni-hagen.de
[2] WSL Institute for Snow and Avalanche Research SLF,
7260 Davos Dorf, Switzerland
bruendl@slf.ch

Abstract. In alpine regions, traffic infrastructure may be endangered by snow avalanches. If not protected by physical structures, roads need to be temporarily closed in order to prevent fatal accidents. For assessing the danger of avalanches, local avalanche services use, amongst others, meteorological data measured on a daily basis as well as expert knowledge about avalanche activity. Based on this data, a system for decision support in avalanche warning has been developed. Feasible models were trained using Balanced Random Forests and Weighted Random Forests, yielding a performance useful for human experts. The results are discussed and options for further improvements are pointed out.

Keywords: Balanced Random Forest, Weighted Random Forest, avalanche warning, decision support system, rare events, class imbalance.

1 Introduction

Snow avalanches pose a serious threat in alpine regions. They may cause significant damages and fatal accidents. Therefore, local avalanche services responsible for avalanche safety in communities and for traffic infrastructure have been established in alpine countries. Their task is to protect people from the impact of snow avalanches by temporary measures, like the closing of roads, ordering people to stay in buildings, evacuation, or artificial avalanche triggering [20]. Thus, assessing the local risk of snow avalanches is of vital importance, and requires expert knowledge, intuition, and process understanding. Decision support systems such as NXD2000 [9,10] based on the method of nearest neighbors [4] help local avalanche forecasters to base their decisions in addition to their knowledge and experience on more objective criteria. Precipitation (new snow or rain), wind, air temperature and solar radiation are the main factors influencing the formation of avalanches. Local avalanche forecasters base their daily judgment of avalanche danger on a careful analysis of meteorological variables and snowpack properties influencing the stability of the snowpack. This assessment relies

G. Agre et al. (Eds.): AIMSA 2014, LNAI 8722, pp. 80–91, 2014.

heavily on a sound understanding of the physical processes in the snowpack but also on experience and comparisons with similar situations observed in the past. When using NXD2000, the ten days being most similar to the current situation and the avalanche activity that occurred within the corresponding time slots are selected by the program and presented to the user. The output of the model provides additional evidence that a similar avalanche activity might take place in a specific situation.

Classification and regression trees [3] were applied in [18] for forecasting large and infrequent snow avalanches. The work reported in [18] concentrated on one single avalanche path. In [12], classification trees were adopted for forecasting avalanches in coastal Alaska; the region accounted for comprised over 100 avalanche paths, and prior to modeling, variable selection was performed. In [17], wet-snow avalanches were predicted using classification trees and Random Forests [2]. In [12,17,18], the training data were selected using statistical methods.

The purpose of the work reported here is to develop a decision support system for assessing the local hazard of snow avalanches, based on the data collected within NXD2000. As local knowledge is essential and generalization to other locations is very difficult, the system has been developed for a specific area, the Canton of Glarus in Switzerland.

A second objective of our work was to investigate the suitability of Random Forests and variants thereof for modeling a decision support system for snow avalanche warning. Since Random Forests [2], which are used in [17], are suited for classification problems where the dependencies between the variables are unknown and non-linear, they are a candidate for modeling our system. However, in the given data set provided by NXD2000, avalanches represent rare events, and Random Forests are biased towards the majority class [5]. Due to the class imbalance, bootstrap samples drawn for decision tree construction may contain few or no examples from the minority class, hence the resulting decision tree will perform poorly on examples from the minority class [5]. Therefore, we employ two variants of Random Forests, Balanced Random Forest and Weighted Random Forest [5], and study their suitability for our application.

A third goal of our work is the elaboration of quality measures for the obtained models from the point of view of the avalanche service which is in charge of assessing the local avalanche danger. While we also use the quality measures employed in the work cited above, we propose to use positive and negative predictive values as additional quality measures for snow avalanche warning, providing an assessment of the forecast probability. It turned out that these measures are particularly useful for the human experts at the avalanche service of the Canton of Glarus.

The rest of this paper is organized as follows: In Section 2, we briefly address the problem of forecasting rare events. In Section 3, we define six different measures for assessing the quality of models and discuss their relevance for our application scenario. The weather and snow data being used and the variables derived from these data are presented in Section 4. The resulting models obtained by employing

Balanced Random Forest and Weighted Random Forest are described and discussed in Section 5 and Section 6. In Section 7 we conclude and point out further work.

2   Forecasting Rare Events

In the given data covering more than 40 years, avalanches represent rare events (i.e. 53 days with avalanches, but 6889 without an avalanche). A model predicting always the negative class, i.e. non-avalanches, achieves a high overall accuracy. However, it would be of no use as a decision support for avalanche warning, since the important cases are the positive ones, i.e. the avalanche days. Fatal accidents may be the consequence of roads not being closed due to a missed avalanche forecast. Therefore, with rare events, the overall accuracy of a classifier is not an adequate quality criterion [19].

Sampling as well as cost-sensitive learning are possible solutions to the problem of predicting rare events. Undersampling the negative class may result in the loss of important information, while oversampling the positive class may introduce duplicates of positive examples into the training data. This, in turn, bears the risk of learning specific examples [19].

When taking measures such as temporary road closures, resulting costs have to be considered. They consist of business interruption costs for the regional economy due to road closure as well as the efforts for road closure and opening and clearance of avalanche debris on the road. In the case of false negative forecasts, costs may be significant due to damages and fatal accidents. Costs need not only be monetary: With every false prediction, the avalanche service looses credibility. For this reason, it is not only important to achieve a low number of false negatives, but the number of false positives has to be low, too.

With cost-sensitive learning, different forecast types are assigned different costs. An ideal classifier minimizes the associated cost function [6]. Costs are case-specific and their estimation is difficult. Long-time damage statistics allow for a quantitative estimation of costs. Since appropriate data were not available in our application, a qualitative estimation of costs from a regional economic view was conducted. The following assumptions apply: With a positive forecast, the affected road section is closed. Considered costs include damage to persons and property and business interruption costs as well as loss of credibility in case of unnecessary closure; hence, costs for correct predictions were set to 0. According to the remarks made in the last paragraph, the costs for false negatives have to be higher than the costs for false positives. Cost-sensitivity can be achieved by assigning different weights to the positive and negative class [6].

In [5], variants of Random Forests [2] suited for the classification of rare events were proposed.

2.1   Balanced Random Forest (BRF)

Balanced Random Forest approaches the problem of class imbalance by constructing each decision tree from equally-sized bootstrap samples from the negative and the positive class. This ensures that positive and negative examples are both included in the training data set. The trees are grown using the CART algorithm [3] without pruning. The determination of the best split in each node is carried out analogously to Random Forests by testing a previously fixed number of randomly chosen variables.

2.2   Weighted Random Forest (WRF)

Weighted Random Forest implements cost-sensitive learning by assigning weights to classes. By assigning a higher weight to the positive class, misclassification costs for positive examples are higher and positive examples therefore gain weight in the training process.

3   Model Assessment

Forecasting an avalanche can be considered as a classification problem. The positive class contains the avalanche days, and in this work, it is assigned the value 1. The negative class contains the non-avalanche days and is assigned the value 0. The results can be represented in contingency tables as shown in Fig. 3.

		Observed	
		0	1
Predicted	0	TN	FN
	1	FP	TP

Fig. 1. Contingency table for event forecasting: TN denotes the number of true negative forecasts, FN the number of false negatives. The number of false positive and true positive forecasts are denoted by FP and TP, respectively.

The number of true negative forecasts is abbreviated as TN and denotes the number of cases in which neither an avalanche was predicted nor an avalanche occurred. Accordingly, the number of correctly predicted avalanches is abbreviated as TP. The number of false negatives refers to the number of missed avalanches and is abbreviated as FN, the number of false positive forecasts is denoted by FP. They refer to the situations in which an avalanche occurred when there was none predicted and vice versa, respectively. For model assessment, the following quality criterions were applied:

Sensitivity (Probability of Detection):

$$POD = \frac{TP}{TP + FN} \tag{1}$$

Specificity (Probability of Non-Event):

$$PON = \frac{TN}{TN + FP} \tag{2}$$

False Alarm Ratio:

$$FAR = \frac{FP}{FP + TP} \tag{3}$$

True Skill Statistic:

$$TSS = \frac{TP}{TP + FN} - \frac{FP}{FP + TN} \tag{4}$$

Positive Predictive Value:

$$PPV = \frac{TP}{TP + FP} = 1 - FAR \tag{5}$$

Negative Predictive Value:

$$NPV = \frac{TN}{TN + FN} \tag{6}$$

Amongst these, the positive predictive value and the negative predictive value [14] are most informative regarding the operational use of a model used for decision support in our application scenario. Given a negative forecast, the negative predictive value is the probability for this forecast being right. Given a positive forecast, the positive predictive value is the probability for an avalanche event.

The ideal classifier shows a high sensitivity as well as a high specificity. But these quality criterions do not allow for an assessment of the forecast. Obviously, high sensitivity and high specificity result in high positive and negative predictive values.

The probability of detection, the probability of a non-event, the false alarm ratio as well as the true skill statistic are established quality measures for model assessment in avalanche forecasting and are adopted in [12,17,18].

4 Data

The Canton of Glarus is located in the eastern part of Switzerland and is characterized by high mountains and steep slopes. In this work, we focused on the alpine valley Kleintal situated in the southeast of the Canton of Glarus. The valley floor of the Kleintal is gently inclined, its elevation ranging from over 600 m.a.s.l. to over 1000 m.a.s.l. The starting zone of a snow avalanche may be situated up to 1700 m above the valley floor and may therefore endanger the main road leading through the valley. The data consist of meteorological variables measured daily in the early morning as well as avalanche information between January 1st, 1972 and April 30th, 2013.

The measures were collected in Elm at 958 m.a.s.l. and at Risiboden, a location situated 2.5 km from Elm at an elevation of 1690 m.a.s.l. They comprised the maximum and minimum air temperature in the last 24 hours, actual wind speed and actual wind direction, degree of sky cover and precipitation in the last 24 hours in Elm as well as snow depth and new snow depth in the last 24 hours at Risiboden. The air temperature is measured in Celsius degrees and recorded in $^1/_{10}$ Celsius degrees in the NXD2000 database. The wind direction is measured in arc degrees and the rounded value of the measured value divided by 10 is recorded with 0 or 36 indicating wind coming from north, and 9, 18, and 27 indicating wind coming from east, south, and west, respectively. The wind speed is measured in meters per seconds and recorded in knots. For standardization purposes, for this work, the wind direction was set to 0 where either the wind speed was 0 or the wind direction was 36. The degree of sky cover was recorded as follows: 0 indicates a clear sky, 4 a coverage of 50% and 8 a cloudy sky. For the precipitation, the water equivalent was given, i.e. the snow was melted and the water amount was recorded in millimeters. The new snow depth as well as the snow depth are measured in centimeters and recorded unaltered.

Meteorological factors are potentially useful for estimating snowpack instability, but interpretation is uncertain and the evidence less direct than for snowpack factors [16]. Avalanche expert knowledge was taken into account by using the derived variables listed in Table 1, which were defined for NXD2000 for the Canton of Glarus and are documentated for internal use. In the following, we explain how these variables and their range of values are derived from the data described above.

Table 1. Meteorological variables are measured daily. The definition of derived variables allows to consider an expert knowledge about avalanche activity.

	Variable	Unit	Range of values
1	Max. air temperature in the last 24 hours	[$^1/_{10}$ °C]	[-178, 240]
2	Max. air temperature in the last 48 to 24 hours	[$^1/_{10}$ °C]	[-178, 240]
3	Min. air temperature in the last 24 hours	[$^1/_{10}$ °C]	[-251, 157]
4	Min. air temperature in the last 48 to 24 hours	[$^1/_{10}$ °C]	[-251, 157]
5	Actual wind direction		{0, 10, ..., 350}
6	Wind direction of the previous day		{0, 10, ..., 350}
7	Wind speed	[kn]	[0, 206]
8	Wind speed of the previous day	[kn]	[0, 206]
9	Degree of sky cover		{0, 12, ..., 96}
10	Precipitation in the last 24 hours	[mm]	[0, 989]
11	Precipitation in the last 48 to 24 hours	[mm]	[0, 989]
12	New snow fallen in the last 24 hours	[cm]	[0, 550]
13	New snow fallen in the last 72 to 24 hours	[cm]	[0, 575]
14	Snow depth		[0, 432]

The maximum and minimum air temperature in the last 48 to 24 hours (lines 2 and 4 in Table 1) refers to the maximum and minimum air temperature recorded for the previous day. The wind direction (lines 5 and 6 in Table 1) is multiplied by 10, the wind speed (lines 7 and 8) is multiplied by 5.1479, and the degree of sky cover (line 9) is multiplied by 12. The precipitation in the last 48 to 24 hours (line 11) refers to the precipitation recorded for the previous day. The amount of new snow fallen in the last 24 hours (line 12) is multiplied by 5. The amount of new snow fallen in the last 72 to 24 hours (line 13) is defined as the sum of the weighted new snow depths of the last 3 days multiplied by 5. The snow depth (line 14) is divided by the mean of all snow depths in the database and multiplied by 100.

Only the avalanches endangering the main road were recorded. In this work, we included 7 avalanche paths with 7 to 13 avalanches each. We did not discriminate between avalanche paths, and days with at least one avalanche being released in one of these paths were considered as one event. The complete data set contained 53 positive examples, i.e. avalanche days, and 6889 negative examples, i.e. non-avalanche days. The ratio of positive to negative examples therefore was approximately 1:130. The data set was divided into a training and a test set as follows: The test set consisted of all entries from November 1st, 2002 to April 30th, 2013. By this means, the ratio of positive to negative examples matched approximately the ratio observed in the real world. The training set consisted of all avalanche days from January 1st, 1972 to April 30th, 2002 and about 10 times as many non-avalanche days drawn randomly every year. The test data set consisted of 12 positive and 1572 negative examples, the training data set consisted of 41 positive and 560 negative examples.

5   Results

Two BRFs were trained using the size of the positive class as bootstrap sample size for positive and negative examples. For the positive class, the cutoffs were set to 0.5 and 0.6, respectively. The number of variables to be tested for the best split was set to 2. The cutoff as well as the number of variables to be tested for the evaluation of the splits were determined using 10-fold cross-validation. For the cutoff, the following values were tested: 0.3, 0.4, 0.5, 0.6, and 0.7. For the number of variables, the following values were tested: 2, 3, and 4. The contingency tables obtained for the test data are shown in Fig. 2.

For model BRF_0.5 with a cutoff of 0.5, the number of true positives is higher with respect to model BRF_0.6 with a cutoff of 0.6. On the other hand, model

		Observed				Observed	
		0	1			0	1
Predicted	0	1496	6	Predicted	0	1517	7
	1	76	6		1	55	5

Model BRF_0.5   Model BRF_0.6

Fig. 2. For model BRF_0.5, the cutoff for the positive class was set to 0.5. For model BRF_0.6, the cutoff for the positive class was set to 0.6. It is observed that a decrease in the cutoff leads to an increase in the number of true positives and false positives.

BRF_0.6 achieved a lower number of false positives. Using WRF, model WRF_5 was trained using a class weight of 5 for the positive class and a class weight of 1 for the negative class. The class weights were determined using 10-fold cross-validation. For the positive class, the following weights were tested: 1, 2, 3, 5, 10, 15, 20, 25, 30, 50, 100, 110, 120, 130, and 150. The class weight for the negative class was set to 1. The contingency table obtained for the test data is shown in Fig. 3. In Table 2, the quality measures for the generated models for the test data are listed.

Observed

		0	1
Predicted	0	1516	7
	1	56	5

Model WRF_5

Fig. 3. For model WRF_5, the class weights for the positive and negative class were set to 5 and 1, respectively. The results are very similar to the ones obtained for model BRF_0.6.

Table 2. In the quality measures obtained for the test data, the analogy between models BRF_0.6 and WRF_5 becomes evident. No WRF similar to model BRF_0.5 could be trained with an acceptable number of false positive forecasts.

Model	TN	FN	FP	TP	POD	PON	FAR	TSS	PPV	NPV
BRF_0.5	1496	6	76	6	0.500	0.952	0.927	0.452	0.073	0.996
BRF_0.6	1517	7	55	5	0.417	0.965	0.917	0.382	0.083	0.995
WRF_5	1516	7	56	5	0.417	0.964	0.918	0.381	0.082	0.995

The performance of models BRF_0.6 and WRF_5 was almost the same. The sensitivity was quite low with only 41.7%, hence the number of false positives was low too compared to model BRF_0.5. No WRF with a sensitivity of 50% and an acceptable number of false positive forecasts could be trained. While in the case of a positive forecast an avalanche occurred only with a probability of 7.3% to 8.3%, depending on the model used, negative forecasts were very reliable with a negative predictive value of 99.5% and 99.6%, respectively. Compared to the quality measures obtained for the training data shown in Table 3, the sensitivity, the true skill score and the positive predictive value were considerably lower.

The BRF models showed a sensitivity of 100% for the training data while for model WRF_5, a sensitivity of 65.9% was obtained. For all models, the sensitivity, the true skill statistic and the positive predictive value were noticeably higher for the training data than for the test data. Accordingly, the false alarm ratio was lower for the training data than for the test data. These differences were more pronounced with the BRF models respect to the WRF model.

Table 3. The quality measures obtained for the training data showed differences between BRF and WRF. The sensitivity for the BRFs was 100%, while for model WRF_5 it was 65.9%.

Model	TN	FN	FP	TP	POD	PON	FAR	TSS	PPV	NPV
BRF_0.5	525	0	35	41	1.000	0.938	0.461	0.938	0.539	1.000
BRF_0.6	538	0	22	41	1.000	0.961	0.349	0.961	0.651	1.000
WRF_5	531	14	29	27	0.659	0.948	0.518	0.607	0.482	0.974

The similarity of the two BRF models became visible in the misclassified examples: For the test data, the false negatives for model BRF_0.5 were a subset of the false negatives for model BRF_0.6. On the other hand, the false positives for model BRF_0.6 were a subset of the false positives of model BRF_0.5. When comparing the misclassified examples for the BRF and WRF models, it was noticed that both models showed the same false negative predictions. 46 false positives showed up in BRF_0.6 as well as in WRF_5. The comparison of all three models can be summarized as follows: 52 examples were misclassified by all three models, consisting of 6 false negatives and 46 false positives. This makes up 63.4% of the misclassifications of model BRF_0.5, 83.9% of the misclassifications of model BRF_0.6 and 82.5% of the misclassifications of model WRF_5.

Compared to the two decision trees described in [12], our models achieved a lower sensitivity but a higher specificity. With different test sets, the models described in [12] had a sensitivity of 61% and 100% with a corresponding specificity of 83% and 21%, respectively. The true skill score was 21% for the second model and therefore lower than in our models. The first model achieved a true skill score of 44% which was lower than in our model BRF_0.5 only. The false alarm ratio in our models is noticeably higher for the test data. With the training data, our models achieved a higher true skill score, specificity and false alarm ratio than did the models in [12]. It has to be remarked that in [12] the ratio of positive to negative examples in the test data was about 1:6. Furthermore, the presented models were trained considering two or three variables while in our models we applied no variable selection.

In [18], large and infrequent snow avalanches are predicted. Considering new snow depth only, a sensitivity of about 65% was achieved which seems favourable compared to our models. The presented models showed a false alarm ration of about 90% which is similar to the false alarm ration of our models on the test data.

For the purpose of comparison, the following classifiers were employed using the default settings in WEKA [11]: AdaBoost.M1 [8] using DecisionStump [13] as base classifier; bagging [1] using DecisionStump [13] and REPTree [7] as base classifier; logistic regression [15]. The resulting models show a significantly lower sensitivity compared to our models and therefore are not applicable for decision support in our case.

6   Discussion

Two types of models proved to be feasible: On the one hand, two models with a sensitivity of slightly more than 40% were trained. On the other hand, one model with a sensitivity of 50% was trained. The latter showed a considerably higher number of false positive forecasts and a slightly lower positive predictive value. No WRF with a sensitivity of 50% and an acceptable number of false positives could be trained. Therefore, with this data, BRF could be chosen for modeling a system for decision support in avalanche warning.

The trained models are feasible as a decision support in avalanche warning: The testing period comprises 11 winter seasons consisting of approximately 181 days each. For this period, 55 to 76 false positive forecasts are acceptable. The misclassification rate is comparable to that of an human expert. Accepting a higher number of false negative forecasts for a higher sensitivity may make sense: Not all avalanches contained in the database reached the road. Danger does not occur always with a false negative forecast and therefore 6 to 7 false negatives are acceptable. It would be interesting to differentiate between avalanches that reached the road and avalanches that did not. In this work, due to the lack of data, this differentiation was not made.

The models are developed for the Kleintal in the Canton of Glarus in Switzerland based on the data contained in NXD2000. In contrast to NXD2000, no comparison with previous similar situations can be made, but the models allow for probabilistic forecasts. Therefore, the cutoff for the positive class for BRF could also be determined using ROC curves and the corresponding weight adopted in WRF for the positive class derived from this cutoff according to the procedure described in [6].

The positive and negative predictive values present valuable information for assessing a given forecast. While sensitivity and specificity are important quality measures and their values have to be as high as possible, they do not allow for an assessment of a given prediction. However, this is an important information for the avalanche service using the system as a decision support in avalanche warning.

In BRF_0.5, BRF_0.6 and WRF_5, mostly the same examples were misclassified. Considering the fact that particularly for the negative examples the misclassified examples comprised less than 5% of all examples, it can be supposed that these misclassifications are due to the data. The training examples were chosen randomly on a yearly basis and therefore few consecutive days are present in the training data set. The training examples could as well be chosen using statistical methods analogous to the approaches employed in [12,17,18]. The test data set contains time series of meteorological variables for up to 178 consecutive days. Differentiation of two consecutive days belonging to different classes poses a major challenge and cannot be made by analyzing the meteorological values only. The definition of additional meaningful variables could improve the differentiation between positive and negative examples.

The model performance with the training data is significantly superior to the model performance with the test data. Generalization seems to be an issue.

One possible reason may be that the test data contains a high percentage of consecutive days, thus the recommendations given in the last paragraph apply.

7 Conclusions and Further Work

Based on meteorological data measured on a daily basis as well as avalanche data, a system for the decision support in avalanche warning has been modeled. In this data, avalanche days are rare events. All trained models have a maximum sensitivity of 50% and a high false alarm ratio. Nevertheless, the trained models are feasible as decision support in avalanche warning. The number of false negative and false positive forecasts are acceptable with respect to the period considered, and approximately match the performance of a human expert.

Compared to the models described in [12,17,18], the following aspects of our work should be noted: First, the quality measures were chosen from the point of view of the designated user, the members of the avalanche service which is in charge of assessing the local avalanche danger. The positive and negative predictive values, which are not presented in the cited approaches, provide an assessment of the forecast reliability. From an operational point of view, these are the most important quality measures. Second, the models allow for probabilistic forecasts and therefore for the characterization of the probability of an event. Third, BRF and WRF proved an adequate starting point for obtaining a feasible system for decision support in snow avalanche warning with rare events.

There are several directions in which the work presented in this paper should be extended. In order to achieve a higher performance, additional meaningful variables should be defined; these may be quantitative as well as qualitative variables. A more sophisticated variable selection could also prove beneficial. Depending on the weather situation, the importance of meteorological variables varies. This can be taken into account by defining variables describing the weather situation. Since the snowpack develops with time, defining variables characterizing weather trends could prove advantageous. The influence of the training data on the generated model should be investigated. Training examples may be chosen according to statistical criteria, or all data not assigned to the test data set may be used for training. The possibility to additionally predict which avalanche path is in danger of being released would be advantageous for the avalanche service. However, this requires an appropriate amount of data.

References

1. Breiman, L.: Bagging predictors. Machine Learning 24(2), 123–140 (1996)
2. Breiman, L.: Random Forests. Machine Learning 45(1), 5–32 (2001)
3. Breiman, L., Friedman, J., Stone, C.J., Olshen, R.A.: Classification and regression trees. CRC Press (1984)
4. Buser, O., Bütler, M., Good, W.: Avalanche forecast by the nearest neighbour method. International Association of Hydrological Sciences 162, 557–570 (1987)

5. Chen, C., Liaw, A., Breiman, L.: Using Random Forest to Learn Imbalanced Data. Tech. rep., University of California, Berkeley (2004), http://www.stat-www.berkeley.edu/tech-reports/666.pdf

6. Elkan, C.: The foundations of cost-sensitive learning. In: Proceedings of the Seventeenth International Joint Conference on Artificial Intelligence, pp. 973–978 (2001)

7. Elomaa, T., Kääriäinen, M.: An Analysis of Reduced Error Pruning. Journal of Artificial Intelligence Research 15, 163–187 (2001)

8. Freund, Y., Schapire, R.E.: Experiments with a New Boosting Algorithm. In: International Conference on Machine Learning, vol. 96, pp. 148–156 (1996)

9. Gassner, M., Brabec, B.: Nearest neighbour models for local and regional avalanche forecasting. Natural Hazards and Earth System Sciences 2, 247–253 (2002)

10. Gassner, M., Etter, H.J., Birkeland, K., Leonard, T.: NXD2000: An improved avalanche forecasting program based on the nearest neighbor method. In: Proceedings of the 2000 International Snow Science Workshop, pp. 52–59 (2000)

11. Hall, M., Frank, E., Holmes, G., Pfahringer, B., Reutemann, P., Witten, I.H.: The WEKA Data Mining Software: An Update. SIGKDD Explorations 11(1) (2009)

12. Hendrikx, J., Murphy, M., Onslow, T.: Classification trees as a tool for operational avalanche forecasting on the Seward Highway, Alaska. Cold Regions Science and Technology 97, 113–120 (2014)

13. Iba, W., Langley, P.: Induction of One-Level Decision Trees. In: Proceedings of the Ninth International Conference on Machine Learning, pp. 233–240 (1992)

14. Lalkhen, A.G., McCluskey, A.: Clinical tests: sensitivity and specificity. Continuing Education in Anaesthesia, Critical Care & Pain 8(6), 221–223 (2008)

15. Le Cessie, S., Van Houwelingen, J.: Ridge Estimators in Logistic Regression. Applied Statistics 41(1), 191–201 (1992)

16. McClung, D., Schaerer, P.A.: The Avalanche Handbook. The Mountaineers Books (2006)

17. Mitterer, C., Schweizer, J.: Analyzing the atmosphere-snow energy balance for wet-snow avalanche prediction. In: Proceedings of the International Snow Science Workshop, pp. 77–83 (September 2012)

18. Schweizer, J., Mitterer, C., Stoffel, L.: On forecasting large and infrequent snow avalanches. Cold Regions Science and Technology 59(2), 234–241 (2009)

19. Weiss, G.M.: Mining with Rarity: A Unifying Framework. ACM SIGKDD Explorations Newsletter 6(1), 7–19 (2004)

20. WSL-Institut für Schnee- und Lawinenforschung SLF, Bundesamt für Umwelt BAFU, Schweizerische Interessengemeinschaft Lawinenwarnsysteme (SILS): Praxishilfe. Arbeit im Lawinendienst: Organisation, Beurteilung lokale Gefährdung und Dokumentation. Information sheet (in German), http://www.slf.ch/dienstleistungen/merkblaetter/praxishilfe_lawdienst_deutsch.pdf

Applying Language Technologies
on Healthcare Patient Records
for Better Treatment of Bulgarian Diabetic Patients

Ivelina Nikolova[1], Dimitar Tcharaktchiev[2],
Svetla Boytcheva[3], Zhivko Angelov[4], and Galia Angelova[1]

[1] Institute of Information and Communication Technologies,
Bulgarian Academy of Sciences (IICT-BAS), Sofia, Bulgaria
[2] Medical University – Sofia, Bulgaria
[3] American University in Bulgaria, Blagoevgrad, Bulgaria
[4] Adiss Lab Ltd. Sofia, Bulgaria
{iva,galia}@lml.bas.bg, dimitardt@gmail.com, sboytcheva@aubg.bg,
angelov@adiss-bg.com

Abstract. This paper presents a research project integrating language technologies and a business intelligence tool that help to discover new knowledge in a very large repository of patient records in Bulgarian language. The ultimate project objective is to accelerate the construction of the Register of diabetic patients in Bulgaria. All the information needed for the Register is available in the outpatient records, collected by the Bulgarian National Health Insurance Fund. We extract automatically from the records' free text essential entities related to the drug treatment such as drug names, dosages, modes of admission, frequency and treatment duration with precision 95.2%; we classify the records according to the hypothesis "having diabetes" with precision 91.5% and deliver these findings to decision makers in order to improve the public health policy and the management of Bulgarian healthcare system. The experiments are run on the records of about 436,000 diabetic patients.

Keywords: Biomedical Natural Language Processing, Business Intelligence, Big Data.

1   Introduction

The constant growth of electronic narratives discussing patient-related information implies constant growth of the attempts to process these texts automatically. It is well known that the most important findings about the patients are kept as free texts in various documents and languages but these text descriptions are usually oriented to human readers. In this way Information Extraction (IE) becomes the dominating natural language processing (NLP) approach to biomedical texts. The main IE idea is to extract automatically important entities from free texts, with accuracy as high as possible, and to build software systems operating on these entities (skipping the non-analysed text fragments). NLP in general is viewed as a rather complex Artificial

G. Agre et al. (Eds.): AIMSA 2014, LNAI 8722, pp. 92–103, 2014.
© Springer International Publishing Switzerland 2014

Intelligence task so IE is proposed as a technology at the middle between keyword search and deep text analysis; it focuses on surface linguistic phenomena that can be recognised without deep inference. The NLP performance gradually improves during the last decades but the IE systems are still rarely used outside the research groups where they have been developed [1]. On the other hand it is expected that IE progress would enable radical improvements in the clinical decision support, biomedical research and the healthcare in general [2]. Leading industrial companies claim that NLP is an enabling technology and can be leveraged today for revenue, efficiency and quality but some 10-15 years are needed for mature technological development [3].

Recent Big Data challenges add a new perspective to the complex task of secondary use of electronic health records. Today typical collections of clinical narratives contain millions of records for millions of patients so even procedures for pseudonymisation and anonymisation are problematic. In this paper we present a project dealing with dozens of millions of outpatient records where NLP is carefully applied to specific text sections of the patient records. The extraction components in use are developed several years ago, continuously upgraded, tested and evaluated to deliver entities extracted with high accuracy. We present an integrated system of a business intelligence tool and NLP modules performing big data normalisation, cleaning and knowledge discovery and show a use case on medical data. We demonstrate that the combination of these tools can help to build the Register of diabetic patients by discovering potential diabetic patients which were not formally diagnosed with diabetes. The IE focus is on the patients' medical treatment and patient anamnesis.

Section 2 presents the project objectives and the data repository we use. Section 3 describes the Business Intelligence tool *BITool*. Section 4 presents the knowledge discovery modules, their application to real data and evaluates their performance. Section 5 sketches further work and the conclusion.

2   Building the Bulgarian Diabetic Register

The ultimate project objective is to accelerate the construction of the Register of diabetic patients in Bulgaria by integration of language technologies and business intelligence tools. Advanced information technologies would enable to: *(i)* keep the established practice of patient registration without burdening the medical experts with additional paper work; *(ii)* reuse the existing standard records in compliance with all legal requirements for safety and data protection; *(iii)* save time and resources by avoiding multiple patient registrations and disturbance of the diagnostic and treatment process. Practically, once entered in the healthcare system, the patient data might be reused in multiple aspects. Multiple registrations and growing administrative burdens are seen as a major obstacle for the development of the Register. A web-interface for self-registration to the Bulgarian Diabetic Register is foreseen as well.

The Register contains 28 indicators of diabetic patients' status, including age, sex, ICD-10 codes of diagnoses of diabetes and its complications, diabetes duration, risk factors, data about compensation, laboratory results, hospitalisations and prescribed medication. Manual collection of data proved to be impractical during the last ten years; in addition there are many diabetic patients who are not formally diagnosed and not treated at all. In the case of diabetes, a progressive chronic disease with serious complications, it is highly desirable to develop a system for early alerts that might signal eventual diabetes symptoms.

It turns out that all the information needed for the Register is available in the outpatient records, collected by the Bulgarian National Health Insurance Fund. There are multiple records stored for the same patient along the months and the years. Given that this information is extracted automatically, a Business Intelligence tool can deliver various types of findings to decision makers in order to improve the public health policy and the management of Bulgarian healthcare system. Actually the *BITool* is useful anyway because the data of the Health Insurance Fund contains a lot of information that is structured using codes of medical classifications and nomenclatures. However in this paper we are interested in the analysis of free texts and capturing some essential entities described there. By means of NLP techniques integrated with the *BITool* we discover the potential diabetic patients which were not formally diagnosed with diabetes.

Thanks to the support of the Bulgarian Ministry of Health and the National Health Insurance Fund, the Medical University - Sofia has received for research purposes a large collection of outpatient records. The data repository currently contains more than 37.9 million pseudonymised reimbursement requests (***outpatient records***) submitted to the National Health Insurance Fund (NHIF) in 2013 for more than 5 million patients, including 436,000 diabetic ones. In Bulgaria the outpatient records are produced by the General Practitioners (GPs) and the Specialists from Ambulatory Care for every contact with the patient (in patient home as well as in doctors' office).

The outpatient records are semi-structured files with predefined XML-format. Despite their primary accounting purpose they contain sufficient text explanations to summarise the case and to motivate the requested reimbursement. The most important indicators like *Age*, *Gender*, *Location*, *Diagnoses* are easily seen since they are stored with explicit tags. The Case history is presented quite briefly in the *Anamnesis* as free text with description of previous treatments, including drugs taken by the patient beyond the ones that are to be reimbursed by the Insurance Fund. *Family history* and *Risk factors* are often included in the *Anamnesis* of diabetic patients. *Patient status* is another section containing free text. It includes a summary of the patient state, symptoms, syndromes, patients' height and weight, body mass index, blood pressure and other clinical descriptions. The values of *Clinical tests and lab data* are enumerated in arbitrary order as free text in another section. A special section is dedicated to the *Prescribed treatment*. Only the drugs prescribed by the GPs and reimbursed by the NHIF are coded, using the specific NHIF nomenclatures. All the other medications and treatment procedures are described as free text. In contrast to clinical discharge letters that might discuss treatments in longer past and future periods, the *Prescribed treatment* section in the outpatient records is more focused to the context at the moment when the record is composed.

The repository given to the Medical University – Sofia is pseudonymised by NHIF which has the keys for mapping the records to the original patients. Our experiments use a completely anonymised data set. Fortunately, using the pseudonymised patient identifier, it is possible to track automatically the multiple visits of the same patient to GPs and Ambulatory Care, which is important in the case of a chronicle disease like diabetes.

An outpatient record might include about 160 tags. The average length of the files is about 1 MB. For our purposes, we work with about 20-30 tags and consider the unstructured content of four sections.

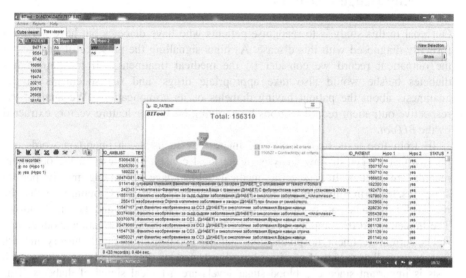

Fig. 1. A *BITool*-constructed concordancer for the word DIABETES in the *Anamnesis* of outpatient records for patients who are not formally diagnosed with Diabetes Mellitus (E11)

3   *BITool* and Data Integration

BITool serves as an integration platform of the analyses performed on the medical data. It offers means to operate on the original data, such as search facilities and means to extract subsets of semi-structured data, as well as to enrich the stored resources with newly extracted features (e.g. from the NLP or statistical modules).

BITool offers powerful functionalities for online analytic processing (OLAP) of large data repositories, organised as a *n*-dimensional cube. The cube is easy to construct by drag-and-drop interface after selection of the desired attributes and their values (see 3 selected dimensions of Fig. 1 upper left). Fig. 1 shows the result of an intermediate step of the task to investigate whether some patients without formal diagnosis of diabetes (code E11) have symptoms typical for this disease. This can be discovered after finding text descriptions of typical events: e.g. mentioning the phrase *"diabetic foot"* in a positive context. Concordancers of text fragments using a 12-words window (6 words before and 6 words after) are constructed around the words of interest. In Fig. 1 the records of 156,310 patients, who are not formally diagnosed with diabetes mellitus, are checked for occurrences of the word *"diabetes"* in the

Anamnesis section. The records of 5,783 patients contain the word "*diabetes*". These are later used in the knowledge discovery phase by the NLP modules. After the processing the results are re-integrated back to *BITool* to be used in further analyses and research or health management tasks such as to help the construction of the diabetic register.

4   Knowledge Discovery

Our goal in this study is to recognise patients who have diabetes but have not been formally diagnosed with this disease. As hints signalling the presence of diabetes in the outpatient record we consider (*i*) the medical treatment – if the patient has diabetes he/she would also take appropriate drugs, and (*ii*) statements in the anamnesis about the patient having diabetes or its complications. We analyse the respective outpatient record sections by applying NLP over feature vectors extracted by the *BITool*.

IE from free texts finds entities of interest by focusing on important words and phrases that trigger shallow analysis in the local context. The texts in the outpatient records are written in a specific medical sublanguage containing mostly phrasal structures, terms in Bulgarian and Latin, typical abbreviations etc. Grouping together the records for the same patient we can track the progress of diabetes and its complications. The implemented IE components extract with high accuracy information about patient status, current treatment, hospitalisation, diabetes compensation, family history and risk factors, as well as values of specific clinical tests and lab data. Identifying values of lab tests is important since e.g. blood sugar levels are a typical signal of diabetes. This extractor uses a rule-based approach for recognition of linguistic patterns corresponding to the entities of interest. Major difficulties encountered in the development are due to the large variety of expressions describing the laboratory tests and clinical examinations. Here we consider in more details the drug extractor [4] that has been extended recently to cope with the NHIF outpatient records.

4.1   Structuring Drug Treatment Information

An automatic procedure analyses the free texts in the *Prescribed treatment* section in order to extract information about: drug names; dosages; modes of admission; frequency and treatment duration. It assigns the corresponding ATC[1] and NHIF codes to each medication event. The list of registered drugs in Bulgaria is provided by the Bulgarian Drug Agency; it contains about 4,000 drug names and their ATC codes.

The extraction is based on algorithms using regular expressions to describe linguistic patterns. There are more than 80 different patterns for matching text units which deal with the ATC and NHIF code, medication name, dosage and frequency. Some regular expressions are illustrated at Fig. 2.

---

[1]   Anatomical Therapeutic Chemical (ATC) Classification System for the classification of drugs, see http://www.who.int/classifications/atcddd/en/

For diabetic patients, currently the extractor handles 2,239 drugs names included in the NHIF nomenclatures. Recent extraction evaluation has been performed with large-scale analysis of the outpatient records of 33,641 diabetic patients for 2013. The precision is 95.2% and the sensitivity - 93.7%. This result is slightly better than the accuracy reported in 2011 [5] when the extractor was a (research) prototype dealing with less than 500 drugs. The performance of the module is evaluated manually. The labelled data is split to 20 equal subsets and randomly selected records are evaluated by an expert (about 40% of each subset). The average of the subset evaluation is the final score of the module.

Fig. 2. Regular expressions of linguistic patterns for analysis of Dosage

The major reasons for incorrect recognition of drug events are: *(i)* misspelling of drug names; *(ii)* drug names occurring in the contexts of other descriptions; *(iii)* undetected descriptions of drug allergies, sensibility, intolerance and side effects; *(iv)* drug treatment described by (exclusive) *OR*; *(v)* negations and temporally interconnected events of various kinds: undetected descriptions of cancelled medication events; of changes or replacements in therapy; of insufficient treatment effect and change of therapy.

About 30% of the medication events in the test corpus were described without any dosage. Lack of explicit descriptions occurs mostly for treatment of accompanying diseases. After applying the recognition algorithm and default daily dosage, the number of records lacking dosage has been reduced to 15.7% in the final result.

4.2   Discovering Potential Diabetic Patients

Now we consider in more detail the discovery of potential diabetic patients that are not formally diagnosed in the NHIF repository for 2013. Medical experts propose criteria for happening of the event "*having diabetes*": e.g. high blood sugar or high glycated hemoglobin in the text of the section *Lab test results*, or admission of drugs used for diabetes treatment mentioned in the *Anamnesis,* or statements in the *Anamnesis* describing diabetes or its symptoms. A concordancer is built on the outpatient records for words related to descriptions of such events. The focused records are subject to further NLP analysis in order to confirm the relevant hypotheses. Iteratively the set of records under consideration can be shrunk by excluding the already checked patients. The *BITool* checks a single hypothesis for about 40 seconds. Checking several hypotheses simultaneously takes approximately the same time for each new one. OLAP is used for checking hypotheses and calculating the support to corresponding associative rules.   The results of text processing at records level are re-integrated back to *BITool* as additional attributes to each outpatient record. These results can be manual annotations (considered "trusted") or automatically generated labels (called "generated"). The process is illustrated at Fig. 3.

4.2.1. Filtering Diabetes Related Records

To confirm/reject the hypothesis of *having diabetes,* we shall apply supervised machine learning (ML) methods for filtering text chunks from the free text zones of the outpatient record. The focus of this example is the word *diabetes* but the same analysis can be performed for any other disease, symptom, word or phrase of interest. The selected outpatient records have no explicit diagnosis of diabetes. We want to show that NLP can help to find contradiction in the repository (thus supporting data cleaning) and/or to extend the records' metadata when matching free text fragments confirming the diabetes diagnosis.

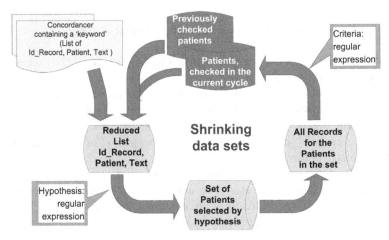

Fig. 3. Iterative shrinking of sets of outpatient records to confirm/reject a hypothesis

4.2.2. Input Data

The input data are text chunks extracted from a concordancer built for the string *диабет* (*diabetes*). The whole data set consists of 67,904 distinct chunks extracted from the records of 156,310 patients who are not formally diagnosed with diabetes. Each chunk contains the word *diabetes* and a 6-token window of its left and right context. The text is only tokenised and stemmed. Our procedure will perform a classification task - confirm/reject the hypothesis without recognising the particular trigger expressions.

Here follow some sample chunks which demonstrate the variety of positive and negative examples:

(*i*) NEG Фамилност- обременен/а-**диабет**ици по майчина линия/.

*Family - heredity - **diabet**ic on maternal line.*

(ii) NEG Необходимо е изключване на стероиден **диабет**; насочва се към ТЕЛК...

*It is necessary to exclude steroid **diabetes**; re-directing to TEMC...*

(iii) POS Покачва кръвно налягане. Има **диабет**. Оплаква се от сърцебиене...

*Raises the blood pressure. Has **diabetes**. Complains of palpitation...*

Examples *(i)* and *(ii)* are negative - the first one shows only heredity but does not confirm the diagnosis *diabetes*, in the second one the diagnosis is negated. However example *(iii)* is positive. Often the positive diabetes statement is given within 3-4 tokens.

4.2.3. Experiments and Results
We apply a combined rule-based and machine-learning approach to solve the problem: *(i)* iterative rule-based rough filtering to reduce the size of the data that potentially confirms our hypothesis; *(ii)* supervised classification of the records in the reduced dataset to train a model which classifies the records as positive or negative to the hypothesis.

Rule-Based Rough Pre-Filtering
The data is extracted from records where *diabetes* is not explicitly mentioned among the diagnoses; therefore, we expect that most of the input chunks are negative examples. We decided first to filter as many of the negative ones as possible in a rule-based approach and reduce the size of the input that will be classified further. Several patterns of the negative examples appeared quite often in our data - e.g. *"no evidence about diabetes"*, *"no diabetes in the family"* etc., which match about 10% of the input records. These are expressions talking about family heredity or rejection of the diabetes diagnosis. Iteratively we created a set of regular expressions on stem level which match such chunks. With a set of 41 expressions in the filter, the number of chunks was reduced to 26,000 (which is about 1/3 of the initial corpus size).

Supervised Classification of Positive/Negative Examples
In the second phase we extracted from the reduced data set two random subsets – one of 282 documents and one of 1,000 documents and we annotated them. The first one was a development set, we used it for selecting the features and make initial tests. It contains 74 positive and 208 negative examples whereas the second one contains 187 positive and 813 negative examples. By using various features and classification algorithms we check the applicability of ML to the automatic extraction of records referring to *"having diabetes"* (similarly to [6, 7, 8]) and set a reasonable baseline for this task. The difference in our approach is that we model and process real world big data.

In the preprocessing phase we stem the chunks to overcome the morphological inflection (Bulgarian is highly inflectional). Due to the large number of abbreviations, dosages, lab test results etc. which contain punctuation marks, the automatic sentence splitting with the available tools is unreliable and we do not apply it.

By achieving comparatively high results with surface and structural features only we prove the applicability of the approach. We experimented with both boolean and nominal feature vectors. The vectors characterising each chunk in the classification task correspond to word stems, bigrams and trigrams which are pre-defined. In the boolean settings the feature is true if the corresponding attribute - stem/bigram/trigram is available in the text chunk and negative otherwise. The list of stem-features is constructed by analysing the set of stems which occur in the datasets. We made experiments with 3 different feature corresponding to the experiments described below.

We tried several algorithms on the same dataset: NaiveBayes, J48, SMO and JRip, all with boolean features: JRip and J48 performed best. We did also classification with nominal features with MaxEnt algorithm and this one outperformed all the rest in means of precision. The features we used were the same - the words' stems, bigrams and trigrams.

In this study we are most interested in the precision on positive examples because in this way we can add to the register (with minimal manual effort/check) with high certainty the correctly recognised positive examples as patients *having diabetes* who were not formally diagnosed with diabetes. By applying several high precision filters like one ones scatched above we could achieve also also good coverage in the end. Here follow details about the experiments. The results are available in Tables 1, 2 and 3.

Experiment 1

We use 93 features which correspond to stems of terms occurring in the text of positive examples (excluding numbers). The results from a 10-fold cross-validation with the best algorithms J48 and JRip are shown in the first column of Table 1. J48 recognised only 63.1 % of the positive examples however the precision achieved by JRip is encouraging – 91.2%. The rules inferred by JRip are only 2 but obviously they fit well the data.

> **Rule 1**: (*family*) = *false*) and (*sugar* = *true*) and (*noninsulindependent*= *true*) =>
>
>   class=pos (30.0/2.0)
>
> **Rule 2**: (*family*) = false) and (*sugar* = true) and (*treat* = true) =>
>
>   class=pos (6.0/0.0)

However on a larger scale these two rules might not suffice for achieving such good performance and we kept elaborating our features.

Experiment 2

We use 112 textual features which correspond to the stems of terms occurring in positive and negative examples. The terms are pre-filtered manually by an expert. Both algorithms - JRip and J48 performed worse than in the first experiment and scored precision under 70% (Table 1, Exp.2).

Experiment 3

In this experiment we took advantage of the automatic feature-selection algorithms. Our initial feature set contained 10,576 attributes corresponding to the stems of all terms in the development dataset. We applied on it the chi-squared attribute evaluator implemented in Weka [9] according to the feature representativeness in the development set. As result 151 features were selected. In Experiments 3a and 3b we used them in combination with bigrams and with bigrams and trigrams respectively (Table 2). The results of both algorithms improve by adding bigrams and by including trigrams rise even more. JRip reaches 65.3% precision and 73.1% when adding trigram features. J48 achieves precision as high as 86.1% after adding bigrams and 87.2 including trigrams in the feature set. In the tree built by J48 one could clearly see the importance of bigrams and trigrams features - out of 17 tree leaves, only 4 are unigrams; the rest are bigrams and trigrams. We explain this with the fact that trigrams capture the order of the tokens and represent concrete expressions signalling the presense/absense of diabetes such as: *"инсулинозависим захарарен диабет"* (*insulindependent diabetes mellitus*), *"неинсулинозависим захарен диабет"* (*noninsulindependent diabetes mellitus*), *"със зах диабет"* (*with diabetes mellitus*), *"майка с диабет"* (*mother with diabetes*). According to our observations the last feature signals the absence of diabetes in the concrete excerpt because normally in the length of one excerpt, if there is a family anamnesis, there is no other description related to the patient which signals diabetes (the length is too small to fit more descriptions along with a family anamnesis). The results suggest that until this point adding more features results in better precision (except for JRip in Experiment 1) and also growing recall when using JRip.

Table 1. Results from Experiment 1 and 2 with manually selected features

	Exp. 1: 93 pos features; stems only; 10-fold cross-validation						Exp. 2: 112 pos/neg features; stems only; 10-fold cross-validation					
	JRip			J48			JRip			J48		
Class	P	R	F	P	R	F	P	R	F	P	R	F
positive	91.2	16.6	28.1	66.7	21.4	32.4	61.1	29.4	39.7	65.8	52.4	58.3
negative	83.9	99.6	91.1	84.4	97.5	90.5	85.5	95.7	90.3	89.5	93.7	91.6
weighted avg	85.2	84.1	84.1	81.1	83.3	79.6	80.9	83.3	80.8	85.1	86	85.4

Table 2. Experiments 3a and 3b - automatic feature selection, bigrams and trigrams

	Exp. 3a: 151 automatically selected stem features + bigrams; 10-fold cross-validation						Exp. 3b: 151 automatically selected stem features + bigrams + trigrams; 10-fold cross-validation					
	JRip			J48			JRip			J48		
Class	P	R	F	P	R	F	P	R	F	P	R	F
positive	65.3	41.2	50.5	86.1	36.4	51.1	73.1	40.6	52.2	87.2	36.4	51.3
negative	87.5	95	91.1	87.1	98.6	92.5	87.6	96.6	91.9	87.1	98.8	92.6
weighted avg	83.4	84.9	83.5	86.9	87	84.8	84.9	86.1	84.5	87.1	87.1	84.9

Table 3. Experiments 4a, 4b – MaxEnt with all textual features, bigrams and trigrams

	MaxEnt					
	Exp. 4a: all stems + bigrams; 10-fold cross validation			Exp. 4b: all stems+bigrams + trigrams; 10-fold cross validation		
Class	P	R	F	P	R	F
positive	91.3	22.6	36.2	91.5	20	32.8
negative	85.6	84.2	85	85.6	84.2	84.9
weighted avg	88.45	53.4	60.6	88.55	52.1	58.9

Experiment 4

We trained a model with MaxEnt algorithm using all textual features plus bigrams (Exp. 4a) and bigrams and trigrams (Exp. 4b) as nominal values. All strings were first stemmed. These two experiments gave almost the same results and outperformed J48 and JRip in means of precision. The best precision on positive examples we reached is as high as 91.5% when including bigrams and trigrams. Using MaxEnt with nominal features has the advantage that the features are not pre-set. The similar results also suggest that using MaxEnt with these features could be set as a reasonable baseline.

The positive examples in our database are so rare (because in principle diabetic patients *are* formally diagnosed) that the data quantity has major impact on the training. Having a larger corpus (which could happen when records from previous years become available) and having more golden data will help for learning better the patterns of positive examples. Nevertheless the current results show that such a hybrid method combining rule-based and machine learning approach can be used to prove the hypothesis *having diabetes* with high precision.

5   Conclusion and Further Work

The paper presents our first steps towards building a computational platform for tackling Big Data in the medical domain in a real-world application. It is obvious that decisions about medical cases cannot be made completely automatically so the final judgment should be always subject to human considerations. But the automatic processing of texts in the patient records defines a completely new horizon for most tasks related to health analytics.

The IE modules are exploited quite carefully, for extraction of a limited number of entities and events only. They are tested in various scenarios and gradually improve their performance using hybrid rule-based and machine learning approaches. We extract automatically from the records' free text essential entities related to the drug treatment such as drug names, dosages, modes of admission, frequency and treatment duration with precision 95.2%; we classify the records according to the hypothesis "*having diabetes*" with precision 91.5% and deliver these findings to decision makers in order to improve the public health policy and the management of Bulgarian healthcare system. We think that large-scale analysis of medical texts can be viewed as a reliable technology if the input is well-structured into zones (which is the case of the outpatient records) and the extraction task has clear and well-defined target entities.

Acknowledgements. The research work presented in this paper is partially supported by the FP7 grant AComIn No. 316087, funded by the European Commission in the FP7 Capacity Programme in 2012–2016. The team acknowledges also the support of Medical University – Sofia, the Bulgarian Ministry of Health and the Bulgarian National Health Insurance Fund.

References

1. Meystre, S., Savova, G., Kipper-Schuler, K., Hurdle, J.F.: Extracting Information from Textual Documents in the EHR: A Review of Recent Research. In: Geissbuhler, A., Kulikowski, C. (eds.) IMIA Yearbook of Medical Informatics, pp. 138–154 (2008)
2. Demner-Fushman, D., Chapman, W., McDonald, C.: What can NLP do for Clinical Decision Support? J. of Biomedical Informatics 42(5), 760–772 (2009)
3. Health Fidelity: The What, When, Where and How of Natural Language Processing. NLP issue brief (2013), http://healthfidelity.com/technology/issue-briefs/nlp-issue-brief
4. Boytcheva, S.: Shallow Medication Extraction from Hospital Patient Records. In: Koutkias, V., Niès, J., Jensen, S., Maglaveras, N., Beuscart, R. (eds.) Studies in Health Technology and Informatics series, vol. 166, pp. 119–128. IOS Press (2011)
5. Boytcheva, S., Tcharaktchiev, D., Angelova, G.: Contextualization in automatic extraction of drugs from Hospital Patient Records. In: Moen, A., et al. (eds) User Centred Networked Health Case. Studies in Health Technology and Informatics, vol. 169, pp. 527–531. IOS Press (2011)
6. Nikolova, I.: Unified Extraction of Health Condition Descriptions. In: Proceedings of the NAACL HLT 2012 Student Research Workshop, Montreal, Canada, pp. 23–28 (June 2012), http://aclweb.org/anthology//N/N12/N12-2005.pdf
7. Savova, G., Ogren, P., Duffy, P., Buntrock, J., Chute, C.: Mayo Clinic NLP System for Patient Smoking Status Identification. Journal of American Medical Informatics Association 15(1), 25–28 (2008)
8. Chu, C.D., Dowling, J.N., Chapman, W.W.: Evaluating the Effectiveness of Four Contextual Features in Classifying Annotated Clinical Conditions in Emergency Department Reports. In: Proceedings of AMIA Annual Symposium, pp. 141–145 (2006)
9. Weka: Data Mining Software in Java, http://www.cs.waikato.ac.nz/ml/weka/

Incrementally Building Partially Path Consistent Qualitative Constraint Networks*

Michael Sioutis and Jean-François Condotta

Université Lille-Nord de France, Artois, CRIL-CNRS UMR 8188
Lens, France
{sioutis,condotta}@cril.fr

Abstract. The Interval Algebra (IA) and a fragment of the Region Connection Calculus (RCC), namely, RCC-8, are the dominant Artificial Intelligence approaches for representing and reasoning about qualitative temporal and topological relations respectively. In this framework, one of the main tasks is to compute the path consistency of a given Qualitative Constraint Network (QCN). We concentrate on the partial path consistency checking problem problem of a QCN, i.e., the path consistency enforced on an underlying chordal constraint graph of the QCN, and propose an algorithm for maintaining or enforcing partial path consistency for growing constraint networks, i.e., networks that grow with new temporal or spatial entities over time. We evaluate our algorithm experimentally with QCNs of IA and RCC-8 and obtain impressive results.

Keywords: qualitative constraint language, chordal graph, triangulation, qualitative reasoning, BA model, partial path consistency.

1   Introduction

Spatial and temporal reasoning is a major field of study in Artificial Intelligence; particularly in Knowledge Representation. This field is essential for a plethora of areas and domains that include, but are not limited to, ambient intelligence, dynamic GIS, cognitive robotics, and spatiotemporal design [7]. The Interval Algebra (IA) [1] and a fragment of the Region Connection Calculus [16], namely, RCC-8, are the dominant Artificial Intelligence approaches for representing and reasoning about qualitative temporal and topological relations respectively.

The state-of-the-art techniques to enforce *partial path consistency* [8] on a set of IA or RCC-8 relations consider a fixed size constraint network to represent and reason with the relations. However, it may be the case that temporal intervals (the case for IA) or regions (the case for RCC-8) are not known a priori, but arrive continuously within different fragments of time. This is a real problem and has not been addressed before in literature. The term *"incremental"* has been used to describe the problem of maintaining or enforcing partial path

---

* This work was funded by a PhD grant from Université d'Artois and region Nord-Pas-de-Calais.

G. Agre et al. (Eds.): AIMSA 2014, LNAI 8722, pp. 104–116, 2014.
© Springer International Publishing Switzerland 2014

consistency for a fixed size network when new constraints among existing nodes are added or existing constraints are tightened. This approach is well described in the work of Gerevini [10] for qualitative temporal reasoning (where complete underlying constraint graphs are considered and, thus, partial path consistency is identical to path consistency [8, chapt. 6]) and the work of Planken et al. for the Simple Temporal Problem (STP) [15], and differs from our approach in that we consider extensions of a given network with new temporal or spatial entities. In a recent theoretical work, Huang showed that IA and RCC-8 have *canonical solutions* [13], i.e., path consistent IA or RCC-8 networks with relations from some maximal tractable subset of their signatures can be extended arbitrarily with the addition of new temporal or spatial entities respectively. In a more practical view, until recently state-of-the-art techniques made use of a matrix to represent a constraint network [20]. Growing a constraint network represented by an adjacency matrix requires $O(|V|^2)$ for every variable addition.

In this paper, we concentrate on the problem of maintaining or enforcing partial path consistency for growing constraint networks and make the following contributions: (i) we present an algorithm that maintains or enforces partial path consistency for an initial partially path consistent constraint network augmented by a new set of temporal or spatial entities and their accompanying constraints, (ii) we implement our algorithm making use of chordal graphs and a hash table based adjacency list to represent and reason with the QCNs as described in [20], (iii) we evaluate our algorithm experimentally with random and real QCNs of IA and RCC-8 and obtain quite interesting results.

The paper is organized as follows. Section 2 introduces the theoretical background of our work. In Section 3 we present our algorithm that maintains or enforces partial path consistency for an initial partially path consistent QCN augmented by a new set of temporal or spatial entities and their accompanying constraints. In Section 4 we use large QCNs of IA and RCC-8 to experimentally compare our algorithm with the state-of-the-art, one-shot partial path consistency algorithm that considers the whole network size all at once, and, finally, in Section 5 we conclude and give directions for future work.

2   Preliminaries

In this section we formally introduce the IA and RCC-8 constraint languages and partial path consistency, and chordal graphs along with triangulation.

The IA and RCC-8 Constraint Languages. A (binary) qualitative temporal or spatial constraint language [18] is based on a finite set B of *jointly exhaustive and pairwise disjoint* (JEPD) relations defined on a domain D, called the set of base relations. The set of base relations B of a particular qualitative constraint language can be used to represent definite knowledge between any two entities with respect to the given level of granularity. B contains the identity relation Id, and is closed under the converse operation $(^{-1})$. Indefinite knowledge can be specified by unions of possible base relations, and is represented by the set

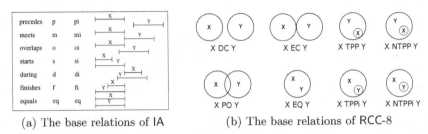

(a) The base relations of IA   (b) The base relations of RCC-8

Fig. 1. IA and RCC-8 constraint languages

containing them. Hence, 2^B will represent the set of relations. 2^B is equipped with the usual set-theoretic operations (union and intersection), the converse operation, and the weak composition operation. The converse of a relation is the union of the converses of its base relations. The weak composition $\diamond$ of two relations s and t for a set of base relations B is defined as the strongest relation $r \in 2^B$ which contains $s \circ t$, or formally, $s \diamond t = \{b \in B \mid b \cap (s \circ t) \neq \emptyset\}$, where $s \circ t = \{(x, y) \mid \exists z : (x, z) \in s \wedge (z, y) \in t\}$ is the relational composition.

The set of base relations of IA [1] is the set $\{eq, p, pi, m, mi, o, oi, s, si, d, di, f, fi\}$. These thirteen relations represent the possible relations between *time intervals*, as depicted in Figure 1a. The set of base relations of RCC-8 [16] is the set $\{dc, ec, po, tpp, ntpp, tppi, ntppi, eq\}$. These eight relations represent the binary topological relations between *regions* that are non-empty regular subsets of some topological space, as depicted in Figure 1b (for the *2D* case). IA and RCC-8 networks are qualitative constraint networks (QCNs), with relation eq being the identity relation in both cases.

Definition 1. *A RCC-8, or IA, network is a pair $\mathcal{N} = (V, C)$ where V is a non empty finite set of variables and C is a mapping that associates a relation $C(v, v') \in 2^B$ to each pair (v, v') of $V \times V$. C is such that $C(v, v) \subseteq \{eq\}$ and $C(v, v') = (C(v', v))^{-1}$.*

Note that we always regard a QCN as a complete network. The underlying (constraint) graph of a QCN $\mathcal{N} = (V, C)$ is a graph $G = (V, E)$, for which we have that $(v, v') \in E$ iff $C(v, v') \neq B$. Given two QCNs $\mathcal{N} = (V, C)$ and $\mathcal{N}' = (V', C')$, and their respective underlying graphs $G = (V, E)$ and $G' = (V', E')$ where $\forall (v, u) \in E$ we have that $v \in V' \backslash V$ or $u \in V' \backslash V$, $\mathcal{N} \uplus \mathcal{N}'$ denotes the QCN $\mathcal{N}'' = (V'', C'')$, where $V'' = V \cup V'$, $C''(v, v') = B$ for all $(v, v') \in (V \backslash V') \times (V' \backslash V)$, $C''(v, v') = C(v, v')$ for all $(v, v') \in (V \times V)$, and $C''(v, v') = C'(v, v')$ for all $(v, v') \in (V' \backslash V) \times V'$. The underlying graph of $\mathcal{N} \uplus \mathcal{N}'$ is graph $G \cup G' = (V \cup V', E \cup E')$. In what follows, $C(v_i, v_j)$ will be also denoted by C_{ij}. Checking the consistency of a QCN of IA or RCC-8 is $\mathcal{NP}$-complete in general [14, 19]. However, there exist large maximal tractable subclasses of IA and RCC-8 for which consistency checking can be done in polynomial time, $O(n^3)$ in particular, with a path consistency algorithm. These maximal tractable subclasses are the classes $\hat{\mathcal{H}}_8, \mathcal{C}_8$, and $\mathcal{Q}_8$ for RCC-8 [17] and $\mathcal{H}_{IA}$ for IA [14]. When path consistency is enforced on the underlying constraint graph of an input QCN, we refer to it as *partial path consistency*. In our case, and throughout this paper, we enforce path consistency on the underlying *chordal* constraint graph of a given QCN, thus, whenever we use the term partial path consistency we implicitly consider underlying chordal constraint graphs. Partial path consistency was originally introduced for finite domain CSPs in [8] and it was most recently used in the case of IA and RCC-8 networks in [9] and [21] respectively.

Chordal Graphs and Triangulation. We begin by introducing the definition of a chordal graph. More results regarding chordal graphs, and graph theory in general, can be found in [11].

Definition 2 ([11]). *Let $G = (V, E)$ be an undirected graph. G is* chordal *or* triangulated *if every cycle of length greater than 3 has a chord, which is an edge connecting two non-adjacent nodes of the cycle.*

Chordality checking can be done in (linear) $O(|V| + |E|)$ time for a given graph $G = (V, E)$ with the maximum cardinality search algorithm which also constructs an elimination ordering α as a byproduct [4]. If a graph is not chordal, it can be made so by the addition of a set of new edges, called *fill edges*. This process is usually called *triangulation* of a given graph $G = (V, E)$ and can run as fast as in $O(|V|+(|E \bigcup F(\alpha)|))$ time, where $F(\alpha)$ is the set of fill edges that result by following the elimination ordering α, eliminating the nodes one by one, and connecting all nodes in the neighborhood of each eliminated node, thus, making it simplicial in the elimination graph. If the graph is already chordal, following the elimination ordering α means that no fill edges are added, i.e., α is actually a *perfect elimination ordering* [11]. In a QCN fill edges correspond to universal relations, i.e., non-restrictive relations that contain all base relations (hence, the universal relation is equivalent to B). Chordal graphs become relevant in the context of qualitative reasoning due to the following result obtained in [2,21] that states that partial path consistency is equivalent to path consistency in terms of consistency checking of tractable QCNs:

Proposition 1 ([2, 21]). *For a given RCC-8, or IA, network $\mathcal{N} = (V, C)$ with relations from the maximal tractable subclasses $\hat{\mathcal{H}}_8, \mathcal{C}_8,$ and $\mathcal{Q}_8,$ or $\mathcal{H}_{IA}$, respectively, and for $G = (V, E)$ its underlying chordal graph, if $\forall (i, j), (i, k), (j, k) \in E$ we have that $C_{ij} \subseteq C_{ik} \diamond C_{kj}$, then $\mathcal{N}$ is consistent.*

In general, partial path consistency can be significantly faster than path consistency as the latter considers much more triangles of relations for a given QCN [9,21]. (A chordal graph has less edges than a complete graph in general.)

3   The iPPC+ Algorithm

In this section we present a new algorithm, viz., iPPC+, that enforces partial path consistency incrementally, together with an auxiliary algorithm, viz., GiPPC, that uses iPPC+ to simulate the construction of a QCN of n entities. Symbol + is only used to differentiate iPPC+ from the iPPC algorithm for the STP of Planken et al. [15], as we do not consider subsequent edge tightenings within a fixed size QCN, but rather extensions of a given QCN with new temporal or spatial entities accompanied by new sets of constraints. iPPC+ is structurally close to the one-shot partial path consistency algorithm presented in [9] and in [21] for IA and RCC-8 respectively. In what follows, we will refer to the one-shot partial path consistency algorithm simply as PPC.

Function iPPC+$(\mathcal{N} \uplus \mathcal{N}', G, G')$

    in   : A QCN $\mathcal{N} \uplus \mathcal{N}' = (V'', C'')$, and two chordal graphs $G = (V, E)$ and $G' = (V', E')$.

  output : False if network $\mathcal{N} \uplus \mathcal{N}'$ results in a trivial inconsistency (contains the empty relation), True if the modified network $\mathcal{N} \uplus \mathcal{N}'$ is partially path consistent.

1   **begin**
2     $Q \leftarrow \{(i,j) \mid (i,j) \in E'\}$;
3   **while** $Q \neq \emptyset$ **do**
4        $(i,j) \leftarrow Q.pop()$;
5   **foreach** k such that $(i,k), (k,j) \in E \cup E'$ **do**
6           $t \leftarrow C''_{ik} \cap (C''_{ij} \diamond C''_{jk})$;
7   **if** $t \neq C''_{ik}$ **then**
8   **if** $t = \emptyset$ **then return** False;
9              $C''_{ik} \leftarrow t; C''_{ki} \leftarrow t^{-1}$;
10             $Q \leftarrow Q \cup \{(i,k)\}$;
11           $t \leftarrow C''_{kj} \cap (C''_{ki} \diamond C''_{ij})$;
12   **if** $t \neq C''_{kj}$ **then**
13   **if** $t = \emptyset$ **then return** False;
14              $C''_{kj} \leftarrow t; C''_{jk} \leftarrow t^{-1}$;
15             $Q \leftarrow Q \cup \{(k,j)\}$;

16   **return** True;

---

iPPC+ receives as input a QCN $\mathcal{N} \uplus \mathcal{N}' = (V'', C'')$, where $\mathcal{N} = (V, C)$ is the initial partially path consistent QCN augmented by a new QCN $\mathcal{N}' = (V', C')$, and $G = (V, E)$ and $G' = (V', E')$ are their respective underlying chordal graphs where $\forall (v, u) \in E'$ we have that $v \in V' \setminus V$ or $u \in V' \setminus V$. The output of algorithm iPPC+ is False if network $\mathcal{N} \uplus \mathcal{N}'$ results in a trivial inconsistency (contains the empty relation), and True if the modified network $\mathcal{N} \uplus \mathcal{N}'$ is partially path consistent and not trivially inconsistent. The queue data structure is instatiated by the set of edges E' (line 2), i.e., the set of edges corresponding to the underlying graph of the new QCN $\mathcal{N}'$. Path consistency is then realised by iteratively performing the following operation until a fixed point $\overline{C''}$ is reached: $\forall i, j, k$ do $C''_{ij} \leftarrow C''_{ij} \cap (C''_{ik} \diamond C''_{kj})$, where edges $(i, k), (k, j) \in E \cup E'$ (line 5). Within the set of edges E' we implicitly consider a small number of edges that maintain chordality of the underlying constraint graph $G \cup G'$ of QCN $\mathcal{N} \uplus \mathcal{N}'$ (as it is possible that E' introduces cycles). These edges can be found before each appliance of the iPPC+ algorithm in time at most linear in the number of vertices [5]. In this paper, we always first construct the graph consisting of all temporal or spatial entities and constraint edges that will be added and triangulate it once, as it is also done in [15, chap. 5], thus reserving incrementally maintaining chordality for future work. The aforementioned path consistency operation will result in a partially path consistent network $\mathcal{N}'$, and a (possibly) modified partially path consistent network $\mathcal{N}$ after constraint tightenings that might occur. Since $G \cup G'$ is a chordal graph, it follows that QCN $\mathcal{N} \uplus \mathcal{N}'$ is partially path consistent with respect to its underlying chordal graph $G \cup G'$, and, thus, due to Proposition 1 we can assert the following theorem:

Theorem 1. *For a given* RCC-8, *or* IA, *network* $\mathcal{N} \uplus \mathcal{N}' = (V'', C'')$ *with relations from classes* $\hat{\mathcal{H}}_8, \mathcal{C}_8,$ *and* $\mathcal{Q}_8,$ *or* $\mathcal{H}_{IA}$, *respectively, where* $\mathcal{N} = (V, C)$ *is the initial partially path consistent* QCN *augmented by a new* QCN $\mathcal{N}' = (V', C')$, *and* $G = (V, E)$ *and* $G' = (V', E')$ *are their respective underlying chordal graphs where* $\forall (v, u) \in E'$ *we have that* $v \in V' \setminus V$ *or* $u \in V' \setminus V$, *function* iPPC+ *decides the consistency of* QCN $\mathcal{N} \uplus \mathcal{N}'$ *with respect to chordal graph* $G \cup G'$.

Regarding data structures, the QCNs are represented by a hash table based adjacency list as described in [20]. This recent technical advancement in qualitative reasoning allows us to extend a QCN with new temporal or spatial entities in constant time.

Function GiPPC($\mathcal{N}, G$)

    in   : A QCN $\mathcal{N} = (V, C)$, and a chordal graph $G = (V, E)$.
    output : False if network $\mathcal{N}$ results in a trivial inconsistency, True if the
               modified network $\mathcal{N}$ is partially path consistent.

1 **begin**
2     $\mathcal{N}_1 \uplus \mathcal{N}_2 \uplus \ldots \uplus \mathcal{N}_i \leftarrow \mathcal{N}; \mathcal{N}' \leftarrow \mathcal{N}_1;$
3   **foreach** $k \leftarrow 2$ *to* i **do**
4   **if** $!$ iPPC+$(\mathcal{N}' \uplus \mathcal{N}_k, G', G_k)$ **then return** False;
5         $\mathcal{N}' \leftarrow \mathcal{N}' \uplus \mathcal{N}_k;$
6     $\mathcal{N} \leftarrow \mathcal{N}';$
7   **return** True;

GiPPC receives as input a QCN $\mathcal{N}$ together with its underlying chordal graph G, and applies iPPC+ iteratively (line 4) on a decomposition of $\mathcal{N}$ (line 2). This decomposition can be any partition of the set of variables. If we start with a single-entity QCN $\mathcal{N}_1$ and extend it with a new QCN $\mathcal{N}_i$ of one new entity at a time applying iPPC+ in total $n - 1$ times (thus, $2 \leq i \leq n$), it follows that we will perform $O(\delta_2 \cdot |E_2| + \ldots + \delta_n \cdot |E_n|)$ intersection and composition operations for constructing a QCN of n temporal or spatial entities, where δ_i is the maximum degree of a vertex of chordal graph $((G_1 \cup G_2) \cup \ldots) \cup G_i$ and $O(|E_i|)$ is the number of constraints that the new QCN $\mathcal{N}_i$ contributes to QCN $((\mathcal{N}_1 \uplus \mathcal{N}_2) \uplus \ldots) \uplus \mathcal{N}_i$. As $\delta_2 \leq \ldots \leq \delta_n$ and $E_2 \cup \ldots \cup E_n = E$ (i.e., the no. of edges in the n entities constraint network after the nth entity is added), it follows that the complexity of iPPC+ is asymptotically upper bounded by $O(\delta_n \cdot |E|)$,

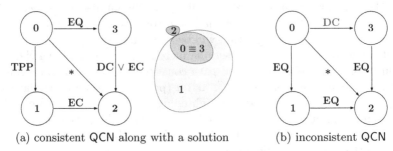

(a) consistent QCN along with a solution   (b) inconsistent QCN

Fig. 2. QCNs with respect to their constraint graphs

which is the complexity of PPC. Thus, we increase on average the performance of applying partial path consistency, as we will also find out in the experimentation to follow, and retain the same worst-case complexity.

3.1   Running Example

Before moving on to our running example, it is important to explain the notions of *processed edges* and *consistency checks*. An edge is processed whenever it is popped out of the queue (line 4), and a consistency check takes place whenever we apply the intersection operator ($\cap$) between two constraints (lines 6 and 11). In our running example we will demonstrate how iPPC+ is able to perform better than the one-shot partial path consistency algorithm (PPC) originally presented in [9] for the case of IA and in [21] for the case of RCC-8. In what follows, we always give the chordal graph $G \cup G'$ of QCN $\mathcal{N} \uplus \mathcal{N}'$ as input to iPPC+ to facilitate description, as the initial network $\mathcal{N}$, along with graph G, and its augmentation $\mathcal{N}'$, along with graph G', are easily identifiable at each step.

Consistent case. Let us consider the consistent RCC-8 network in Figure 2a. We will first build a partial path consistent version of this network incrementally, beginning with node 0 and adding nodes 1, 2, and, 3, one at each step. We always pop edges from the left of the queue and push to the right (FIFO), and whenever needed we use the converse relation corresponding to an edge. First, $(\{0\}, \emptyset)$ is given as input to iPPC+ with no edges whatsoever, the queue is initialized by an empty set, no edges are processed, and, thus, no consistency checks occur. Then, $(\{0, 1\}, \{(0, 1)\})$ is given as input to iPPC+, the queue is initialized with the set of edges $\{(0, 1)\}$, a single edge is processed, and no consistency checks occur as there are no triangles in the network. Then, $(\{0, 1, 2\}, \{(0, 1), (0, 2), (1, 2)\})$ is given as input to iPPC+, the queue is initialized with the set of edges $\{(1, 2)\}$, viz., the constraint edges that accompany the newly inserted spatial entity (region) 2. Edge $(0, 2)$ is not included in the queue as it corresponds to the universal relation $*$ and we do not consider it at all during initialization (this detail is not provided in algorithm iPPC+). (The intersection of $*$ with any other relation leaves the latter relation intact.) Edge $(1, 2)$ is popped out of the queue. Two consistency checks take place among edges $(0, 1)$, $(0, 2)$, and $(1, 2)$, leading to the pruning of the universal relation $*$ for edge $(0, 2)$ into the relation $DC \vee EC$, and, edge $(0, 2)$ is inserted in the queue which now holds the set of edges $\{(0, 2)\}$. Edge $(0, 2)$ is popped out of the queue. Two consistency checks take place among edges $(0, 1)$, $(0, 2)$, and $(1, 2)$, leading to no pruning of relations for edges $(0, 1)$ and $(1, 2)$. Finally, $(\{0, 1, 2, 3\}, \{(0, 1), (0, 2), (0, 3), (1, 2), (2, 3)\})$ is given as input to iPPC+, the queue is initialized with the set of edges $\{(0, 3), (2, 3)\}$. Both edges are popped out of the queue, each one leading to two consistency checks, with no further pruning of relations. In total we have processed 5 edges and performed 8 consistency checks.

We now proceed with PPC which is fairly easier to describe. First, $(\{0, 1, 2, 3\}, \{(0, 1), (0, 2), (0, 3), (1, 2), (2, 3)\})$ is given as input to PPC. The queue is initialized with the set of edges $\{(0, 1), (0, 3), (1, 2), (2, 3)\}$. Edge $(0, 1)$ is popped out of the queue. Two consistency checks take place among edges $(0, 1), (0, 2)$, and $(1, 2)$, leading to the pruning of the universal relation $*$ for edge $(0, 2)$ into the relation $DC \vee EC$. Edge $(0, 2)$ is inserted in the queue which now holds the set of edges $\{(0, 3), (1, 2), (2, 3), (0, 2)\}$. All edges are popped out of the queue with no further pruning of relations. Edges $(0, 3)$, $(1, 2)$, and $(2, 3)$ lead to two consistency checks each, and edge $(0, 2)$ to four, as it is part of two triangles. In total we have processed 5 edges and performed 12 consistency checks.

PPC proccesses the same number of edges as iPPC+, but performs 4 more consistency checks. The numbers may vary a bit depending on the order of the edges in the initialized queue (in our running example we have considered a sorted initial queue of edges), but the trend is that for consistent QCNs, iPPC+ will perform less consistency checks than PPC, and will process only slightly more edges than PPC, depending on whether an edge already exists in the queue or not. (Since PPC works with a large queue, an edge might not have to be popped and pushed often as it may already exist in queue.)

Inconsistent case. Let us consider now the inconsistent QCN in Figure 2b. Regions 0, 1, 2, and 3 are all equal to each other (they are essentially the one same region), thus, regions 0 and 3 can not be disconnected. We will not go into detail as we did with the consistent case, by now the reader should be able to verify (assuming a sorted initial queue of edges in every case) that iPPC+ processes in total 4 edges and performes 5 consistency checks, while PPC processes in total 2 edges and performes 3 consistency checks. In fact, Figure 2b describes the worst case scenario for iPPC+; an inconsistency that occurs when the last temporal or spatial entity is added in the network. By that point iPPC+ will have already fully reasoned with all previous entities and their accompaning constraints. On the other hand, PPC will always do a first iteration of the queue, and might be able to immediately capture the inconsistency, as with our running example. Again, the numbers may vary a bit depending on the order of the edges in the initialized queue, but the trend is that for inconsistent QCNs iPPC+ will perform more consistency checks than PPC, and will process more edges than PPC.

Concluding our running example, we have observed that iPPC+ should perform better than PPC in the case of consistent QCNs and worse than PPC in the case of inconsistent QCNs. Further, iPPC+ works with a very small queue at each step. It is natural that a path consistency algorithm will run faster when there

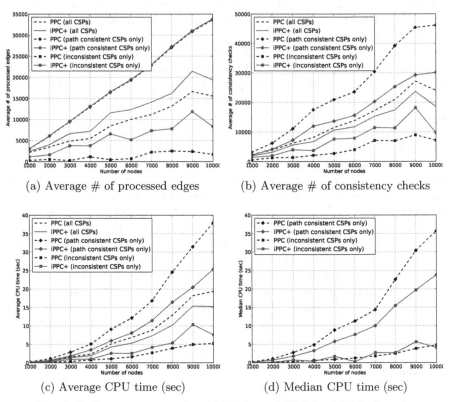

(a) Average # of processed edges   (b) Average # of consistency checks

(c) Average CPU time (sec)   (d) Median CPU time (sec)

Fig. 3. Performance comparison of iPPC+ and PPC for RCC-8 networks

is an inconsistency in the input network, as the inconsistency will not allow the algorithm to reason exhaustively with the network relations. Thus, we expect that the overall performance of iPPC+ should be better than that of PPC in the average case. We are about to experimentally verify this in the next section.

4   Experimental Evaluation

We considered random datasets consisting of *large* IA and RCC-8 networks generated by the BA(n, m) model [3], the use of which is well motivated in [20], and the standard A(n, d, l) model [19], used extensively in literature. In short, BA(n, m) creates random scale-free-like networks of size n and a preferential attachment value m, and A(n, d, l) creates random networks of size n, degree d, and an average number l of IA and RCC-8 relations per edge. For model BA(n, m) the average number of IA or RCC-8 relations per edge defaults to $|B|/2$, where B is the set of base relations of IA or RCC-8 respectively. We also considered real RCC-8 datasets that consist of admingeo [12] and gadm-rdf (http://gadm.geovocab.org/) comprising 11761/77907 nodes/edges and 276728/590865 nodes/edges respectively. In short, admingeo describes the administrative geography of Great Britain using RCC-8 relations, and `gadm-rdf` the world's administrative areas likewise. The experiments were carried out on a computer with a CPU frequency of 2.00 GHz, 4 GB RAM, and the Lucid Lynx x86_64 OS (Ubuntu Linux). The implementations of iPPC+ and PPC were run with the CPython interpreter (`http://www.python.org/`), which implements Python 2.6. Only one of the CPU cores was used for the experiments. Regarding iPPC+, we begin with a single node and grow the network one node at a time. All tools and datasets used in this paper can be found online in the following address: `http://www.cril.fr/~sioutis/work.php`.

Random datasets. BA(n, m) model: For RCC-8 we considered 30 networks for each size between 1000 and 10000 nodes with a 1000 step and a preferential attachment value of $m = 2$. For this specific value of m and for the network sizes considered, the networks lie within the *phase transition* region, where it is equally possible for networks to be consistent or inconsistent, thus, they are harder to solve [20]. We assess the performance of iPPC+ and PPC using the following parameters: average CPU time, median CPU time, average number of processed edges, and average number of consistency checks. On the average case, i.e., when all networks are considered, iPPC+ processes around 26.8% more edges than PPC, as shown in Figure 3a, iPPC+ performs around 15.3% less consistency checks than PPC, as shown in Figure 3b, and, finally, regarding average CPU time, iPPC+ runs around 18.9% faster than PPC, and 21.3% faster in the final step where networks of 10000 nodes are considered, as shown in Figure 3c; for the networks of 10000 nodes iPPC+ runs in an average time of 15.3 sec, and PPC in 19.4 sec. In Figure 3d we can also see the median CPU time for path consistent and inconsistent networks. The interesting thing to note is that in the case of inconsistent networks the median allows us to get rid of some outlying measurements that influence the average CPU time in Figure 3c. As our dataset consists of a little more than 50% inconsistent networks for almost all network sizes, the diagram for the median CPU time for the combined dataset of path consistent and inconsistent networks was very close to that of the inconsistent case, as in almost all cases the median would correspond to the CPU processing time of an inconsistent network. Thus, we did not include this diagram as it was pretty erratic and did not offer any additional information.

For IA we considered 30 networks for each size between 500 and 5000 nodes with a 500 step and a preferential attachment value of $m = 3$. We found that for this specific value of m and for the network sizes considered, the networks lie within the *phase transition* region, as is the case with RCC-8. However, we note that the phase transition for IA occurs for a different value of m (viz., $m = 3$) than the value of m for RCC-8 (viz., $m = 2$). This is probably because IA is a bigger calculus than RCC-8, containing 13 base relations instead of 8 respectively, which allows for consistent networks to be denser as there are more relations to be pruned and more relations that can *support* consistency in the network. On the average case, iPPC+ runs around 27.4% faster than PPC, and 34.4% faster in the final step where networks of 5000 nodes are considered; for the networks of 5000 nodes iPPC+ runs in an average time of 44 sec, and PPC in 67.1 sec.

$\underline{A(n, d, I)}$ model: Due to space constraints it is not possible to give analytical figures regarding this model. However, it should suffice to note that experimentation using this model yielded qualitatively similar results with those using the BA(n, m) model, i.e., we had roughly the same trends and speed-ups for both RCC-8 and IA calculi. In particular, for 50 RCC-8 networks of 1000 nodes and 50 IA networks of 500 nodes in the phase transition region there was a speed-up of around 20% and 30% respectively.

Real datasets. Both `admingeo` and `gadm-rdf` are consistent RCC-8 networks comprising 11761/77907 nodes/edges and 276728/590865 nodes/edges respectively. iPPC+ was able to enforce partial path consistency on `admingeo` in 267.6 sec, and PPC in 349.08 sec. Hence, regarding `admingeo`, iPPC+ runs 22.6% faster than PPC. Regarding `gadm-rdf`, iPPC+ outruns PPC even more significantly. In particular, iPPC+ was able to enforce partial path consistency on `gadm-rdf` in 6.27 sec, and PPC in 10.88 sec. This translates to iPPC+ running 42.4% faster than PPC regarding `gadm-rdf`. We note that our findings concerning real datasets agree with the findings concerning random datasets. Surprisingly, both algorithms run the `gadm-rdf` experiment faster than the `admingeo` one, but this is due to more relations being inferred in the latter case as a result of dataset particularities that affect the reasoning process.

At this point we conclude our experimentation. We have demonstrated that iPPC+ performs better than PPC on average for random QCNs of IA and RCC-8, and real QCNs of RCC-8. It should be noted that since iPPC+ works with a small queue at each incrementation step (as it considers the edges of a subnetwork of the whole network), as opposed to a full queue utilized by PPC (as it considers the edges of the whole network), iPPC+ is also much more memory efficient. To the best of our knowledge, the networks of IA and RCC-8 used in this paper are the biggest ones to date of all others that exist in literature (which scale up to a few hundred nodes only).

5   Conclusion and Future Work

In this paper we presented an algorithm, viz., iPPC+, for maintaining or enforcing partial path consistency for growing constraint networks, i.e., networks that grow with new temporal or spatial entities over time. Through a complexity analysis, and thorough experimental evaluation, we showed that iPPC+ is able to perform better than the state-of-the-art one-shot partial path consistency algorithm (PPC) originally presented in [9] for IA and in [21] for RCC-8, which is an advancement in the field of qualitative reasoning. The importance of our results can be roughly compared with those of Bessiére in [6], in that we also present an improvement of the state-of-the-art path consistency algorithm that on average increases its performance by avoiding redundant consistency checks, and, thus, table lookups. However, in our case, our approach is both more time and memory efficient and appropriate for all network sizes. Thus, state-of-the-art reasoners can immediately gain a performance boost by opting for iPPC+

as the preprocessing step in their backtracking algorithms for general networks, but also for solving tractable QCNs.

Future work consists of exploring if we can also have an approach for incrementally building consistent QCNs by using iPPC+ as the consistency checking step of a backtracking algorithm, i.e., we would like to investigate if in backtracking there is any benefit in reasoning with a smaller network, as a result of reversing the incremental building process. Finally, iPPC+ is tailored for dynamic spatial and temporal reasoning and, in this regard, it can become completely online by implementing a mechanism to incrementally maintain chordality [5].

References

1. Allen, J.F.: Maintaining knowledge about temporal intervals. CACM 26, 832–843 (1983)
2. Amaneddine, N., Condotta, J.F., Sioutis, M.: Efficient Approach to Solve the Minimal Labeling Problem of Temporal and Spatial Qualitative Constraints. In: IJCAI (2013)
3. Barabasi, A.L., Albert, R.: Emergence of scaling in random networks. Science 286, 509–512 (1999)
4. Berry, A., Blair, J.R.S., Heggernes, P.: Maximum Cardinality Search for Computing Minimal Triangulations. In: Kučera, L. (ed.) WG 2002. LNCS, vol. 2573, pp. 1–12. Springer, Heidelberg (2002)
5. Berry, A., Heggernes, P., Villanger, Y.: A vertex incremental approach for maintaining chordality. Discrete Mathematics 306 (2006)
6. Bessière, C.: A Simple Way to Improve Path Consistency Processing in Interval Algebra Networks. In: AAAI/IAAI (1996)
7. Bhatt, M., Guesgen, H., Wölfl, S., Hazarika, S.: Qualitative Spatial and Temporal Reasoning: Emerging Applications, Trends, and Directions. Spatial Cognition & Computation 11, 1–14 (2011)
8. Bliek, C., Sam-Haroud, D.: Path consistency on triangulated constraint graphs. In: IJCAI (1999)
9. Chmeiss, A., Condotta, J.F.: Consistency of Triangulated Temporal Qualitative Constraint Networks. In: ICTAI (2011)
10. Gerevini, A.: Incremental qualitative temporal reasoning: Algorithms for the point algebra and the ord-horn class. Artif. Intell. 166(1-2), 37–80 (2005)
11. Golumbic, M.C.: Algorithmic Graph Theory and Perfect Graphs, 2nd edn. Elsevier Science (2004)
12. Goodwin, J., Dolbear, C., Hart, G.: Geographical Linked Data: The Administrative Geography of Great Britain on the Semantic Web. TGIS 12, 19–30 (2008)
13. Huang, J.: Compactness and its implications for qualitative spatial and temporal reasoning. In: KR (2012)
14. Nebel, B.: Solving Hard Qualitative Temporal Reasoning Problems: Evaluating the Efficiency of Using the ORD-Horn Class. In: ECAI (1996)
15. Planken, L., de Weerdt, M., Yorke-Smith, N.: Incrementally Solving STNs by Enforcing Partial PC. In: ICAPS (2010)
16. Randell, D.A., Cui, Z., Cohn, A.: A Spatial Logic Based on Regions and Connection. In: KR (1992)
17. Renz, J.: Maximal Tractable Fragments of the Region Connection Calculus: A Complete Analysis. In: IJCAI (1999)

116 M. Sioutis and J.-F. Condotta

18. Renz, J., Ligozat, G.: Weak Composition for Qualitative Spatial and Temporal Reasoning. In: van Beek, P. (ed.) CP 2005. LNCS, vol. 3709, pp. 534–548. Springer, Heidelberg (2005)
19. Renz, J., Nebel, B.: Efficient Methods for Qualitative Spatial Reasoning. JAIR 15, 289–318 (2001)
20. Sioutis, M., Condotta, J.-F.: Tackling large Qualitative Spatial Networks of scale-free-like structure. In: Likas, A., Blekas, K., Kalles, D. (eds.) SETN 2014. LNCS, vol. 8445, pp. 178–191. Springer, Heidelberg (2014)
21. Sioutis, M., Koubarakis, M.: Consistency of Chordal RCC-8 Networks. In: ICTAI (2012)

A Qualitative Spatio-Temporal Framework Based on Point Algebra

Michael Sioutis, Jean-François Condotta, Yakoub Salhi, and Bertrand Mazure

Université Lille-Nord de France, Artois
CRIL-CNRS UMR 8188
Lens, France
{sioutis,condotta,salhi,mazure}@cril.fr

Abstract. Knowledge Representation and Reasoning has been quite successfull in dealing with the concepts of *time* and *space* separately. However, not much has been done in designing qualitative spatiotemporal representation formalisms, let alone reasoning systems for that formalisms. We introduce a qualitative constraint-based spatiotemporal framework using Point Algebra (PA), that allows for defining formalisms based on several qualitative spatial constraint languages, such as RCC-8, Cardinal Direction Algebra (CDA), and Rectangle Algebra (RA). We define the notion of a qualitative spatiotemporal constraint network (QSTCN) to capture such formalisms, where pairs of spatial networks are associated to every base relation of the underlying network of PA. Finally, we analyse the computational properties of our framework and provide algorithms for reasoning with the derived formalisms.

Keywords: point algebra, qualitative spatiotemporal reasoning, qualitative spatiotemporal framework, satisfiability, minimality, algorithm.

1 Introduction

Qualitative Reasoning is based on qualitative abstractions of aspects of the common-sense background knowledge, such as *space* and *time*, on which our human perspective on the physical reality is based. Spatiotemporal reasoning has become a significant field of research in Qualitative Reasoning, and, more generally, in Knowledge Representation and Reasoning. This field is essential for a plethora of areas and domains that include dynamic GIS, cognitive robotics, spatiotemporal design, and planning [5,13].

The Point Algebra (PA) [2,3,21] is one of the dominant Artificial Intelligence approaches for representing and reasoning about qualitative temporal relations, and forms the basis of several richer temporal languages, such as Interval Algebra (IA) [1]. In particular, PA encodes temporal relations between two points in the timeline. Likewise, a fragment of the Region Connection Calculus [16], namely, RCC-8, Cardinal Direction Algebra (CDA) [8], and Rectangle Algebra (RA) [12], are among the dominant Artificial Intelligence approaches for representing and reasoning about qualitative spatial relations. In particular, RCC-8 encodes topological relations between two regions that are non-empty regular subsets of some

G. Agre et al. (Eds.): AIMSA 2014, LNAI 8722, pp. 117–128, 2014.
© Springer International Publishing Switzerland 2014

topological space, CDA encodes direction relations between spatial objects, and RA encodes relative position relations between multi-dimensional objects. All these qualitative constraint languages have been extensively studied *separately*, but there has not been a framework so far that allows for combining them in unique formalisms in order to reason about both time and space effectively.

The spectrum of spatiotemporal formalisms has mainly focused on adopting the propositional temporal logic (PTL), and combining it with RCC-8 [22] or even richer fragments than RCC-8, such as the modal logic $S4_u$ [9] interpreted over topological spaces. A study of such formalisms along with their computational properties can be found in [23]. Most of the PTL-based formalisms are very elegant and expressive, but deciding their satisfiability is PSPACE-complete at best [9]. Delving deeper into modal logics, there have been multimodal logic approaches to qualitative spatiotemporal reasoning studied in the works of Burrieza et al. [6,7], Muñoz-Velasco et al. [15], and Golinska-Pilarek et al. [11].

Unfortunately, constraint-based formalisms have not been paid the attention they deserve except in the work of Gerevini et al. [10] where the Interval Algebra (IA) [1] is combined with RCC-8 in a unique spatiotemporal formalism called STCC.

We take a similar approach to that of Gerevini et al. [10] by creating a framework that allows for combining a qualitative temporal constraint language with a spatial one, namely, PA with any spatial language such as RCC-8, CDA, and RA, and make the following contributions: (i) we define our framework in detail and describe the notion of a qualitative spatiotemporal constraint network (QSTCN), (ii) we analyse the computational properties of our framework and provide algorithms for reasoning with the derived formalisms.

Our approach is different to that of Gerevini et al. [10] in that we associate pairs of spatial networks to every base relation of a network of PA, while Gerevini et al. associate a spatial network to every variable of a network of Interval Algebra (IA) [10]. Thus, our approach is more flexible and richer. In particular, Gerevini et al. associate a static spatial configuration to a temporal interval (a variable of IA) which leads to a very rigid framework; every time two temporal intervals overlap in any way, it is clear that the associated spatial configurations must be identical for both intervals, as we can only have a unique spatial configuration within a period of time. This leads to $\mathcal{NP}$-completeness even in trivial cases where one only uses base relations and the two universal[1] relations of the qualitative constraint languages considered (Theorem 2 in [10]). On the other hand, and as we will explore later in the paper, our framework allows for many tractability cases that include large fragments of the relations of the qualitative constraint languages considered. Further, Gerevini et al. handle a maximum of $O(n)$ spatial configurations for a IA network of n variables. Since we associate spatial configurations to every base relation of a network of PA, and every such network of n variables can have $O(n^2)$ relations, we consider in total $O(n^2)$

---

[1] The universal relation of a qualitative constraint language is the non-restrictive relation that contains all base relations. It signifies the lack of knowledge between two entities in a qualitative constraint network.

pairs of spatial configurations. Moreover, taking into account the semantics of the base relations $\{<, =, >\}$ of PA, that is, their natural interpretation over time points in $\mathbb{Q}$, a pair allows us to capture both the past and the future of a spatial configuration with a particular base relation. For example, we can think of base relation $<$ as a relation that associates the past of a spatial configuration with its future. Therefore, we have the ability to define general laws about qualitative change, which the formalization proposed by Gerevini et al. lacks [10].

The paper is organized as follows. Section 2 introduces the notion of qualitative constraint languages. In Section 3 we define our framework and the concept of a QSTCN, analyse its computational properties, and present algorithms for reasoning with derived formalisms. Finally, in Section 4 we conclude and discuss future work.

2   Preliminaries

A (binary) qualitative temporal or spatial constraint language [18] is based on a finite set B of *jointly exhaustive and pairwise disjoint* (JEPD) relations defined on a domain D, called the set of base relations. The set of base relations B of a particular qualitative constraint language can be used to represent definite knowledge between any two entities with respect to the given level of granularity. B contains the identity relation Id, and is closed under the converse operation ($^{-1}$). Indefinite knowledge can be specified by unions of possible base relations, and is represented by the set containing them. Hence, 2^B will represent the set of relations. 2^B is equipped with the usual set-theoretic operations (union and intersection), the converse operation, and the weak composition operation. The converse of a relation is the union of the converses of its base relations. The weak composition $\diamond$ of two relations s and t for a set of base relations B is defined as the strongest relation $r \in 2^B$ which contains $s \circ t$, or formally, $s \diamond t = \{b \in B \mid b \cap (s \circ t) \neq \emptyset\}$, where $s \circ t = \{(x, y) \mid \exists z : (x, z) \in s \land (z, y) \in t\}$ is the relational composition.

The qualitative temporal constraint language PA [2, 3, 21] consists of the set of base relations $\{<, =, >\}$, where the relation symbols display the natural interpretation over time points in $\mathbb{Q}$. We denote the set of base relations of PA by B_{PA}. Thus, $2^{B_{PA}}$ represents the set of relations $\{\emptyset, <, =, >, \leq, \geq, \neq, = \lor \neq\}$, with $=$ being the identity relation. (Note that $\neq$ is an abbreviation for $> \lor <$, $\geq$ an abbreviation for $> \lor =$, and $\leq$ an abbreviation for $< \lor =$.) Likewise, qualitative spatial constraint languages RCC-8 [16], CDA [8], and RA [12] have their own set of base relations. As an example, RCC-8 consists of the set of base relations $B_{RCC8} = \{DC$ (disconnected), EC (externally connected), PO (partially overlaps), TPP (tangential proper part), $NTPP$ (non-tangential proper part), $TPPi$ (tangential proper part inverse), $NTPPi$ (non-tangential proper part inverse), EQ (equals)$\}$, with EQ being the identity relation, and $2^{B_{RCC8}}$ enumerates a total of 256 relations.

Qualitative temporal or spatial constraint languages can be formulated as qualitative constraint networks (QCNs) as follows:

Definition 1. *A* QCN *is a pair* $\mathcal{N} = (V, C)$ *where* V *is a finite set of variables and* C *a mapping associating a relation* $C(v, v') \in 2^{\mathsf{B}}$, *to each pair* (v, v') *of* $V \times V$. C *is such that* $C(v, v) \subseteq \mathsf{Id}$ *and* $C(v, v') = (C(v', v))^{-1}$ *for every* $v, v' \in V$.

Note that we always regard a QCN as a complete network. Given two QCNs $\mathcal{N} = (V, C)$ and $\mathcal{N}' = (V, C')$, $\mathcal{N} \cap \mathcal{N}'$ denotes the QCN $\mathcal{N}'' = (V, C'')$ where $C'' = C(v, v') \cap C'(v, v')$ for every $v, v' \in V$.

Definition 2. *A* solution *of a* QCN $\mathcal{N} = (V, C)$ *is a mapping* σ *defined from* V *to the domain* D, *such that* $\forall (v, v') \in V \times V$, $(\sigma(v), \sigma(v'))$ *satisfies* $C(v, v')$, *i.e., the base relation* b *defined by* $(\sigma(v), \sigma(v'))$ *exists in* $C(v, v')$. $\mathcal{N}$ *is* consistent *or* satisfiable *iff it admits a solution. A* sub-QCN $\mathcal{N}'$ *of* $\mathcal{N}$ *is a* QCN (V, C') *such that* $C'(v, v') \subseteq C(v, v')$ *for every* $v, v' \in V$. *An* atomic QCN *is a* QCN *where each constraint is defined by a base relation. A* scenario *of* $\mathcal{N}$ *is an atomic consistent sub-QCN of* $\mathcal{N}$. $\mathcal{N}$ *admits a solution iff it admits a scenario. Given a* QCN $\mathcal{N} = (V, C)$, *base relation* r *is* feasible *iff there exists a scenario* $\mathcal{N}_{\mathsf{atomic}} = (V, C_{\mathsf{atomic}})$ *of* $\mathcal{N}$ *such that* $C_{\mathsf{atomic}}(v, v') = \{r\}$. *A* QCN $\mathcal{N}$ *is* minimal *iff it comprises only feasible relations.*

Checking the consistency of a QCN of PA can be done in polynomial time, $O(n^3)$, using a path consistency algorithm [2].[2] It follows that the whole set of relations of PA, viz., $2^{\mathsf{B_{PA}}}$, is tractable. On the other hand, QCNs of RCC-8, CDA, and RA are intractable in the general case. However, there exist large maximal tractable subclasses of their relations, for which the satisfiability problem is tractable. As an example, checking the consistency of a QCN of RCC-8 is $\mathcal{NP}$-complete in general [19], but there exist the maximal tractable subclasses $\hat{\mathcal{H}}_8, \mathcal{C}_8$, and $\mathcal{Q}_8$ [17] for which the satisfibility problem is tractable. Checking the consistency of a QCN of RCC-8, CDA, or RA comprising only relations from maximal tractable subclasses can be done in polynomial time, $O(n^3)$ in particular, using a path consistency algorithm.

3   A Spatio-Temporal Framework Based on Point Algebra

We obtain a spatiotemporal framework by defining the concept of a qualitative spatiotemporal constraint network (QSTCN) that builds on PA and allows plugging in any spatial constraint language, such as RCC-8, CDA, and RA. In particular, in a QSTCN we assign a pair of spatial QCNs to every base relation of the underlying QCN of PA. We formally define a QSTCN as follows.

Definition 3. *A* QSTCN *is a tuple* $\mathcal{N} = (V_{\mathsf{T}}, V_{\mathsf{S}}, C, \alpha)$, *where* V_{T} *is a finite set of temporal variables,* V_{S} *is a finite set of spatial variables,* C *a mapping associating a relation* $C(v, v') \in 2^{\mathsf{B_{PA}}}$ *to each pair* $(v, v') \in V_{\mathsf{T}} \times V_{\mathsf{T}}$, *and* α *a mapping associating a pair of spatial* QCNs $(\mathcal{N}_v^{r(v, v')}, \mathcal{N}_{v'}^{r(v, v')})$ *to each base*

---

[2] Actually, in [2, chap. 3] there exists an even faster, but very particular algorithm, that checks the consistency of a QCN of PA in $O(n^2)$ time. The path consistency algorithm is a more general approach that applies to most qualitative calculi.

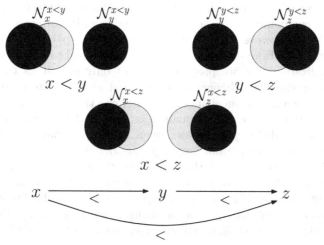

Fig. 1. Example of a QSTCN

relation $r(v,v') \in C(v,v')$, i.e., $\alpha(r(v,v')) = (\mathcal{N}_v^{r(v,v')}, \mathcal{N}_{v'}^{r(v,v')})$. By overloading notation $\alpha(C(v,v')) = \{\alpha(r(v,v')) \mid r(v,v') \in C(v,v')\}$. C is such that $C(v,v) \subseteq \{=\}$ and $C(v,v') = (C(v',v))^{-1}$ $\forall v,v' \in V_{\mathsf{T}}$, and α is such that all spatial QCNs $\mathcal{N}_v$, where $v \in V_{\mathsf{T}}$, share the same set of spatial variables V_{S}, $\mathcal{N}_v^{v=v'} \equiv \mathcal{N}_{v'}^{v=v'}$, and $\alpha(C(v,v')) = \{(\text{second entry of 2-tuple } t, \text{ first entry of 2-tuple } t) \mid t \in \alpha(C(v',v))\}$ for all $v,v' \in V_{\mathsf{T}}$.

Example 1. An example of a QSTCN $\mathcal{N}$ is presented in Figure 1. To begin with, $\mathcal{N}$ builds upon a QCN of PA that comprises the set of temporal variables $\{x,y,z\}$ and the set of constraints $\{x < y, y < z, x < z\}$ (using infix notation). In a sense, a scenario of a QCN of PA yields a totally ordered set of points in a finite timeline ranging over values in $\mathbb{Q}$. Thus, it allows us to a acquire a description of how a spatial configuration evolves over time. For every base relation between every pair of variables of the QCN of PA, a pair of spatial configurations is attached. For the sake of our example, we can view these configurations as QCNs of RCC-8. All QCNs of RCC-8 share the same set of spatial variables V_{S}, which in our case comprises the image of the moon and the sun. In fact, our example describes an eclipse. The pair of QCNs of RCC-8 attached to the base relation $<$ between temporal variables x and y is the pair $(\mathcal{N}_x^{x<y}, \mathcal{N}_y^{x<y})$. $\mathcal{N}_x^{x<y}$ comprises the set of constraints $\{PO(\text{moon}, \text{sun})\}$, and $\mathcal{N}_y^{x<y}$ comprises the set of constraints $\{EQ(\text{moon}, \text{sun})\}$. Upon instatiation of the base relation $<$ between variables x and y, variable x acquires QCN $\mathcal{N}_x^{x<y}$ and variable y acquires QCN $\mathcal{N}_y^{x<y}$.

In the above example, we can think of relation $<$ as a relation that associates the past of a spatial configuration with its future. Likewise, relation $>$ associates the future of a spatial configuration with its past, and relation $=$ describes only one unique spatial configuration at a point of time, i.e., $\mathcal{N}_v^{v=v'} \equiv \mathcal{N}_{v'}^{v=v'}$ for all $v,v' \in V_{\mathsf{T}}$. Another way to think of a base relation of a QCN of PA is like the command c of a Hoare triple $\{p\}\ c\ \{q\}$, where p and q are its assertions, p is named the precondition and q the postcondition. When the precondition is met, executing the command establishes the postcondition. In our example, $\mathcal{N}_y^{x<y}$ is identical to $\mathcal{N}_y^{y<z}$. However, in general we do not require spatial QCNs associated to a variable $v \in V_T$ from different sources to be identical to one another.

The formalization that we propose does not only permit us to describe a spatial configuration that changes over time, but also to state general laws of *how* the spatial configuration changes by describing the transition from its past to its future within a pair. The formalization proposed by Gerevini et al. [10], lacks this ability to define general laws about qualitative change. Depending on the qualitative spatial constraint language considered, one can use the pairs of spatial configurations to restrict the movement, the direction, the position, the topology, or any other sort of property of a spatial configuration between two different points in time. This makes our approach very favorable for applications in many fields that deal with qualitative change, such as dynamic GIS, cognitive robotics, spatiotemporal design, and spatiotemporal planning [5, 13].

Definition 4. *A solution of a QSTCN $\mathcal{N} = (V_T, V_S, C, \alpha)$ is a solution of the underlying QCN of PA, that is compatible with a solution of every spatial QCN associated to every $v \in V_T$. $\mathcal{N}$ is consistent or satisfiable iff it admits a solution. A sub-QSTCN $\mathcal{N}'$ of $\mathcal{N}$ is a QSTCN where the underlying QCN of PA is a sub-QCN of the underlying QCN of PA of $\mathcal{N}$ and all associated spatial QCNs are sub-QCNs of the corresponding spatial QCNs of $\mathcal{N}$. An atomic QSTCN is a QSTCN where the underlying QCN of PA is atomic and all associated spatial QCNs are atomic. A scenario of $\mathcal{N}$ is an atomic consistent sub-QSTCN of $\mathcal{N}$. $\mathcal{N}$ admits a solution iff it admits a scenario.*

Theorem 1. *Checking the consistency of a given QSTCN $\mathcal{N}$ is $\mathcal{NP}$-complete.*

Proof. Suppose that you are provided with a candidate scenario of a given QSTCN $\mathcal{N} = (V_T, V_S, C, \alpha)$. To check whether the candidate scenario is indeed a scenario of the given QSTCN, we must first use the path consistency algorithm to check if the temporal scenario is consistent, which takes $O(n^3)$ time, where $n = |V_T|$. Then, we can check the consistency of all the linear in the number n of temporal variables spatial QCNs in $O(n \cdot m^3)$ time, where $m = |V_S|$, again with the path consistency algorithm. $\mathcal{NP}$-hardness follows from the fact that checking the consistency of a single spatial QCN is $\mathcal{NP}$-complete (intractable). Hence, checking the consistency of a given QSTCN $\mathcal{N}$ is $\mathcal{NP}$-complete.   □

The rest of this section is devoted to characterizing cases for which we have tractability.

We first define a classical constraint satisfaction problem (CSP [14]) $\mathcal{N}_S$ that corresponds to the spatial aspect of a QSTCN $\mathcal{N}$ where the spatial QCNs represent definite knowledge between entities, i.e., they are atomic. Inconsistent atomic QCNs are filtered out with a path consistency algorithm. An atomic QCN can be seen as a constant value, as it yields a unique minimal signature of itself.

Definition 5. *Given a QSTCN $\mathcal{N} = (V_T, V_S, C, \alpha)$ where the associated spatial QCNs are atomic, $\mathcal{N}_S = \langle X, D, R \rangle$ will correspond to the constraint satisfaction problem where:*

- *the set of variables $X = \{x_1, \ldots, x_n\}$ with $n = |V_T|$,*
- *the set of domains $D = \{d_1, \ldots, d_n\}$ where $d_i = \{\bigcap(\{first\ entry\ of\ 2\text{-}tuple\ t \mid t \in \alpha(C(v_i, v_j))\}\ or\ \{second\ entry\ of\ 2\text{-}tuple\ t \mid t \in \alpha(C(v_j, v_i))\}) \mid v_j \in V_T\}$ for all i with $0 \leq i \leq n$,*
- *and the set of binary constraints $R = \{r(x_i, x_j) \mid 0 \leq i, j \leq n\}$ where each $r(x_i, x_j) = \{first\ entry\ of\ 2\text{-}tuple\ t \in d_i\ and\ second\ entry\ of\ 2\text{-}tuple\ t \in d_j \mid t \in \alpha(C(v_i, v_j))\}$.*

The observant reader will note that constructing the set of variables X, the set of domains D, and the set of constraints R for $\mathcal{N}_S$ in that particular order, as defined, will result in a node and arc consistent network.

Example 2. Let us consider the QSTCN $\mathcal{N} = (V_T, V_S, C, \alpha)$, where $V_T = \{x, y, z\}$, $C(x, y) \subseteq \{<\}$, $C(y, z) \subseteq \{>\}$, $C(x, z) \subseteq \{<, =\}$, $\alpha(C(x, y)) = \{(a_x^{x<y}, b_y^{x<y})\}$, $\alpha(C(y, z)) = \{(b_y^{y>z}, c_z^{y>z})\}$, and $\alpha(C(x, z)) = \{(a_x^{x<z}, c_z^{x<z}), (b_x^{x=z}, b_z^{x=z})\}$. Note that a, b, and c correspond to different atomic spatial QCNs over a set of variables V_S, i.e., we can consider them as constant domain values. We can view the CSP $\mathcal{N}_S$ with a set of variables $\{x_s, y_s, z_s\}$, value domain $\{a\}$ for variable x_s, value domain $\{b\}$ for variable y_s, value domain $\{c\}$ for variable z_s, and constraints $r(x_s, y_s) = \{(a_x^{x<y}, b_y^{x<y})\}$, $r(y_s, z_s) = \{(b_y^{y>z}, c_z^{y>z})\}$, and $r(x_s, z_s) = \{(a_x^{x<z}, c_z^{x<z})\}$. (Note that $r(x_s, z_s) \subset \alpha(C(x, z))$.)

Proposition 1. *Given a QSTCN $\mathcal{N} = (V_T, V_S, C, \alpha)$ where the associated spatial QCNs are atomic, we have that for $\mathcal{N}_S = \langle X, D, R \rangle$ the size of domain d_i for each variable $x_i \in X$ is at most 3, and each constraint $r \in R$ contains at most three 2-tuples.*

Proof. It is easy to see that each $v \in V_T$ can be associated with at most three different atomic QCNs, as there can be at most three PA base relations in any constraint $C(v, v')$, where $v' \in V_T$, and each PA base relation contributes a single atomic spatial QCN to v (and a single atomic spatial QCN to v'). Thus, the possible atomic spatial QCNs for each $v \in V_T$ will be the intersection of all atomic spatial QCNs contributed from each constraint $C(v, v')$, which can be at most three. Further, each PA base relation can contribute a single unique 2-tuple, thus, each constraint $C(v, v')$ can contribute at most three.   □

We note that in a CSP a binary constraint $r(x_i, x_j)$ between variables x_i and x_j can be represented as a $(0, 1)$-matrix with $|d_i|$ rows and $|d_j|$ columns by imposing an ordering on the domains of the variables [14]. The entries that correspond to the 2-tuples of a binary constraint have value 1, and all others have value 0.

Definition 6 ([4]). *A binary relation $r(x_i, x_j)$ represented as a $(0, 1)$-matrix is row convex iff in each row all of the 1s are consecutive; that is, no two 1s within a single row are separated by a 0 in that same row.*

Based on the definition of the notion of row convexity we can obtain the definition of the weaker notion of directional row convexity which is sufficient for the results in our paper.

Definition 7 ([4]). *Given a binary* CSP $\mathcal{N} = \langle X, D, R \rangle$ *and an ordering* $x_i \ldots x_n$ *of its variables, network* $\mathcal{N}$ *is directionally row convex if each of the binary relations* $r(x_i, x_j)$ *represented as a* $(0,1)$*-matrix, where* x_i *occurs before variable* x_j *in the ordering, is row convex.*

Then we have the following result from literature:

Theorem 2 ([4]). *Let* $\mathcal{N} = \langle X, D, R \rangle$ *be a path consistent binary* CSP. *If there exists an ordering of the domains* $d_i, \ldots, d_n$ *of* D *such that the constraints of* R *are directionally row convex then a solution for* $\mathcal{N}$ *can be found without backtracking.*

By Proposition 1 we can deduce that for a CSP $\mathcal{N}_S = \langle X, D, R \rangle$ that corresponds to a QSTCN $\mathcal{N} = (V_T, V_S, C, \alpha)$ where the associated spatial QCNs are atomic, all binary constraints in the set of constraints R can be represented by a $i \times j$ $(0,1)$-matrix, with $i, j \leq 3$, with at most three entries of 1s. By exhaustive enumeration of all the possible constraints of a CSP $\mathcal{N}_S$ it can be found that such a network always is, or can be made, directionally row convex. Therefore, by Theorem 2 we can have the following result:

Theorem 3. *Given a* QSTCN $\mathcal{N} = (V_T, V_S, C, \alpha)$ *where the associated spatial* QCN*s are atomic, applying path consistency on the derived* CSP $\mathcal{N}_S = \langle X, D, R \rangle$ *is sufficient to guarantee a backtrack-free solution.*

Proof. As noted earlier, given a CSP $\mathcal{N}_S = \langle X, D, R \rangle$ that corresponds to a QSTCN $\mathcal{N} = (V_T, V_S, C, \alpha)$ where the associated spatial QCNs are atomic, all binary constraints in the set of constraints R can be represented by a $i \times j$ $(0,1)$-matrix, with $i, j \leq 3$, with at most three entries of 1s. This property will obviously hold even after path consistency is applied on $\mathcal{N}_S$ as the size of the domains can only decrease. Thus, we will obtain a path consistent CSP $\overline{\mathcal{N}_S}$ where the maximum domain size will be at most 3. Then, we can sort the variables of $\overline{\mathcal{N}_S}$ according to their domain size in decreasing order and we will obtain $i \times j$ $(0,1)$-matrices, with $i, j \leq 3$ and $i \geq j$. All 3×3 $(0,1)$-matrices are row convex since they can have at most three entries of 1s and, thus, only a single 1 can exist at each row and column, otherwise it would be a $i \times j$ $(0,1)$-matrix with $i < 3$ or $j < 3$. The rest of the matrices of $\overline{\mathcal{N}_S}$, always assuming the aforementioned ordering, will be $i \times j$ $(0,1)$-matrices with $i \leq 3$ and $j \leq 2$ and $j \leq i$. It is clear that for a number of columns less than or equal to 2 ($j \leq 2$) the corresponding matrix is row convex, as we need at least three colummns for a 0 to exist between two 1s in a single row. Thus, there exists an ordering for which $\overline{\mathcal{N}_S}$ is directionally row convex. The result follows directly from the implication of Theorem 2.   □

It is time to introduce our path consistency algorithm, that operates both on the temporal and the spatial aspect of a QSTCN $\mathcal{N}$. We note that the composition of two constraints for the corresponding CSP $\mathcal{N}_S$ is the standard relational

---

Algorithm 1. stPC($\mathcal{N}$, $\mathcal{N}_\mathsf{S}$)

in	: A QSTCN $\mathcal{N} = (V_\mathsf{T}, V_\mathsf{S}, C, \alpha)$, and CSP $\mathcal{N}_\mathsf{S} = \langle X, D, R \rangle$.
output	: False if network $\mathcal{N}$ results in a trivial inconsistency (contains the empty relation), True if the modified network $\mathcal{N}$ is path consistent.

```
1  begin
2  │  Q ← {(i,j) | (i,j) ∈ V_T × V_T};
3  │  while Q ≠ ∅ do
4  │  │  (i,j) ← Q.pop();
5  │  │  foreach k ← 1 to V_T, (i ≠ k ≠ j) do
6  │  │  │  t ← C(i,k) ∩ ((C(i,j) ◇ C(j,k)) ∩ α⁻¹(r(i_s,j_s) ∘ r(j_s,k_s)));
7  │  │  │  if t ≠ C(i,k) then
8  │  │  │  │  if t = ∅ then return False;
9  │  │  │  │  C(i,k) ← t; C(k,i) ← t⁻¹;
10 │  │  │  └  Q ← Q ∪ {(i,k)};
11 │  │  │  t ← C(k,j) ∩ ((C(k,i) ◇ C(i,j)) ∩ α⁻¹(r(k_s,i_s) ∘ r(i_s,j_s)));
12 │  │  │  if t ≠ C(k,j) then
13 │  │  │  │  if t = ∅ then return False;
14 │  │  │  │  C(k,j) ← t; C(j,k) ← t⁻¹;
15 │  │  │  └  Q ← Q ∪ {(k,j)};
16 │  return True;
```

composition. Given for example two 2-tuples of atomic QCNs (a,b) and (b,c), $(a,b) \circ (b,c)$ yields 2-tuple (a,c).

Algorithm stPC presented in Algorithm 1 receives as input a QSTCN $\mathcal{N}$ and its spatial CSP $\mathcal{N}_\mathsf{S}$, and performs path consistency on both the underlying PA network and network $\mathcal{N}_\mathsf{S}$. This is achieved by iteratively performing a composition operation both on the temporal and the spatial aspect of network $\mathcal{N}$ in lines 6 and 11. Let us refer to Example 2. We have a QSTCN $\mathcal{N} = (V_\mathsf{T}, V_\mathsf{S}, C, \alpha)$, where $V_\mathsf{T} = \{x, y, z\}$, $C(x,y) \subseteq \{<\}$, $C(y,z) \subseteq \{>\}$, $C(x,z) \subseteq \{<, =\}$, $\alpha(C(x,y)) = \{(a_x^{x<y}, b_y^{x<y})\}$, $\alpha(C(y,z)) = \{(b_y^{y>z}, c_z^{y>z})\}$, and $\alpha(C(x,z)) = \{(a_x^{x<z}, c_z^{x<z}), (b_x^{x=z}, b_z^{x=z})\}$, and CSP $\mathcal{N}_\mathsf{S}$ with a set of variables $\{x_s, y_s, z_s\}$, value domain $\{a\}$ for variable x_s, value domain $\{b\}$ for variable y_s, value domain $\{c\}$ for variable z_s, and constraints $r(x_s, y_s) = \{(a_x^{x<y}, b_y^{x<y})\}$, $r(y_s, z_s) = \{(b_y^{y>z}, c_z^{y>z})\}$, and $r(x_s, z_s) = \{(a_x^{x<z}, c_z^{x<z})\}$. The composition $C(x,y) \diamond C(y,z)$ regarding PA, yields the set of relations $\{<, =, >\}$ which is the universal relation B_{PA}. Since $C(x,z) \subseteq \{<, =\} \subset \mathsf{B}_{\mathsf{PA}}$, the underlying PA network of QSTCN $\mathcal{N}$ is path consistent. However, the composition $r(x_s, y_s) \circ r(y_s, z_s)$ yields the set $\{(a_x^{x<z}, c_z^{x<z})\}$. Therefore, since $\alpha^{-1}(r(x_s, y_s) \circ r(y_s, z_s)) = \{<\}$, we must intersect $\{<\}$ with $\{<, =\}$ to acquire the set of relations $\{<\}$ for $C(x,z)$. As a result, network $\mathcal{N}$ will be path consistent for both its temporal and spatial aspect.

We can assert the following proposition for the case of an atomic QSTCN:

Proposition 2. *Given an atomic QSTCN $\mathcal{N} = (V_\mathsf{T}, V_\mathsf{S}, C, \alpha)$ algorithm stPC enforces path consistency on $\mathcal{N}$ and is able to correctly decide its consistency in $O(n^3)$ time, where $n = |V_\mathsf{T}|$.*

Proof. It is clear that algorithm stPC enforces path consistency on the underlying QCN of PA and the corresponding CSP $\mathcal{N}_S$. Suppose though that the path consistency of the temporal aspect is not interdependent to the path consistency of the spatial aspect, and vice versa. Then, there should exist a triple of variables i, j, and k for which we have that $(C(i,j) \diamond C(j,k)) \cap \alpha^{-1}(r(i_s, j_s) \circ C(j_s, k_s))$ $= \emptyset$. Because of line 6 in the algorithm this is a contradiction, as stPC would have returned False if this was the case. Since path consistency decides the consistency of an atomic QCN of PA, and it also decides the consistency of CSP $\mathcal{N}_S$ by Theorem 3, it holds that it is able to decide the consistency of QSTCN $\mathcal{N}$. Algorithm stPC is a standard path consistency algorithm as the one described in [20] for qualitative spatial reasoning which runs in $O(n^3)$ time. In our case we only extend the usual composition operation with an additional check on the spatial aspect of a given QSTCN $\mathcal{N}$ which can be done in constant time.   □

Let us now consider the more complicated case, where a QSTCN $\mathcal{N} = (V_T, V_S, C, \alpha)$ comprises atomic spatial QCNs and an underlying QCN of PA with relations from the convex class of relations $\{\emptyset, <, =, >, \leq, \geq, = \vee \neq\}$, i.e., relation $\neq$ is not premitted [2]. Then we have the following result from literature:

Theorem 4 ([2]). *Let $\mathcal{N}$ be a path consistent QCN of PA. If $\mathcal{N}$ comprises relations from the convex class of relations $\{\emptyset, <, =, >, \leq, \geq, = \vee \neq\}$ then $\mathcal{N}$ is minimal and globally consistent and a solution is found with no backtracking.*

By Proposition 2, Theorem 3, and Theorem 4, we have the following result:

Theorem 5. *Given a QSTCN $\mathcal{N} = (V_T, V_S, C, \alpha)$ that comprises atomic spatial QCNs and an underlying QCN of PA with relations from the convex class of relations $\{\emptyset, <, =, >, \leq, \geq, = \vee \neq\}$, algorithm stPC enforces path consistency on $\mathcal{N}$ and is able to correctly decide its consistency in $O(n^3)$ time, where $n = |V_T|$.*

Proof. Enforcing path consistency with stPC on QSTCN $\mathcal{N}$ will result in a globally consistent underlying QCN of PA by Theorem 4, denoted by $\mathcal{N}_{PA}$, and a path consistent corresponding spatial CSP $\mathcal{N}_S$, in a total of $O(n^3)$ time, where $n = |V_T|$. All the scenarios (path consistent atomic networks) that exist for $\mathcal{N}_{PA}$ are interdependent to respective scenarios of $\mathcal{N}_S$ due to Proposition 2, and vice versa. Thus, all scenarios of $\mathcal{N}$, are both scenarios of $\mathcal{N}_{PA}$ and $\mathcal{N}_S$. Due to global consistency for $\mathcal{N}_{PA}$ and the implication of Theorem 3 for $\mathcal{N}_S$, a solution of $\mathcal{N}$ can be obtained by instantiating a single base relation of $\mathcal{N}$, and consistently extending it to a scenario of $\mathcal{N}$ in a backtrack-free manner.   □

Up to this point, and as long as tractability was the issue, we have been concerned with a QSTCN $\mathcal{N} = (V_T, V_S, C, \alpha)$ that comprises atomic spatial QCNs. If the associated spatial QCNs are not atomic, it is not possible to construct the corresponding spatial CSP $\mathcal{N}_S$ as provided by Definition 5. This is mainly because there is no way to know the possible values of the spatial QCNs that we will obtain in a scenario of QSTCN $\mathcal{N}$, i.e., it is no longer the case that spatial QCNs yield unique constant values of themselves. Two non-atomic QCNs can intersect and yield a different value, not just the empty relation $\emptyset$. A possible approach would be to enumerate all the scenarios for each non-atomic QCN and use stPC in the way that we described so far. However, for a QSTCN that comprises an atomic underlying QCN of PA we can have the following result and a simple algorithm sketched in its proof.

Theorem 6. *Checking the consistency of a QSTCN $\mathcal{N} = (V_T, V_S, C, \alpha)$ that comprises an atomic underlying QCN of PA, has the same complexity with checking the consistency of the associated spatial QCNs.*

Proof. We can check the consistency of the underlying QCN of PA in $O(n^3)$ time, where $n = |V_T|$, with a path consistency algorithm. We then have to obtain the set of spatial QCNs that correspond to each $v \in V_T$. This would be the set $S = \{\mathcal{N}_1, \ldots, \mathcal{N}_{|V_T|}\}$ where $\mathcal{N}_i = \{\bigcap(\{\text{first entry of 2-tuple } t \mid t \in \alpha(C(v_i, v_j))\}$ or $\{\text{second entry of 2-tuple } t \mid t \in \alpha(C(v_j, v_i))\}) \mid v_j \in V_T\}$ for all i with $0 \leq i \leq |V_T|$. Set S can be constructed in $O(n^2 \cdot m^2)$ time, where $m = |V_S|$. In the case of atomic QCNs we could create constant values out of them, hash values, and compare them in constant time. In this case, we need to go over the $O(m^2)$ constraints for each spatial QCN and intersect them with the constraints of another QCN. After that, checking the consistency of the spatial QCNs is in $\mathcal{P}$ if they are tractable (in $O(n \cdot m^3)$ time with a path consistency algorithm that will go over n spatial QCNs), and in $\mathcal{NP}$ if the they are not tractable. $\square$

4   Conclusion and Future Work

In this paper, we defined a qualitative constraint-based spatiotemporal framework using Point Algebra (PA), that allows for defining formalisms based on several qualitative spatial constraint languages, such as RCC-8, Cardinal Direction Algebra (CDA), and Rectangle Algebra (RA). We formally defined the notion of a qualitative spatiotemporal constraint network (QSTCN), studied its computational properties for the consistency checking problem, and presented algorithms for reasoning with derived formalisms.

Future work consists of further exploring cases of tractability, especially for QSTCNs that comprise non-atomic spatial QCNs. Then, we would like to formally define algorithms for these general QSTCNs, explore heuristics, introduce random and real datasets, identify the phase transition region for such datasets, and create and experiment with a benchmark of QSTCN instances for evaluation. Further, we would like to extend our framework to pointisable IA [2].

Acknowledgments. This work was funded by a PhD grant from Université d'Artois and region Nord-Pas-de-Calais.

References

1. Allen, J.F.: Maintaining Knowledge about Temporal Intervals. Commun. ACM 26, 832–843 (1983)
2. van Beek, P.: Reasoning About Qualitative Temporal Information. Artif. Intell. 58, 297–326 (1992)

3. van Beek, P., Cohen, R.: Exact and approximate reasoning about temporal relations. Computational Intelligence 6, 132–144 (1990)
4. van Beek, P., Dechter, R.: On the Minimality and Decomposability of Row-Convex Constraint Networks. JACM 42, 543–561 (1995)
5. Bhatt, M., Guesgen, H., Wölfl, S., Hazarika, S.: Qualitative Spatial and Temporal Reasoning: Emerging Applications, Trends, and Directions. Spatial Cognition & Computation 11, 1–14 (2011)
6. Burrieza, A., Muñoz-Velasco, E., Ojeda-Aciego, M.: A PDL Approach for Qualitative Velocity. International Journal of Uncertainty, Fuzziness and Knowledge-Based Systems 19(1), 11–26 (2011)
7. Burrieza, A., Ojeda-Aciego, M.: A Multimodal Logic Approach to Order of Magnitude Qualitative Reasoning with Comparability and Negligibility Relations. Fundam. Inform. 68, 21–46 (2005)
8. Frank, A.U.: Qualitative Spatial Reasoning with Cardinal Directions. In: ÖGAI (1991)
9. Gabelaia, D., Kontchakov, R., Kurucz, A., Wolter, F., Zakharyaschev, M.: On the Computational Complexity of Spatio-Temporal Logics. In: FLAIRS (2003)
10. Gerevini, A., Nebel, B.: Qualitative Spatio-Temporal Reasoning with RCC-8 and Allen's Interval Calculus: Computational Complexity. In: ECAI (2002)
11. Golińska-Pilarek, J., Muñoz-Velasco, E.: Reasoning with Qualitative Velocity: Towards a Hybrid Approach. In: Corchado, E., Snášel, V., Abraham, A., Woźniak, M., Graña, M., Cho, S.-B. (eds.) HAIS 2012, Part I. LNCS, vol. 7208, pp. 635–646. Springer, Heidelberg (2012)
12. Guesgen, H.W.: Spatial Reasoning Based on Allen's Temporal Logic. Tech. rep., International Computer Science Institute (1989)
13. Hazarika, S.: Qualitative Spatio-Temporal Representation and Reasoning: Trends and Future Directions. IGI Global (2012)
14. Montanari, U.: Networks of constraints: Fundamental properties and applications to picture processing. Inf. Sci. 7, 95–132 (1974)
15. Muñoz-Velasco, E., Burrieza, A., Ojeda-Aciego, M.: A logic framework for reasoning with movement based on fuzzy qualitative representation. Fuzzy Sets and Systems 242, 114–131 (2014)
16. Randell, D.A., Cui, Z., Cohn, A.: A Spatial Logic Based on Regions and Connection. In: KR (1992)
17. Renz, J.: Maximal Tractable Fragments of the Region Connection Calculus: A Complete Analysis. In: IJCAI (1999)
18. Renz, J., Ligozat, G.: Weak Composition for Qualitative Spatial and Temporal Reasoning. In: van Beek, P. (ed.) CP 2005. LNCS, vol. 3709, pp. 534–548. Springer, Heidelberg (2005)
19. Renz, J., Nebel, B.: On the Complexity of Qualitative Spatial Reasoning: A Maximal Tractable Fragment of the Region Connection Calculus. AI 108, 69–123 (1999)
20. Renz, J., Nebel, B.: Efficient Methods for Qualitative Spatial Reasoning. JAIR 15, 289–318 (2001)
21. Vilain, M., Kautz, H., van Beek, P.: Constraint Propagation Algorithms for Temporal Reasoning: A Revised Report. In: Readings in Qualitative Reasoning about Physical Systems, pp. 373–381. Morgan Kaufmann Publishers Inc. (1990)
22. Wolter, F., Zakharyaschev, M.: Spatio-temporal representation and reasoning based on RCC-8. In: KR (2000)
23. Wolter, F., Zakharyaschev, M.: Qualitative Spatiotemporal Representation and Reasoning: A Computational Perspective. In: Exploring Artificial Intelligence in the New Millennium. Morgan Kaufmann Publishers Inc. (2003)

Training Datasets Collection and Evaluation of Feature Selection Methods for Web Content Filtering

Roman Suvorov, Ilya Sochenkov, and Ilya Tikhomirov

Institute for Systems Analysis of Russian Academy of Sciences
{rsuvorov,sochenkov,tih}@isa.ru

Abstract. This paper focuses on the main aspects of development of a qualitative system for dynamic content filtering. These aspects include collection of meaningful training data and the feature selection techniques. The Web changes rapidly so the classifier needs to be regularly re-trained. The problem of training data collection is treated as a special case of the focused crawling. A simple and easy-to-tune technique was proposed, implemented and tested. The proposed feature selection technique tends to minimize the feature set size without loss of accuracy and to consider interlinked nature of the Web. This is essential to make a content filtering solution fast and non-burdensome for end users, especially when content filtering is performed using a restricted hardware. Evaluation and comparison of various classifiers and techniques are provided.

Keywords: Dynamic content filtering, text classification, automatic topic identification, active content recognition, feature selection, TF-IDF, thematic importance characteristic, information gain, focused crawling.

1 Introduction

The problem of improper use of the Web has been worrying rather broad categories of people such as employers and parents since the Web came to each house. A number of various attempts to solve this problem have been proposed by the society: FOSI (former ICRA) content labeling initiative, thematic catalogs of resources, lists of URLs and regular expressions, methods for dynamic content filtering. Due to the nature of the Web only the dynamic content filtering can be considered as an adequate solution: all other approaches require hard and conscientious labor to keep databases up-to-date. The latter task is hardly solvable because of fast growth of the Web and existence of Web-anonymizers.

A good dynamic content filtering system must classify content on-the-fly with high quality. In most cases it is a compromise: the faster method is, the more often it makes errors. There are a number of commercial systems that declare use of dynamic classification: PureSight Owl ©, Blue Coat WebFilter ©, NetNanny © etc. Most systems target web resources grouped not only by theme but also by type (forums, shops etc). In this paper we will refer to such groups as categories.

G. Agre et al. (Eds.): AIMSA 2014, LNAI 8722, pp. 129–138, 2014.
© Springer International Publishing Switzerland 2014

In most cases content filters detect the following categories of content: purchase of alcohol, tobacco and drugs; web-anonymizers and proxies; chats, forums, instant messengers, dating sites and social networks; materials with cruelty and criminal information; suicide methods and stories; religious sects; news sites; file sharing sites, warez, video, image and music hostings, torrent trackers; travelling and entertainment; health and beauty; gambling and online games; popularization of various kinds of discrimination; job search websites; adult content; online shops; hobbies: sports, cars, pets etc; sites about weapons purchase and construction.

To summarize, the problem of content filtering differs from the text classification in the following aspects:

- Processing time and memory consumption is crucial (web filters perform in real time and often run on restricted hardware).
- Rates of various classification errors may vary depending on the situation (filtering may be configured more or less strict).
- The target data constantly changes (new lexis can be introduced in order to bypass filters; such resources as chats and forums don't have any specific fixed lexis).

These differences restrict the classification method that can be used: it cannot use complex techniques for feature extraction and the procedure of retraining of the classifier must be simplified.

The research presented in this paper continues the work [1]. We introduce some extensions to the original method, evaluate them in near-real conditions and compare with other classification methods. Such conditions were established using a simple thematic web crawler. The thematic web crawler is a particular case of a focused crawler that aims at collecting web pages on a certain topic [2].

The rest of the paper is organized as follows: in Chapter 2 we review available information about the dynamic content filtering and the focused crawling; in Chapter 3 we describe the modifications proposed to the original method; in Chapter 4 we discuss difficulties in collecting data for training and evaluation and ways to overcome these difficulties; Chapter 5 presents the experiment setup and the results; in Chapter 6 we sum up the work done and discuss the future research.

2 Related Work

Text categorization is a well explored area and many surveys and comparisons have been published [3, 4]. As mentioned above, the problem of the dynamic content filtering is similar to the text categorization but has several significant distinctions that have not yet been considered in the existing comparisons.

The focused crawler is a program that collects data from the Web according to some complex criteria (e.g. only scientific publications with no regard to their knowledge domain or pages on a certain topic). Rather extensive research was done in this field. According to [5] the following major approaches were developed: ontology-based, metadata abstraction, user modeling-based and other.

Most of them require resources (ontologies or models) that are expensive to generate. Babouk [2] is a keyword-based thematic web crawler that needs nearly no pre-training. It starts from a small number of keywords or URLs and crawls documents on the same topic. It uses BootCaT [6] procedure to expand the set of keywords describing the topic of interest. BootCaT itself can be used for crawling but it's not as effective as specialized crawlers are because it retrieves pages only through global search engines (search engines ban clients that try to use them intensively).

3 Method

The method discussed and improved in this paper was described in detail in [1]. According to this method a document is considered to belong to a category if its normalized Thematic Importance Characteristic (nTIC) exceeds the corresponding threshold derived during the training stage. Thematic Importance Characteristic estimates how specific the term to a particular collection of documents is comparing to another collection. To achieve this TIC relies on principles that are similar to ones that information gain relies on. nTIC of a document is estimated as a weighted sum of words TICs in it. We will refer to this method as nTIC.

Let's clarify the terms used. Category is a label that is assigned to a group of documents sharing similar features (e.g. web pages downloaded from a forum about cars have labels "forums" and "cars" assigned to them). To belong to a category is to have the corresponding label. A web site is a group of web pages that are accessible through URLs which have the same host domain name (including subdomains). Web sites and web pages can belong to multiple categories.

In this paper we propose two modifications that take into account interlinked nature of the Web. These modifications are:

- Take into account categories of the Web-sites that the currently classified page refers to.
- Tokenize URLs found on the classified page and treat these tokens as the usual lexis.

The first modification roots in the so-called thematic isolation: web pages often refer to pages from the same website or to thematically similar resources. This principle can be generalized by introducing a frequency distribution of topics of the referred resources. We propose to treat topics of resources as usual lexical features in context of the TIC-based classifier [1]. The corresponding part of the feature set is generated according to the following algorithm.

1. Extract URLs from the body of a page.
2. Extract server domain names from the retrieved URLs.
3. Map domains to categories labels using a gazetteer.
4. Calculate weights as if these labels were usual words.

The most crucial part of this algorithm is the domain-category mapping. It can be initially constructed from a catalog such as Open Directory Project [7] (e.g. an online tobacco shop would get a label "url_shopping_tobacco"). Later on it can be iteratively expanded with domains of pages that got class label with high confidence (margin between the rating of a page and the corresponding threshold is large). Each domain may correspond to multiple topics. Each topic is represented by a unique label constructed from a prefix "url_" and a title of the topic (e.g. adult, chats etc.). The goal of using the prefix is to distinguish topics of the referred pages from the usual lexis.

The second modification makes sense because human-readable URLs become more widespread (e.g. http://example.com/catalog/pages-on-some-topic instead of http://example.com/catalog.php?topicId=31415). Similar ideas were developed in [8,9]. We propose to tokenize URLs found in the body of the page and to treat all the extracted tokens as a part of usual lexis. By tokenization we mean splitting the URL using a set of delimiters, e.g. an URL "http://mega-news/catalog/news?q=tech" is converted to the following list of tokens: url_mega, url_news, url_example, url_catalog, url_news, url_tech. Short tokens and numbers are ignored.

4   Data Sets

A good evaluation of a method for content filtering is not a trivial task because of absence of reliably marked data sets. Such standard data sets for text categorization and clustering as 20 Newsgroups and Reuters-21578 don't fit the task because they contain only textual features (no hyperlinks and markup) and sets of labels used in these corpora differ from ones that make sense for content filtering (they are less thematic and more associative). Public access lists are updated rarely and contain addresses of pages that have disappeared or have been sold to another owner. Therefore, it is useless to collect pages from such lists.

There are a couple of ways to overcome this issue:

- Use unsupervised methods of machine learning (datasets marking is not necessary).
- Use methods that require few examples to learn (in this case datasets can be marked manually).
- Introduce a technique of collecting datasets that does not require much additional manual marking.

In modern conditions such corpora cannot be easily created manually because the Web changes rapidly and it is necessary to retrain the filter periodically in order to fit it to the current state of the Web. Moreover, when training on a small dataset one cannot guarantee that recall of the lexis-based filter in real life will be the same as during the experimental evaluation.

Therefore, we have chosen the third way. We created a special web crawler to collect web pages only on the topics of interest. This crawler addresses the focused crawling problem [5,10,11]. However, this problem in general is very difficult. Our goal was to create a rather simple system that collects pages on the specified topic and needs no or almost no manual configuration. Babouk [2] is a thematic web crawler that is similar to the one proposed in this paper. Our system differs from it in the web resouces walk order and the decision rule.

The general idea of data collection was to reproduce behavior of the end users. To achieve this, experts tried to find pages of each category of interest using global search engines (Google, Bing etc). Then the system automatically addressed other search engines using the most productive queries. Finally, the found pages were crawled. During the crawling, the topic of the downloaded page is compared with topics of the previous pages and only the pages on similar topics are added to the dataset.

Seed URLs are addresses of pages retrieved from the global search engines. Seed pages are the pages the seed URLs refer to. Root pages are main (home) pages of web sites.

General algorithm of this crawler contains two major steps: seed URLs collection and recursive crawling.

To collect the seed URLs, we applied the following approach. An expert tried to simulate behavior of a user. To accomplish this, the expert looked for web pages on the topic of interest using the global search engines and wrote down the most productive queries. Then the system automatically sent these queries to the other global search engines and collected addresses of the found pages (seeds URLs). Duplicate URLs were then removed from the resulting list.

Before describing the crawling algorithm let us make some definitions. Let K be the number of keywords that should be extracted from the analyzed page. Let T be the list of keywords describing the subject of crawling. It consists of pairs $(word, weight)$ where $weight$ is the number of pages that contained $word$ since the last reduction of T. T is periodically reduced to increase recall of crawling. P is the list of keywords of the currently processed page.

The system recursively crawls the specified number of pages starting from the seed list according to the following rules:

- Web sites are traversed in the breadth-first order.
- If a seed URL points to a root page (path and query string of its URL are empty) then the system will crawl recursively the referred pages.
- If a seed URL points to a non-root page (path or query are not empty) then the system will download it but will not proceed recursively. We decided to do that to reduce amount of candidate pages.
- If the current page is a seed page then the system will extract K keywords from it and add them to the list T containing keywords that describe the topic of interest.
- If the current page is not a seed page then:
  1. Extract K keywords from the current page and put them to the list P.
  2. If $|P \cap T| < M$, where M is a minimal number of keywords set by an expert, then stop processing the current page.
  3. Update T using P (details of this step will be described below).

References found in the current page are enqueued for crawling with no regard to the topic of the page. It is done so because there is no general topic-relevant order of the web site traversal and thus we cannot guarantee that non-relevant pages do not refer to the relevant ones and vice versa. There are works on more advanced focused crawling techniques [2, 10, 11] that try to reduce amount of considered pages.

The list of keywords of a page consists of K stems of tokens that have the greatest TF-IDF rating [12]. We used Snowball algorithm [13] to extract stems. IDFs were calculated over rather large subsets of English and Russian Wikipedia. If the table of IDFs does not contain a stem, the corresponding token will be ignored. The table of IDFs was built using POS-tagging and contains only stems of verbs, nouns and adjectives. Thus, the stopwords, numbers and misspelled words are naturally filtered out from the list of keywords of a page.

If the current page belongs to the topic of interest and is not a seed page, list of topic keywords T will be updated according to the following algorithm.

1. For each keyword w in the list P of the current page do
   (a) if w in T then increment its weight by 1;
   (b) otherwise add $(w, 1)$ to T.
2. If reduction of T has not been performed for R times, reduce it. The reduction consists of two steps:
   (a) remove from T $|T| - Max_T$ entries that have the smallest weights;
   (b) decrease weights of the rest entries by the maximal weight of the removed ones.

The crawling algorithm has the following configuration parameters:

- K - the number of keywords to extract from each page;
- Max_T - maximal number of terms representing the topic of interest (maximal size of T);
- M - minimal number of keywords of a page that must be in T for this page to be added to the dataset;
- R - the number of pages to process before next reduction of T.

This algorithm allows collecting web pages on the restricted topic that is specified by a set of initial pages (seeds). Periodical updates and reductions of T give some freedom to the crawler: it can slightly diverge from initial topic. The degree of the divergence depends on the configuration parameters. We do not have a technique to estimate these parameters automatically at the moment. Furthermore, we doubt that such technique can exist because of the vicious circle: to build a classifier we already need a classifier. Fortunately, these parameters seem to be rather easy for an expert to set.

5   Evaluation

The aim of the evaluation within this work is to estimate quality of the content filtering in conditions that are rather close to the real life. To achieve this goal, we collected about 170000 web pages on 17 topics (10000 pages per topic) in English and Russian (5000 pages in each language) according to the technique described in Chapter 3. These topics are:

- Adult content - resources with pornographic and erotic videos, images and animations.
- Chats and forums - resources for chatting on non-professional topics and dating including most popular social networks.
- Criminals - resources exposing violent and related to criminal materials.
- Drugs - resources on how to purchase drugs including legal ones (spice, alcohol, tobacco) or how to make them at home.
- Entertainment - websites with advertisements and recommendations on how to spend the spare time.
- Online games - browser games, MMORPG, online shooters, racing games, online casino.
- Health - resources "for women", articles on wellness, fitness, health etc.
- Hostings - sites for sharing files, videos and images (including torrents).
- Jobs - resources with job descriptions and advertisements.
- Pets - resources on pets care.
- Proxy - Web-anonymizers, lists of proxy servers and virtual private networks.
- Sect - materials containing information on possibly dangerous religious sect stories, meetings, ceremonies etc.
- Shopping - online shops, and auctions.
- Sports - news about sports events, articles for sports lovers.
- Suicide - suicide methods, stories of self-murderers.
- Tech - news about modern technologies (IT, cars etc).
- Weapons - information on how to buy, use or create weapons at home.

The topics above are considered as "bad". The system must block documents of these topics. We also added about 20000 "good" pages to the dataset. These pages were collected from Wikipedia and other informational resources and should not be blocked.

We treat this problem as a multi-class and multi-label classification problem. It means that the system must assign zero or more labels to each document. If no labels are assigned to a document, it is considered to be "good" and is not blocked. We reduce this problem to a set of binary problems using "one-vs-all" technique.

Quality of the classification is measured using macro-averaged precision, recall and F1-measure.

The following combinations of features were used during the experiment:

- *Base* - only usual lexis.
- *Cat&Tok* - *Base* feature set extended with categories and tokens of links.

The results are present in table 1.

The most difficult categories were "Jobs", "Adult" and "Sect", probably because of their breadth and fuzzy boundaries. The easiest ones were "Sports", "Games" and "Drugs".

Table 1. Comparison of classifiers and feature extraction techniques

Classifier	Feature set	Precision			Recall			F1		
		Min	Max	Avg	Min	Max	Avg	Min	Max	Avg
nTIC	Base	0.739	0.972	0.895	0.918	0.994	**0.968**	0.819	0.983	0.929
nTIC	Cat&Tok	0.812	0.986	**0.934**	0.909	0.988	0.963	0.878	0.986	**0.948**
SVM	Base	0.98	0.999	**0.996**	0.962	0.996	**0.988**	0.971	0.997	**0.992**
SVM	Cat&Tok	0.98	0.999	0.996	0.953	0.996	0.985	0.973	0.997	0.991

Table 2. Results of experiments on feature selection

Technique	IDF			nTIC			IG		
N	P	R	F1	P	R	F1	P	R	F1
5 000	0.977	**0.948**	**0.962**	1	0.852	0.921	0.99	0.83	0.908
10 000	0.983	**0.962**	**0.972**	0.981	0.955	0.968	**0.99**	0.908	0.951
100 000	0.992	**0.975**	**0.984**	0.995	0.951	0.972	**0.997**	0.958	0.977

As one can see, the extended feature set yields 7% increase in precision and 2% increase in F1. SVM performs better than nTIC on most categories, but it trains about 3 times slower and requires 2 times more memory in production. Memory consumption can become a bottleneck when working with many categories.

Often content filtering solutions work on hardware that is far from state-of-the-art (e.g. in schools, on smartphones). It means that consumption of the computational resources should be reduced as much as possible. Each category needs memory for feature selection and classifier model: e.g. if we have about 100000 features, we need about 500KB for classifier model and about 100KB to represent a document. This means that if we have 50 categories (as most modern content filters do), we need about 25MB of working memory only to classify a page (besides additional memory needed for preprocessing). Memory is a very scarce resource on most mobile devices. Furthermore, smaller feature set allows the system to work faster.

The second series of experiments addresses this issue: its goal is to determine how quality of classification depends on the feature selection technique used. During this series quality indices were evaluated over four-dimensional grid using SVM classifier with linear kernel. Dimensions were:

- Category (the same as above).
- Technique for feature selection (Inverse Document Frequency, Thematic Importance Characteristic, Information Gain) [14, 15].
- N - the number of top-rated features that must be included into the resulting feature set;
- Threshold - minimal number of documents that must use a feature in order to the feature to be significant (not considered as noise).

Results of the second series of evaluation are present in table 2. The present quality indices are macro-averaged over categories. Threshold values are also omitted for brevity (only the best values are taken into account). The numbers present in Table 2 are slightly smaller than in Table 1 because the original feature set contained about 600000 features.

6   Conclusion

In this work we evaluated and compared two classifiers, two techniques for feature extraction and three techniques for feature set reduction (feature selection) in near-real conditions. Techniques for feature extraction address the idea of using interlinked nature of the Web and the thematic isolation to improve the classification quality. We think that adding categories and tokens of URLs to the feature set does not improve the SVM accuracy because of optimization (it is already rather good). Taking into account tokens and categories of URLs may be useful to filter results of image search like Google Images. Such pages usually contain almost no text and refer to the search engine itself, but references often include addresses of the found web resources as parameters. Without the URL tokenization there are no other text-based features to classify such page.

Methods for feature selection address the need of deploying content filtering systems on any hardware (including old servers and mobile devices). Thematic Importance Characteristic gains better accuracy with small feature sets. Feature selection technique is not that important with middle-scale and large feature sets (all techniques show similar performance). The difference in quality between various feature sets originates in the way the technique resolves the tradeoff between frequent but non-special lexis and very special but rare. The more special the used lexis is the better precision is but the worse recall is.

Furthermore, a simple and easy-to-tune method for thematic web crawling is proposed and applied for collection of training data. The method is based on metasearch, keyword extraction and IDF term weighting. It is similar to Babouk [2] but due to the dynamic topic representation (i.e. reduction of the list of keywords) should provide higher control over the area of crawling. Evaluation and comparison of such web crawlers should be a subject of another research.

The main direction of the future work is the development of a heterogeneous filter. It should include a functional classifier, an analyzer of graphical content and take into account user behavior. Functional classification should improve detection of forums, shops and other types of resources that do not have very special lexis. It also probably will allow stripping out only parts of pages that contain, e.g. "dirty" advertisements. Analysis of graphics should improve detection of violent materials and pornography resources that do not use text to describe images or videos.

Acknowledgments. The project is supported by Russian Foundation for Basic Research grant 12-07-33012. The described work is performed within the Exactus Expert [16,17] project and the results are used in the TSA WebFilter software.

References

1. Suvorov, R., Sochenkov, I., Tikhomirov, I.: Method for pornography filtering in the WEB based on automatic classification and natural language processing. In: Železný, M., Habernal, I., Ronzhin, A. (eds.) SPECOM 2013. LNCS, vol. 8113, pp. 233–240. Springer, Heidelberg (2013)
2. de Groc, C.: Babouk: Focused web crawling for corpus compilation and automatic terminology extraction. In: 2011 IEEE/WIC/ACM International Conference on Web Intelligence and Intelligent Agent Technology (WI-IAT), vol. 1, pp. 497–498 (August 2011)
3. Yang, Y.: An evaluation of statistical approaches to text categorization. Information Retrieval 1(1-2), 69–90 (1999)
4. Sebastiani, F.: Machine learning in automated text categorization. ACM Comput. Surv. 34(1), 1–47 (2002)
5. Dong, H., Hussain, F.K., Chang, E.: State of the art in semantic focused crawlers. In: Gervasi, O., Taniar, D., Murgante, B., Laganà, A., Mun, Y., Gavrilova, M.L. (eds.) ICCSA 2009, Part II. LNCS, vol. 5593, pp. 910–924. Springer, Heidelberg (2009)
6. Baroni, M., Bernardini, S.: Bootcat: Bootstrapping corpora and terms from the web. In: LREC (2004)
7. AOL Inc.: Open directory project, http://www.dmoz.org
8. Baykan, E., Henzinger, M., Marian, L., Weber, I.: Purely url-based topic classification. In: Proceedings of the 18th International Conference on World Wide Web, WWW 2009, pp. 1109–1110. ACM, New York (2009)
9. Shih, L.K., Karger, D.R.: Using urls and table layout for web classification tasks. In: Proceedings of the 13th International Conference on World Wide Web, WWW 2004, pp. 193–202. ACM, New York (2004)
10. Aggarwal, C.C., Al-Garawi, F., Yu, P.S.: Intelligent crawling on the world wide web with arbitrary predicates. In: Proceedings of the 10th International Conference on World Wide Web, pp. 96–105. ACM (2001)
11. Jamali, M., Sayyadi, H., Hariri, B.B., Abolhassani, H.: A method for focused crawling using combination of link structure and content similarity. In: IEEE/WIC/ACM International Conference on Web Intelligence, WI 2006, pp. 753–756. IEEE (2006)
12. Salton, G., McGill, M.J.: Introduction to modern information retrieval. McGraw-Hill, New York (1983)
13. Porter, M.F.: Snowball: A language for stemming algorithms (2001)
14. Liu, T., Liu, S., Chen, Z., Ma, W.Y.: An evaluation on feature selection for text clustering. In: ICML, vol. 3, pp. 488–495 (2003)
15. Mitchell, T.: Machine Learning. McGraw Hill (1997)
16. Osipov, G., Smirnov, I., Tikhomirov, I., Vybornova, O.: Technologies for semantic analysis of scientific publications. In: 2012 6th IEEE International Conference on Intelligent Systems (IS), pp. 058–062 (September 2012)
17. Osipov, G., Smirnov, I., Tikhomirov, I., Shelmanov, A.: Relational-situational method for intelligent search and analysis of scientific publications. In: Proceedings of the Integrating IR Technologies for Professional Search Workshop, pp. 57–64 (2013)

Feature Selection by Distributions Contrasting

Varvara V. Tsurko[1] and Anatoly I. Michalski[1,2]

[1] Institute of Control Sciences Russian Academy of Sciences,
Profsouznaya str. 65, 117997, Moscow, Russian Federation
v.tsurko@gmail.com
[2] National Research University Higher School of Economics,
Bolshoy Tryokhsvyatitelsky Pereulok 3, 109028, Moscow, Russian Federation
amikhalsky@hse.ru

Abstract. We consider the problem of selection the set of features that are the most significant for partitioning two given data sets. The criterion for selection which is to be maximized is the symmetric information distance between distributions of the features subset in the two classes. These distributions are estimated using Bayesian approach for uniform priors, the symmetric information distance is given by the lower estimate for corresponding average risk functional using Rademacher penalty and inequalities from the empirical processes theory. The approach was applied to a real example for selection a set of manufacture process parameters to predict one of two states of the process. It was found that only 2 parameters from 10 were enough to recognize the true state of the process with error level 8%. The set of parameters was found on the base of 550 independent observations in training sample. Performance of the approach was evaluated using 270 independent observations in test sample.

Keywords: classification, features selection, information distance between in-class distributions, Rademacher penalty, set of manufacture process parameters.

1 Introduction

The problem of features selection is very important due to three reasons [16]. First, most machine learning algorithms don't operate well on the big amount of features, the number of training examples can grow exponentially with the number of inconclusive features. Second, as the number of features increases the algorithm run-time grows dramatically. And the last but not the least, accuracy of the algorithm decreases and the over-fitting problem can occur [6].

The benefits of feature selection include better algorithm accuracy, improved stability, better understanding and possibly better visualization of the data, reduced measurement and storage requirements, reduced training and inference time [16]. And the main point is that algorithm worked on the selected features has better generalization ability.

There are methods of feature selection that are performed as a preprocessing step: they are independent of the main data analysis algorithm [2]. These methods reduce the space dimension by using space transformations (e.g. Fourier

G. Agre et al. (Eds.): AIMSA 2014, LNAI 8722, pp. 139–149, 2014.
© Springer International Publishing Switzerland 2014

decomposition, principal component analysis) or ranking features by level of relevance to classes (e.g. RELIEF [5]).

Another feature selection methodology proposes to create sets of features that can be useful in the main machine learning algorithm. These methods optimize the accuracy of the algorithm by exhaustive or partial search for features subsets. For example, finding the resulting subset of features that optimizes Support Vector Machine accuracy. But the exponential growth of the number of subsets in the exhaustive enumeration process makes it impractical for tasks with more than 20-30 features. Often subsets search methods, such as sequential features adding or removing, depth- or breadth-first search, genetic algorithms, are used in the present-day complex problems of data mining.

A frequent practice in feature selection is to maximize the divergence between features distributions in classes. According to the method from [1] frequencies of features occurring in the data were calculated and features with the maximal difference of these frequencies in different classes were selected . In [12] the method of features selection was focused on the maximization of dependence between selected features and the target variable.

Ideas from Information Theory [4] are commonly used for feature selection. In [6] the algorithm is based on selection of such features subset that minimize modified Kullback-Leibler divergence [9] between classes distributions estimated for all features set and a certain subset. In [3] the main idea of the approach is to select such features subset that maximize the Kullback-Leibler divergence or its modifications between the resulting class conditional probability densities.

Since we don't know the real distributions of random variables Kullback-Leibler divergence is estimated on the empirical data. The accuracy of this estimation depends on the sample size and on the complexity of the class of distributions [14].

This article is devoted to the description of the feature selection procedure that considers the complexity of the distributions class in which the Kullback-Leibler divergence is estimated. We consider the features selection problem in case of two classes, which are described by the conditional distributions of the features. The goal of the method is to find the set of features for which conditional distributions in classes have the maximal difference. This difference is characterized by a functional of average risk. Maximization of it is equivalent to the Kullback-Leibler divergence maximization between the two in-class distributions.

The form of the proposed average risk functional allows us to implement the results of the empirical process theory and obtain statistically reliable estimations. The estimation of the average risk functional relies on the empirical risk corrected by the Rademacher penalty term, which takes into account the complexity of distribution functions in the features subset. The proposed method selects a set of features that provides the average risk maximum with the guaranteed probability. The method was applied to empirical data of a manufacturing process.

2   Distributions Contrasting Algorithm

2.1   Average Risk

A lot of data analysis problems like classification, regression, probability density reconstruction could be formulated in terms of average risk minimization. Problem of distribution contrasting can be set in the same way. Let x be a random vector of continuous features, $\varphi_0(x)$ and $\varphi_1(x)$ be probability density functions (pdfs) which estimate the conditional distribution of x under hypotheses H_0 and H_1 respectively, y be a class label variable which takes values 0 or 1 and states the number of hypothesis. The loss function is defined in following form

$$f_{\varphi_0,\varphi_1}(x,y) = -y \ln \varphi_0(x) - (1-y) \ln \varphi_1(x).$$

The functional of average risk is defined as an expectation of the loss function [14]:

$$M(\varphi_0, \varphi_1) = -E_{x,y}(y \ln \varphi_0(x) + (1-y) \ln \varphi_1(x)), \tag{1}$$

where expectation is taken by joint distribution of x and y.

The functional has the meaning of $\varphi_0(x)$ and $\varphi_1(x)$ pdfs crossentropy weighted by a priori probabilities of the hypotheses H_0 and H_1. Minimization of it by $\varphi_0(x)$ and $\varphi_1(x)$ is equivalent to probability density reconstruction under different hypotheses. If the class label y is unobserved in the data then minimization of (1) is equivalent to a clusterization problem [11].

In this paper we consider a different problem of average risk maximization. Rewrite the functional of average risk gives

$$M(\varphi_0, \varphi_1) = I(\varphi_0, \varphi_1) - E_{x,y}(y \ln p(x|H_1) + (1-y) \ln p(x|H_0)),$$

where

$$I(\varphi_0, \varphi_1) = -E_{x,y}\left(y \ln \frac{\varphi_0(x)}{p(x|H_1)} + (1-y) \ln \frac{\varphi_1(x)}{p(x|H_0)}\right).$$

Maximization of the average risk functional by $\varphi_0(x)$ and $\varphi_1(x)$ is equivalent to maximization of $I(\varphi_0, \varphi_1)$ which is close to the Kullback-Leibler divergence between two pairs of distributions $\varphi_0(x), p(x|H_1)$ and $\varphi_1(x), p(x|H_0)$ [9]. Functional $I(\varphi_0, \varphi_1)$ equals half of J-divergence functional if prior probabilities of the hypothesis H_0 and H_1 are equal. In other words, optimally selected functions $\hat{\varphi}_0, \hat{\varphi}_1$ should maximally differ from the true pdfs of x under the hypothesis H_1 and H_0 respectively.

In the paper we maximize the functional of average risk over the sets of features. In each set for probability density functions $\varphi_0(x), \varphi_1(x)$ we use Bayesian estimates of conditional distributions $p(x|H_0)$ and $p(x|H_1)$ in form of histogram [15]

$$\varphi_y^b(x) = \sum_{i=1}^{k} I(x \in \sigma_i) \frac{(1-y)n_i + ym_i + 1}{(1-y)l_0 + yl_1 + k}, \tag{2}$$

where k is the number of bins in the histogram used in estimations, $I(x \in \sigma_i)$ is indicator which equals to 1 if vector x belongs to bin σ_i, l_0 and l_1 denotes the sizes of independent samples obtained under hypotheses H_0 and H_1 respectively, n_i and m_i are the numbers of observations from independent samples obtained under hypotheses H_0 and H_1 respectively that put into the bin σ_i. Given formula represents Bayesian probability estimation in case of uniform prior distribution on a k-fold simplex. We use Bayesian estimates to avoid zero values of probabilities under the logarithm function.

With defined estimations the average risk for $\varphi_0^b(x)$ and $\varphi_1^b(x)$ is obtained by substituting them in (1). Maximization of it over the sets of features in the way explained further gives the set of features for which the Bayesian estimates of conditional distributions under different hypotheses maximally differ. It allows us to conclude that the constructed set includes features for which the hypothesis can be tested with less error than in case of using all features in the data sets.

2.2   Empirical Risk

Average risk (1) can not be calculated directly because pdfs $p(x|H_0)$ and $p(x|H_1)$ are unknown. We will maximize a sample estimate for average risk - empirical risk with penalization term to incorporate difference between mathematical expectation and sample mean.

Let $x_1^0, \ldots, x_{l_0}^0$ be a sample with the pdf $p(x|H_0)$ and $x_1^1, \ldots, x_{l_1}^1$ be a sample with the pdf $p(x|H_1)$. The formula for pdf (2) is based on Bayesian estimates of histograms with k bins for first and the second classes. We denote these estimates [15] of probability density of the bin i as

$$\varphi_0^b(i) = \frac{n_i + 1}{\sum\limits_{j=1}^{k} n_j + k}, \quad \varphi_1^b(i) = \frac{m_i + 1}{\sum\limits_{j=1}^{k} m_j + k}. \tag{3}$$

It is clear, that

$$0 < c \le \varphi_y^b(i), \; y = 0, 1, \; i = 1, \ldots, k, \; c = \frac{1}{k + l_0 + l_1}; \tag{4}$$

$$\sum\limits_{i=1}^{k} \varphi_y^b(i) = 1, \; y = 0, 1, \tag{5}$$

and the vector $\varphi_y^b = (\varphi_y^b(1), \ldots, \varphi_y^b(k))$ which satisfies conditions (4), (5) defines a histogram. Finally, let F denote a set of histograms pairs $(\varphi_0^b, \varphi_1^b)$ calculated from empirical data using (3) for all subsets of features set.

The average risk $M(\varphi_0^b, \varphi_1^b)$ reflects the divergence between Bayesian estimates and real densities. In order to evaluate the average risk we calculate an empirical risk by using average of empirical data instead of expectation. Applying $E_y(y) = l_1/(l_0 + l_1)$ we get a formula for the functional of empirical risk

$$M_e(\varphi_0^b, \varphi_1^b) = -\frac{1}{l_0 + l_1} \left(\sum\limits_{i=1}^{k} m_i \ln \varphi_0^b(i) + \sum\limits_{i=1}^{k} n_i \ln \varphi_1^b(i) \right). \tag{6}$$

The relation between the average risk and the empirical risk was discussed in [14]. Consideration of this relation allows us to switch from the maximization of the average risk problem to the maximization of the empirical risk corrected by a penalty term. A form of the penalty term will be discussed below.

2.3   Rademacher Penalization

The functional of empirical risk (6) can be presented in the form

$$
M_e(\varphi_0^b, \varphi_1^b) = -\frac{1}{l_0 + l_1} \left(\sum_{i=1}^{l_1} \ln \varphi_{0,x_i^1}^b + \sum_{i=1}^{l_0} \ln \varphi_{1,x_i^0}^b \right),
$$

where $\varphi_{y,x_i^t}^b$ – the Bayesian estimate of conditional probability under the hypothesis H_y for the bin to which x_i^t belongs.

Let $\delta_1^0, \ldots, \delta_{l_0}^0, \delta_1^1, \ldots, \delta_{l_1}^1$ be a sequence of independent and identically distributed random variables which take values $+1$ and -1 with probability $1/2$ each independently of $(x_1^0, \ldots, x_{l_0}^0, x_1^1, \ldots, x_{l_1}^1)$. The Rademacher penalty term [8,10] is defined then as

$$
R = \sup_{\varphi_0^b, \varphi_1^b \in F} \left| \frac{1}{l_0 + l_1} \left(\sum_{i=1}^{l_1} \delta_i^1 \ln \varphi_{0,x_i^1}^b + \sum_{i=1}^{l_0} \delta_i^0 \ln \varphi_{1,x_i^0}^b \right) \right|. \tag{7}
$$

With Δ_i^y denoting the sum of δ_t^y that correspond to the same bin i it could be represented as

$$
R = \sup_{\varphi_0^b, \varphi_1^b \in F} \left| \frac{1}{l_0 + l_1} \sum_{i=1}^{k} \left(\Delta_i^1 \ln \varphi_0^b(i) + \Delta_i^0 \ln \varphi_1^b(i) \right) \right|.
$$

In order to solve the optimization problem we remove the modulus be representing the penalty term in form

$$
R = \frac{1}{l_0 + l_1} \max \{A, -A\}, \tag{8}
$$

where

$$
A = \sup_{\varphi_0^b \in F} \sum_{i=1}^{k} \Delta_i^1 \ln \varphi_0^b(i) + \sup_{\varphi_1^b \in F} \sum_{i=1}^{k} \Delta_i^0 \ln \varphi_1^b(i).
$$

Thus, in order to find the optimal solution of initial problem we consider the optimization subproblem, which is to find

$$
R' = \sup_{\varphi^b \in F} \sum_{i=1}^{k} \Delta_i \ln \varphi^b(i). \tag{9}
$$

Mind that $\varphi^b(i)$ satisfy constraints (4) and (5).
The solution is given by the rules 1-3:

1. If $\Delta_i > 0$, $i = 1\ldots,k$ then

$$R' = \sum_{i=1}^{k} \Delta_i \ln \frac{\Delta_i}{\sum_{t=1}^{k} \Delta_t}$$

2. If $\Delta_i \leq 0$, $i = 1,\ldots,k$ then

$$R' = \sum_{i=1}^{k} \Delta_i \ln c + \Delta_m \ln \frac{1 - c(1 - k)}{c},$$

where $m = \arg\max_i \Delta_i$

3. If $\Delta_i > 0$, $i = 1,\ldots,s$ and $\Delta_i \leq 0$, $i = s+1,\ldots,k$ then

$$R' = \sum_{i=1}^{s} \Delta_i \ln \frac{\Delta_i(1 - c(k - s))}{\sum_{t=1}^{s} \Delta_t} + \sum_{i=s+1}^{k} \Delta_i \ln c.$$

It's important to notice that particular relation between Δ_i, c and size of empirical data set is significant. The solution for more general case, when Δ_i can take any values, is quite similar, but the rules become more complex.

By substitution of extremal values of (9) into (8) we obtain the value for the Rademacher penalty term.

2.4   Average Risk Evaluation

Values of the penalty term and the empirical risk can be used for estimation of the average risk using the symmetrization inequality [7,8]. For the class of functions uniformly bounded by a constant U and for all $t > 0$ the following holds

$$P\left\{\sup_{\varphi \in F} |M(\varphi) - M_e(\varphi)| \geq 2R + \frac{3tU}{\sqrt{l_0 + l_1}}\right\} \leq \exp\left(-\frac{t^2}{2}\right). \quad (10)$$

For Bayesian estimates (3) it is valid that $\frac{1}{l_0 + l_1 + k} \leq \varphi_y^b < 1$, from which we obtain $0 < |\ln \varphi_y^b| \leq \ln(l_0 + l_1 + k) = U$.

Using (10) and fixing the probability $\eta = \exp\left(-\frac{t^2}{2}\right)$ we derive the following inequality

$$P\left\{\sup_{\varphi \in F} |M(\varphi) - M_e(\varphi)| < 2R + \frac{3\sqrt{-2\ln\eta}\ln(l_0 + l_1 + k)}{\sqrt{l_0 + l_1}}\right\} \geq 1 - \eta.$$

Hence, with the probability not less than $1 - \eta$ the lower bound of the functional of average risk is

$$M(\varphi) > M_e(\varphi) - 2R - \frac{3\sqrt{-2\ln\eta}\ln(l_0 + l_1 + k)}{\sqrt{l_0 + l_1}}. \quad (11)$$

2.5   Distributions Contrasting Algorithm

We consider a set of features $X = (f_1, f_2, \ldots, f_n)$ measured in two different classes. The goal is to find such subset X_j of X for which two classes maximally differ in terms of the conditional distributions divergence. There are two stages: first task is to form the sequence of features subsets, second is to define the subset satisfied the goal.

We start with building histograms of k bins for each feature in a class. Then the value of the empirical risk (6) is calculated and the feature with maximum value of the empirical risk forms the beginning of sequence. Without restricting the generality let the feature f_1 be the one with a maximum value of the empirical risk functional, so $X_1 = f_1 \in X$.

Then all possible pairs of features are constructed with one feature f_1 obtained in the first step and two-dimensional histograms for each pair in two classes are built. Selecting the pair with the maximum value of the empirical risk, e.g. the pair will be (f_1, f_2), leads to the second subset in the target sequence $X_2 = (X_1, f_2) = (f_1, f_2) \in X$ which is a superset for X_1.

At the third step we create all possible triplets of features with two features fixed on the previous step obtaining X_3 in a similar way. The process continues until all features are put in the sequence. As the result of the first stage we obtain the sorted subsets of the features.

At the second stage of the algorithm we evaluate the average risk value (1) using the estimation (11) for the features in X_j for each $j \in \{1, \ldots, n\}$. The result is a set of estimates of the average risk : $M_{X_1}, M_{X_2}, \ldots, M_X$. Finally, we select the average risk M_{X_j} with the maximum value and corresponding subset of features X_j.

Proposed algorithm of distributions contrasting has $O(n^2)$ complexity.

3   Experimental Results

The algorithm of distributions contrasting was evaluated using the empirical data from a real manufacturing process presented by time records of 10 parameters. The two states of the process were matter of interest and data were labeled by an expert with the class label for each time point: 562 points in the first class and 258 points in the second class. For evaluation purpose the data were divided into a test sample and a training sample in proportion $1 : 2$.

Then the distributions contrasting algorithm was applied to the training sample. We sorted the sequence of ten parameters and evaluated lower bound of the average risk. The value of the empirical risk and the 90% lower bound of the average risk for the features sequences are shown in the Table 1.

The lower bound of average risk riches its maximum on the pair of features. With increasing number of features in the set the lower bound of the average risk goes down. So the optimal number of parameters to distinguish two given states of the system is two and it includes parameters #10 and #1.

To verify the results both training and test samples were classified using different sets of parameters described above. The Naive Bayes Classifier [11] was

Table 1. Results of the distributions contrasting

Number of parameters in the set	Parameters included in the set	Empirical risk	Lower bound of average risk
1	10	4.799	2.1275
2	10, 1	6.884	**2.9038**
3	10, 1, 4	10.1013	1.9058
4	10, 1, 4, 5	13.4695	0.6662
5	10, 1, 4, 5, 2	16.8365	-0.4554
6	10, 1, 4, 5, 2, 7	20.2038	-1.5773
7	10, 1, 4, 5, 2, 7, 3	23.5711	-2.6551
8	10, 1, 4, 5, 2, 7, 3, 6	26.9384	-3.738
9	10, 1, 4, 5, 2, 7, 3, 6, 8	30.3057	-4.8208
10	10, 1, 4, 5, 2, 7, 3, 6, 8, 9	33.673	-5.9037

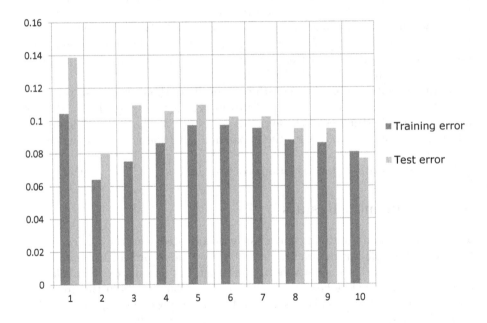

Fig. 1. Naive Bayes Classifier errors

used for classification. Figure 1 illustrates errors of classification for ten classifiers. Each pair of columns reflects classification errors on the training sample and on the test sample. The vertical axis describes the absolute value of error and horizontal axis shows the number of features in the set.

The figure 1 shows that the error on the training sample, which was used for features selection, is minimal for the set composed by parameters #10 and #1. Exactly the same set was selected by the algorithm of distributions contrasting as optimal one to distinguish the two classes. The training error is about 6%,

Table 2. Results of the algorithms comparison

Number of parameters in the set	Parameters included in the set	J-divergence	Γ
1	10	1.68	0.70822
2	10, 1	1.5357	0.82617
3	10, 1, 4	3.6693	1.0676
4	10, 1, 4, 5	6.9196	1.0762
5	10, 1, 4, 5, 2	10.3001	1.07664
6	10, 1, 4, 5, 2, 7	13.6979	**1.07673**
7	10, 1, 4, 5, 2, 7, 3	17.0991	1.07673
8	10, 1, 4, 5, 2, 7, 3, 6	20.5003	1.07673
9	10, 1, 4, 5, 2, 7, 3, 6, 8	23.9015	1.07673
10	10, 1, 4, 5, 2, 7, 3, 6, 8, 9	**27.3027**	1.07673

the test error is minimal on the same set of features and equals 8%. For sets with more features both errors are greater because of higher dimension of feature space. For the one parameter set both errors are greater because single parameter is not enough for good classification of the considering process states. Results of Naive Bayes classification show that the set of features obtained by the algorithm of distributions contrasting gives precise and stable results of classification.

To evaluate the performance of the distributions contrasting algorithm we compared it with feature selection algorithm from [3]. The author of the paper [3] proposes several functionals to evaluate the distance between two densities and makes an assumption that class-conditioned likelihood ratio is approximately Gaussian. We calculated the J-divergence for two conditional distributions and its modification Γ defined for the features sequences obtained by distribution contrasting. Mentioned functionals defined in [3] using the Kullback-Leibler divergence which is

$$D(p_0 : p_1) = \sum_x p_0(x) \ln \frac{p_0(x)}{p_1(x)},$$

where $p_i = p_i(x) = p(x|H_i), i = 0, 1$. The J-divergence then defined as the symmetric Kullback-Leibler divergence

$$J(p_0, p_1) = D(p_0 : p_1) + D(p_1 : p_0),$$

and the functional Γ is the modification of J-divergence

$$\Gamma(p_0, p_1) = \frac{J(p_0, p_1)}{[\Psi(p_0 : p_1) - D^2(p_0 : p_1)]^{1/2} + [\Psi(p_1 : p_0) - D^2(p_1 : p_0)]^{1/2}},$$

where $\Psi(p_0 : p_1)$ is given by following equation

$$\Psi(p_0 : p_1) = \sum_x p_0(x) \left(\ln \frac{p_0(x)}{p_1(x)} \right)^2.$$

The values of J-divergence and the functional Γ for ten feature subsets are presented in Table 2. The functional achieves the maximum on the set of six features. For this set both the training error and the test error are higher than for the set selected by the distributions contrasting algorithm. In the presented example the distributions contrasting algorithm demonstrates better generalization ability and doesn't make additional assumptions about in-class likelihood ratio pdfs.

4   Conclusion

In this paper two modern directions in data analysis are combined. One direction is an Information Theoretical approach to data analysis and the other is an Average Risk estimation on small samples of empirical data. It is shown that J-divergence functional between conditional in-class distributions is a particular case of average risk functional defined as mathematical expectation of a logarithm of a probability density function. This fact allows to implement well developed techniques for estimation of average risk by empirical risk calculated on data. We use Rademacher penalization to incorporate both sample size and complexity in the reliable estimation of the average risk. In the paper the average risk functional is used for feature selection problem. Presented approach aims to find a feature set for which divergence between distributions in two classes is maximal. This method is kind of opposite to often considered problem of approximation of "real" distribution, which implies minimization of divergence between distributions.

Proposed algorithm of feature selection was applied to select a set of parameters to classify two states on real data. Two considered classes were related to the two different states of the manufacture process. The constructed features subset contains two parameters which allows to predict the state of the process with high enough precision. This fact was confirmed by test sample classification with Naive Bayes approach. Implementation of alternative approach for feature selection using the information divergence criterion resulted with more features and lower percent of correct classification of as training, so test samples.

In other applications classes can be formed differently and features set can have more specific meaning. One of example is described in [13]. The features space was formed by ICD-10 codes of diseases that a person had at the end of his life. The class label in that case was formed by the logical condition "did the person have a cancer?". By selection the features set for which lower bound estimates for distance between distributions of diseases in the two classes had maximal value we obtained a list of diseases related to the cancer. All those diseases act the cancer stimulation role and should be cured at the initial state to lower the risk of cancer incidence.

References

1. Bay, S.D., Pazzani, M.J.: Detecting group differences: mining contrast sets. Data Mining and Knowledge Discovery 5, 213–246 (2001)
2. Blum, A., Langley, P.: Selection of relevant features and examples in machine learning. AI 97(1-2), 245–271 (1997)
3. Coetzee, F.M.: Correcting Kullback-Leibler Distance for Feature Selection. Pattern Recognition Letters 26(11), 1675–1683 (2005)
4. Cover, T., Thomas, J.: Elements of Infornation Theory. Wiley (1991)
5. Kira, K., Rendell, L.: The feature selection problem: Traditional methods and a new algorithm. In: Tenth National Conference on Artificail Intelligence, pp. 129–134. MIT Press (1992)
6. Koller, D., Sahami, M.: Toward Optimal Feature Selection. In: Proceedings of the Thirteenth International Conference on Machine Learning, pp. 284–292. Morgan Kaufmann Publishers (1996)
7. Koltchinskii, V.: Rademacher penalties and structural risk minimization. IEEE Transactions on Information Theory (1999)
8. Koltchinskii, V.: Oracle Inequalities in Empirical Risk Minimization and Sparse Recovery Problems. LNM, vol. 2033. Springer, Heidelberg (2008)
9. Kullback, S., Leibler, R.A.: On information and sufficiency. The Annals of Mathematical Statistics 22(1), 79–86 (1951)
10. Lozano, F.: Model selection using Rademacher Penalization. In: The Second ICSC Symposia on Neural Computation (NC 2000). ICSC Adademic (2000)
11. Manning, C., Raghavan, P., Schutze, H.: An Introduction to Information Retrieval. Cambridge University Press, Cambridge (2009)
12. Song, L., Smola, A., Gretton, A., Bedo, J., Borgwardt, K.: Feature selection via dependence maximization. Journal of Machine Learning Research 13, 1393–1434 (2012)
13. Tsurko, V., Michalski, A.: Statistical Analysis of Links between Cancer and Associated Diseases (in Russian). Adv. Geront. 26(4), 766–774 (2013)
14. Vapnik, V.: Statitical Learning Theory. Wiley Interscience (1998)
15. Vapnik, V., Chervonenkis, A.: Pattern Recognition Theory (in Russian). Nauka, Moscow (1974)
16. Wolf, L., Shashua, A.: Features Selection for Unsupervised and Supervised Inference: The Emergence of Sparsity in a Weight-Based Approach. Journal of Machine Learning Research 6, 1855–1887 (2005)

Educational Data Mining
for Analysis of Students' Solutions

Karel Vaculík, Leona Nezvalová, and Luboš Popelínský

KD Lab, FI MU Brno
{xvaculi4,popel}@fi.muni.cz, xnezva36@mail.muni.cz

Abstract. We introduce a novel method for analysis of logical proofs constructed by undergraduate students that employs sequence mining for manipulation with temporal information about all actions that a student performed, and also graph mining for finding frequent subgraphs on different levels of generalisation. We show that this representation allows one to find interesting subgroups of similar solutions and also to detect outlying solutions. Specifically, distribution of errors is not independent of behavioural patterns and we are able to find clusters of erroneous solutions. We also observed significant dependence between time duration and an appearance of the most serious error.

Keywords: educational data mining, logical proofs, clustering, outlier detection, sequence mining.

1 Introduction

Teaching constructive tasks, i.e. tasks that a student has to build in several steps, like tasks in descriptive geometry or logical and math proofs, requires advanced evaluation techniques. For example, in the case of resolution proofs in logic, see examples in Fig. 1, it is not sufficient to assign the mark based only on the conclusion that the student reached. To evaluate a student solution properly, a teacher needs not only to check the final result of a solution (the set of clauses is or is not contradictory) but also to analyse the sequence of steps that a student performed, with respect to correctness of each step and with respect to correctness of that sequence. We show that novel machine learning methods, such as graph mining and sequence mining, can be very helpful in that situation because of their capability to process structural and temporal information. Specifically, graph mining methods work with data represented as graphs, in our case one graph for each instance, and take into account the structural information of the graphs. An overview of graph mining methods can be found in [5]. Sequence mining is another topic of data mining oriented to structured data. In comparison to graph mining, data are arranged in sequences, usually ordered by time, and they are assumed to be discrete. More information on sequence mining can be found in [6].

Up to our knowledge, there is no work on analysis of student solutions of logical proofs by means of graph mining. Definitely, solving logical proofs, especially

G. Agre et al. (Eds.): AIMSA 2014, LNAI 8722, pp. 150–161, 2014.
© Springer International Publishing Switzerland 2014

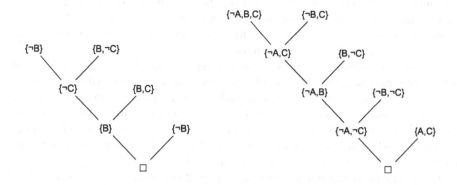

Fig. 1. An example of a correct and an incorrect resolution proof

by means of resolution principle, is one of basic graph-based models of problem solving in logic. In problem-solving processes, graph mining has been used in [21] for mining concept maps, i.e. structures that model knowledge and behaviour patterns of a student, for finding commonly observed subconcept structures. The combination of multivariate pattern analysis and hidden Markov models for the discovery of major phases that students go through in solving complex problems in algebra is introduced in [1]. Markov decision processes for generating hints to students in logical proof tutoring from historical data has been solved in [2,3,18]. Authors of [11] analyzed students' ordinary handwritten coursework with a digital pen by means of sequence mining techniques to identify patterns of actions that are more frequently exhibited by either good- or poor-performing students. In [14] sequential pattern mining was used for analysis of activity around an interactive tabletop and finding frequent sequences that differentiate high achieving from low achieving groups.

In our previous work [19] we presented a method for analysis of students' solutions of resolution proofs that employed graph mining. It used frequent subgraph mining algorithm Sleuth [23] for finding frequently occurring subgraphs. These subgraphs were then generalized and used as new features for representing the data. In [20] we extended this procedure and left out the frequent subgraph mining algorithm. More information can be found in Section 5.1. Although the method displayed very high accuracy in classification, it was unable to exploit temporal information about how the students solve the task, neither was appropriate for finding outliers – anomalous solutions.

In this paper we propose a novel method that is much more robust and is actually independent of a particular student task. In addition, it processes also information about the sequence of steps that a student performed, like adding/deleting a node or edge in the proof and also about text (i.e. a formula) modification. It also exploits information about the time a particular operation was performed and uses temporal information for finding outlying solutions. We use three different feature extraction methods and show that by using those new temporal and structural features we are able to find clusters of erroneous solutions. We also show that there is a significant dependence between time duration and appearance of errors in a student solution. Moreover, by means of class outlier detection we are able to find solutions that are anomalous and for that reason difficult to detect automatically.

It has to be stressed at the very beginning that we do not aim at building a tool for automatic classification of students' solutions. The goal that we cope with here is different: to find typical and abnormal patterns in data that are strongly correlated with wrong solutions and with particular errors. We show that with the proposed feature extraction methods we are able to find students that are at risk of producing wrong solutions. We are also able to detect outliers, i.e. solutions that are hard to detect as incorrect.

The paper is structured as follows. Section 2 contains a description of data and two representations of frequent subsequences (episodes). In Section 3 we use clustering on those two representations and show our first results about the distribution of errors in the clustered data. In Section 4 we analyse explicit time stamps and show that there is a significant dependence between time intervals of particular operations (as well as total time duration) and an appearance of the most serious error. Outlier detection in resolution proofs is solved in Section 5. A discussion and conclusion are presented in the last two sections.

2   Data and Data Pre-Processing

The data, 873 solutions altogether, was obtained in the course on Introduction to logic. Via a web-based tool, each of the 351 students solved at least three tasks randomly chosen from 19 exercises. The data set contained the resolution tree and also dynamics of the solutions, i.e. all the actions performed together with temporal information. Among these 873 different students' solutions of resolution proofs in propositional calculus, 101 of them were classified as incorrect and 772 as correct.

The most serious error in resolution is resolving on two literals. In this text we denote this error as E3. Other common errors in resolution proofs are the following: repetition of the same literal in the clause, a literal is missing in the resolved clause, resolving on the same literals (not on one positive and one negative), resolving within one clause, resolved literal is not removed, the clause is incorrectly copied, switching the order of literals in the clause, proof is not finished, intentional negation of literals in a clause. Information about the error that appeared in the logical proof is also part of the data.

All actions that a student performed, such as adding/deleting a node, drawing/removing an edge, writing/deleting a text into a node, were saved into a database. The collected data contains also a timestamp for each action performed by a student. The timestamps also allow us to get the order of actions and use techniques for sequence mining as discussed in the following sections.

2.1   Types of Sequences

For processing temporal information, we used two sequence representations of the data. Each resolution proof can be represented as a sequence in either representation. The first one is composed of two events: addition of a node (clause) into a proof and addition of an edge. An example of such sequence is CCCCCEEEE, where C denotes node addition and E denotes edge addition. The second representation uses the same events as the first one, but it also contains events of text modification[1]. An example of a sequence in the second representation is CTCTCTCTCTEEEE, where elements C, T and E denote node addition, text modification and edge addition, respectively. From now on, we will use CE and CET abbreviations for the representations.

2.2 Frequent Subsequences

As sequences cannot be processed by commonly used machine learning algorithms, such as classification, clustering or outlier detection algorithms, we need to transform the data. Simple and common practice is to use subsequences as features [6]. We considered only subsequences consisting of elements that are consecutive in original sequences, i.e. without *gaps*, because subsequences with gaps are not descriptive in case of our sequences. Formally, we say that a sequence $\alpha = \alpha_1\alpha_2...\alpha_n$ is a *subsequence* of another sequence $\beta = \beta_1\beta_2...\beta_m$ with $m \geq n$, if there exists an integer $1 \leq k \leq m - n + 1$ such that $\alpha_j = \beta_{k+j-1}$ for each $1 \leq j \leq n$.

To find all potentially useful subsequences, we employed cSpade [22] algorithm for frequent sequence mining. For a given value $min\_support \in [0, 1]$, this algorithm finds all subsequences whose $support \geq min\_support$. Support of a subsequence α is a fraction of input sequences which contain α as a subsequence. Specifically, we set $min\_support = 0.1$ to get only subsequences that occur at least in 10% of all input sequences. We obtained 121 frequent subsequences from sequences in CE representation, and 242 subsequences in case of CET representation. Each frequent subsequence is used as a new feature with value equal to 1 if the subsequence appears in the given sequence, and 0 otherwise.

3   Sequence Clustering

3.1 Method

Having the resolution proofs represented by features constructed from the two representations of sequences, we performed clustering on features of each representation. For the purpose of clustering, we set the values of features as follows: if the sequence contains the corresponding subsequence, we set the value as the squared length of the subsequence. Otherwise we set the value to 0. The rationale for this is that long subsequences should be more explanatory so they carry more weight.

On this representation of data we performed cluster analysis using the AGNES (AGglomerative NESting) hierarchical clustering and PAM (Partitioning Around

---

[1] We also tried sequences with node- and edge-deletion events, but these events did not affect the results due to their sparse occurrence.

Table 1. Internal evaluation of CE-2, CE-8, CET-2, and CET-8 clusterings by Dunn index (DI) and avg. silhouette width (SIL)

	AGNES		PAM	
clustering	DI	SIL	DI	SIL
CE-2	0.14	0.60	0.01	0.60
CE-8	0.35	0.78	0.05	0.76
CET-2	0.09	0.53	0.14	0.57
CET-8	0.16	0.64	0.02	0.61

Medoids) algorithms, description of both algorithms can be found in [13]. In case of AGNES we used average linkage method and for both algorithms we used Manhattan distance metric.

To evaluate different numbers of clusters, i.e. different cuts in AGNES dendrogram and different number of PAM medoids, we utilized two metrics, Dunn index [7] and average silhouette width [17]. Higher value of either metric indicates better clustering. For each algorithm we performed clustering with different numbers of clusters, specifically we used all integers from the interval [2, 12]. To select the most appropriate number of clusters, we ranked the values of the two metrics for each algorithm. The higher the value of a metric, the lower the rank. Then we calculated the total rank by summing over all four ranks. For the CE representation, the value of total rank decreased with larger number of clusters, but from 8 clusters onward, the change was not substantial, so we selected 8 clusters as sufficient. Specifically, the values of total rank from 2 to 12 clusters were following: 60, 56, 50, 43, 46, 31, 25, 31, 24, 21, and 21. In case of CET representation, the lowest value of total rank was calculated for 8 clusters. Results for both cases are depicted in Table 1, clustering is encoded as [*sequence representation*]–[*# of clusters*]. We also included results for CE-2 and CET-2, cases with the smallest number of clusters, for comparison. These two clustering divisions are also considered in statistical tests described later. In Table 1 we can see that the Dunn index was generally quite low, especially for PAM algorithm.

For each cluster we also looked for the most representative sequence. In case of PAM algorithm, it was enough to take the medoids. However, hierarchical clustering algorithms does not use medoids, so we designed and used the following procedure. First, we computed the average value for each feature on the set of sequences from a specific cluster. The features were the same as for clustering. Then we used the same distance metric, Manhattan distance, to find a sequence most similar to the average.

3.2   Clustering Results

Resulting representative sequences for the above mentioned clusterings are shown in Table 2. By comparing both algorithms, we can see that they share a lot of similar representatives. Simple division of the proofs can be seen in case of the

CE-2 clustering for both algorithms, see the first two rows of the table. The first cluster groups proofs solved in a step-by-step fashion, where a step means an application of the resolution rule and relevant edges are added immediately after clauses (nodes) in each step. The second cluster groups proofs solved in such a way that the nodes are added first and all the edges afterwards.

Table 2. Cluster representatives of CE-2, CE-8, CET-2, and CET-8 clusterings

Clustering	AGNES	PAM
CE-2	CCCEECCCCEECEECEE	CCCEECCEECCCCEEE
	CCCCCCCCCEEEEEEEE	CCCCCCCCCEEEEEEEE
CE-8	CCCEECCEECCEECCEE	CCCEECCEECCEECCEE
	CCCCEECEECCCCEEEE	CCCCCEEEECCCCEEEE
	CCCCCCEEEEEE	CCCCCCEEEEEE
	CCCEECCEECCEE	CCCEECCEECCEE
	CCCCCCCCCEEEEEEEE	CCCCCCCCCCCEEEEEEEEEE
	CCCEECCCECEEE	CCCEECCEECCEECCEECCEE
	CCCCCEEEE	CCCCCEEEE
	CCCEECCEE	CCCCCCCCCEEEEEEEE
CET-2	CTTCTCTCTCTCTCTCTCTEEEEEECTEE	CTCTCTCTEECEETCTCEECTTCTEE
	CCCCCCTTTTTTCTEEEEEE	CTCTCTCTCTCTCTCTCTEEEEEEEE
CET-8	CTCTCTCTCTCTCTCTCTEEEEEEEE	CTCTCTCTCTCTCTCTCTEEEEEEEE
	CCCCCCTTTTTTCCCTTTEEEEEEEE	CCCCCTTTTTCCCTTTCTEEEEEEEE
	CCCCCCCTTTTTTTEEEEEE	CCCCCCCTTTTTTTEEEEEE
	CTCTCTCTCTEEEE	CTCTCTCTCTEEEE
	CTTCTCTEETCTCEETCTCTEECTCTEE	CTTCTCTEECTCEETCTCTECTE
	CTCTCTEECTCTEE	CTCTTCTCTCTCTCTEEEEEE
	CCCCTTEETTTCCCEETTEETCTCTEET	CTCTCTCTEECEETCTCEECTTCTEE
	CTCTCTEE	CCCCCCCCTTTTTTTTTTTTCTEEEEEEEE

On the other hand, CET-8 clustering of AGNES, for example, is slightly more difficult to analyse than CE-2. Nevertheless, several distinctive characteristics can be seen from the representatives. In the first half of sequences, all edges are added last, and in the second half, they are added approximately after each step. Similar phenomenon can be observed for the T events with respect to the C events – in some cases, most of the C events are added first, and in some cases, these events are alternating with the T events.

In addition, the first and the fifth clusters from CET-8 of AGNES contained most of the instances, precisely 696 out of 873 instances. This means that for most of the proofs, it should hold that they are not short, node and text addition is alternating, and there is no prevailing way of edge addition. The last cluster in the table, represented by the CTCTCTEE sequence, was also interesting, as its 12 out of 16 instances contained E3 error. Let us remind that there were only 53 instances with E3 error in total. It means that almost 1/4 of all erroneous solutions appeared in that cluster. We can exploit the information from the representative sequence for detecting potential error and maybe warn a student or offer them a hint even before finishing their solution.

3.3 Analysis of Sequence Clusters

As we want to find whether some behavioural patterns of students are connected with errors in solutions, we analysed our sequential data with respect to solution errors. From the set of common errors, we considered the most serious error – the error of resolving on two or more literals at the same time, i.e. the E3 error. This error is the most common and also the only one with occurrence rate greater than 5%. We performed Fisher's exact test [9] to compare the occurrence of the E3 error and each of the four sequence clusterings, taken for both clustering algorithms. Considering the 5% significance level, we can conclude from the test results that the data provides convincing evidence that the occurrence of E3 error is not independent of any of those four clusterings of any of the two algorithms. Moreover, by analysing clusters as in the previous section, we can discover more useful patterns, such that for CET-8 there was a cluster with majority of wrong solutions. Such information may help early detect students that are at risk of performing the E3 error.

4   Time Duration

4.1   Time Features

We also analysed explicit time information. For each resolution proof, we computed several time characteristics. They are expressed in seconds and can be divided into two groups. The first group reflects time duration between simple events such as node addition and deletion, edge addition and deletion, and text modification. As there was no time limit for solving a resolution proof, students could keep the web tool opened for a long time without actually solving the exercise. Therefore we marked 5% of the longest durations as outliers and replaced each such outlier by a mean value of non-outlying durations of the corresponding proof. Specifically, the threshold for outliers was 15 seconds in this case. From these new values of durations we calculated the mean and the maximum value[2] and also the sum of durations, which represents the total duration of proof solving.

The second group was derived from durations of resolution rule application. In particular, for each application of resolution rule, we calculated time interval between text modification of parents and text modification of the corresponding resolvent. We omitted cases in which a parent was modified after its resolvent and, again, 5% of the longest durations were replaced by mean values. In this case, the outlier threshold was 60 seconds exactly. At the end, the maximum, the minimum and the mean values were calculated as well as the sum of the values.

4.2   Analysis of Time Duration

Then we investigated the relation between errors and time duration of a resolution proof solution. Although the duration is a continuous variable, we did

---

[2] We omitted the minimum value as it was almost always equal to zero because some actions were saved simultaneously.

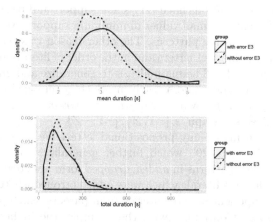

Fig. 2. Density plots for mean duration (top) and total duration (bottom) with respect to E3 error

not use the Kolmogorov-Smirnov test because the data contains mostly integer values with many ties between groups. Instead, we performed a permutation test which is based on a comparison of means. More precisely, we used two-sample exact permutation test estimated by Monte Carlo [8] on data consisting of duration summary statistics of simple events. For each summary statistic, this test compared distributions corresponding to erroneous and correct solutions for a given type of error.

Significant results were obtained only for the E3 error and for mean values of durations and sums of durations, i.e. total durations of resolution proofs. The significance level was 0.05. In both cases, the test even yielded the p-value of 0.002. This means that the distribution of mean values of durations for solutions without the E3 error is different from the distribution for solutions with the error. The same holds for the two distributions of sums of durations. Distributions of mean values can be found in the density plot at the top of Fig. 2. At the bottom of the same figure, the density plot of the distributions of total durations is depicted. We can conclude that solutions built in short time but on the basis of slow actions are more likely to contain the E3 error.

5   Outlier Detection in Resolution Graphs

5.1   Generalized Subgraphs

As in the case of sequences, graph features may be constructed from graph substructures found by an algorithm for frequent subgraph mining [5]. The task of frequent subgraph mining is very similar to the frequent sequence mining, but now the frequently occurring subgraphs are looked up in a set of graphs. Next, boolean features are used and the value of a feature depends on whether the corresponding substructure occurs in the given instance or not. Because the node labels of incorrect solutions have very small support and frequent subgraph mining can be inefficient for small values of minimum support, we created a new method for finding interesting patterns. The other reason was that similar subgraphs differed only in alphabet letters and in the order of literals in text labels or in the order of parent nodes. The main idea is to generalize and unify subgraphs consisting of two parent nodes and a resolvent. Such generalized subgraphs may be expressed shortly in the following form: $parent1; parent2 -- > resolvent$. An example of generalized subgraph is $\{\neg Y, Z\}; \{\neg Y, \neg Z\} -- > \{\neg Y\}$, where Y, Z are variables that can match any propositional letter. We also created a new, higher-level, generalization [20], which further generalizes the patterns found earlier. As a result, new patterns in *added; dropped* form are created. The *added* component simply denotes literals which were added erroneously to the resolvent and the *dropped* component denotes literals from parents which participated in the resolution process. Continuing with the example, the higher-level pattern will be $\{\}; \{\neg Z, Z\}$. The detailed description can be found in [20].

5.2   Method

The data we process has been labeled. We focused only on the E3 error as it was the most common and the most serious error. Specifically, there were two values of the class attribute, corresponding to the occurrence or nonoccurrence of the error. Unlike in common outlier detection, where we look for outliers that differ from the rest of "normal" data, we need to exploit information about a class. That is why we used weka-peka [16] that looks for class outliers [15] using Random Forests (RF) [4]. It extends the outlier detection method in RF implemented in Weka [10] that actually works for classical settings – normal vs. anomalous data to manage class information. The main idea lies in different computation of the proximity matrix that exploits also information about a class label [16].

5.3   Results

When analysing the strongest outliers that weka-peka discovered, we found three groups according to the outlier score. The two most outlying examples, with outlier factor overcoming 130, significantly differ from the others. The second cluster consists of four examples with the outlier score between 50 and 100, and the last group is comprised of instances with the lowest score of 15.91. Analysis of individual outliers let us draw several conclusions. Two most outlying instances contain one specific pattern, *looping*. This pattern represents the ellipsis in a resolution tree, which is used for tree termination if the tree cannot lead to a refutation. Both instances contain this pattern, but neither of them contains the pattern of correct usage of the resolution rule. The important thing is that these two instances contain neither the E3 error nor other errors. This shows that it is not sufficient to find all errors and check the termination of proofs, but we should also check whether the student performed at least few steps by using the resolution rule. Otherwise we are not able to evaluate the student's skills. Instances with the outlier score less than 100 are less different from other instances.

6   Discussion

The other method that we used for outlier detection was CODB [12]. However, when compared with weka-peka, CODB returned much worse results mainly because of using density and distances (to nearest neighbours and to all members of the class) for outlier detection. Such poor results may be caused by the fact that those metrics are too rough for our task. Moreover, it is much more difficult to obtain a comprehensive explanation of why a particular solution is an outlier.

As we stressed in the introduction, this method has not been developed for recognition of correct or incorrect solutions. However, to verify that the feature construction is appropriate, we also learned various classifiers of that kind. Best result was achieved by SMO (SVM implementation in Weka) – 96.9% accuracy[3]. Similar results were obtained when only the higher level of subgraph generalization was used, again with SMO.

7   Conclusion

We proposed three different feature extraction methods for temporal as well as structural information in logical proof solutions. We showed that by using those new features we are able to find clusters of erroneous solutions, dependence between time duration and errors in a student solution, and also solutions that are anomalous and for that reason difficult to detect automatically.

We believe that this method can be used even by non-expert in machine learning. There are only few parameters to be set: the minimum support for frequent subsequences and also frequent subgraphs, maximum number of clusters in sequence clustering, and the length of delay that indicates outliers in time duration analysis.

This method is general. It can be used also for other logical proofs, such as tableaux proofs, and for any construction tasks that are based on graphs. The main idea, frequent subgraph mining and sequential frequent mining, can be used without a change. However, a new way of subgraph pattern generalization must be developed for a specific construction task.

There is a big potential of the results displayed above in practical education. By using the detection and the analysis of clusters with higher frequency of erroneous solutions a teacher can detect potential reasons of errors and find shortcomings in tutoring. Even in the process of solving the task, it is possible to detect behavioural patterns before completing the proof and warn the student. Outlier detection particularly helps to discover picturesque students' solutions and fix drawbacks in automatic evaluation.

---

[3] The rate of correct predictions made by the model.

In future we will use this method for resolution in predicate logic and also for tableaux proofs. This may require an extraction of different structural features (frequent subgraphs) but the rest can be used without changes. A challenge is also the use of inductive logic programming that can better cope with domain knowledge.

Acknowledgments. This work has been supported by Faculty of Informatics, Masaryk University and the grant CZ.1.07/2.2.00/28.0209 Computer-aided-teaching for computational and constructional exercises. We would like to thank Alex Popa and members of KD lab for their help.

References

1. Anderson, J.R.: Discovering the Structure of Mathematical Problem Solving. In: Proceedings of EDM (2013)
2. Barnes, T., Stamper, J.: Toward automatic hint generation for logic proof tutoring using historical student data. In: Woolf, B.P., Aïmeur, E., Nkambou, R., Lajoie, S. (eds.) ITS 2008. LNCS, vol. 5091, pp. 373–382. Springer, Heidelberg (2008)
3. Barnes, T., Stamper, J.: Automatic Hint Generation for Logic Proof Tutoring Using Historical Data. Educational Technology and Society 13(1), 3–12 (2010)
4. Breiman, L.: Random Forests. Machine Learning 45(1), 5–32 (2001)
5. Cook, D.J., Holder, L.B.: Mining graph data. Wiley-Interscience, Hoboken (2007)
6. Dong, G., Pei, J.: Sequence data mining. Springer, New York (2007)
7. Dunn, J.C.: A Fuzzy Relative of the ISODATA Process and Its Use in Detecting Compact Well-Separated Clusters. Journal of Cybernetics 3(3), 32–57 (1973)
8. Fay, M.P., Shaw, P.A.: Exact and Asymptotic Weighted Logrank Tests for Interval Censored Data: The interval R package. Journal of Statistical Software 36(2), 1–34 (2010), http://www.jstatsoft.org/v36/i02/
9. Fisher, R.A.: Statistical Methods for Research Workers. Oliver & Boyd (1970)
10. Hall, M., Frank, E., Holmes, G., Pfahringer, B., Reutemann, P., Witten, I.H.: The WEKA Data Mining Software: An Update. SIGKDD Explor. Newsl. 11(1), 10–18 (2009)
11. Herold, J., Zundal, A., Stahovich, T.F.: Mining Meaningful Patterns from Students' Handwritten Coursework. In: Proceedings of EDM (2013)
12. Hewahi, N., Saad, M.: Class outliers mining: Distance-based approach. International Journal of Intelligent Technology 2(1), 55–68 (2007)
13. Kaufman, L., Rousseeuw, P.J.: Finding groups in data: an introduction to cluster analysis. Wiley, Hoboken (2005)
14. Martinez, R., Yacef, K., Kay, J., Al-Qaraghuli, A., Kharrufa, A.: Analysing frequent sequential patterns of collaborative learning activity around an interactive tabletop. In: Proceedings of EDM (2011)
15. Papadimitriou, S., Faloutsos, C.: Cross-outlier detection. In: Hadzilacos, T., Manolopoulos, Y., Roddick, J., Theodoridis, Y. (eds.) SSTD 2003. LNCS, vol. 2750, pp. 199–213. Springer, Heidelberg (2003)
16. Pekarčíková, Z.: Supervised outlier detection. Master's thesis (in Czech). Masaryk University (2013), http://is.muni.cz/th/207719/fi_m/diplomova_praca_pekarcikova.pdf

17. Rousseeuw, P.J.: Silhouettes: a Graphical Aid to the Interpretation and Validation of Cluster Analysis. Computational and Applied Mathematics 20, 53–65 (1987)
18. Stamper, J.C., Eagle, M., Barnes, T., Croy, M.J.: Experimental Evaluation of Automatic Hint Generation for a Logic Tutor. I. J. Artificial Intelligence in Education 22(1-2), 3–17 (2013)
19. Vaculík, K., Popelínský, L.: Graph Mining for Automatic Classification of Logical Proofs. In: CSEDU (2014)
20. Vaculík, K., Popelínský, L., Nezvalová, L.: Graph mining and outlier detection meet logic proof tutoring. Submitted to Graph-based Educational Datamining 2014 (2014)
21. Yoo, J.S., Cho, M.H.: Mining Concept Maps to Understand University Students' Learning. In: Proceedings of EDM (2012)
22. Zaki, M.J.: Sequences Mining in Categorical Domains: Incorporating Constraints. In: 9th ACM CIKM, pp. 422–429 (2000)
23. Zaki, M.J.: Efficiently Mining Frequent Embedded Unordered Trees. Fundamenta Informaticae 66(1-2), 33–52 (2005)

Differentiation of the Script
Using Adjacent Local Binary Patterns

Darko Brodić[1], Čedomir A. Maluckov[1],
Zoran N. Milivojević[2], and Ivo R. Draganov[3]

[1] University of Belgrade, Technical Faculty in Bor, V.J. 12, 19210 Bor, Serbia
[2] College of Applied Technical Sciences, Aleksandra Medvedeva 20, 18000 Niš, Serbia
[3] Technical University Sofia, Department of Radio Communications and Video
Technologies, Boulevard Kliment Ohridsky 8, 1000 Sofia, Bulgaria
dbrodic@tf.bor.ac.rs

Abstract. The paper proposed an algorithm for script discrimination using adjacent local binary patterns (ALBP). In the first stage, each letter is modeled according to its height. The real data are extracted from the probability distribution of the letter heights. Then, the gray scale co-occurrence matrix is computed. It is used as a starting point for the feature extraction. The extracted features are classified according to ALBP. Because of the variety in script characteristics, the statistical analysis shows the differences between scripts. Accordingly, the linear discrimination function is proposed to distinct the scripts. The proposed method is tested on the samples of the printed documents, which include Cyrillic and Glagolitic script. The results of experiments are encouraging.

Keywords: Cryptography, Optical character recognition, Script recognition, Statistical analysis, Typographical features.

1 Introduction

Script identification represents a key step in multi script and multilingual document image analysis [1]. The existing approaches can be classified in two categories: global and local [2]. Global approaches utilize the statistical feature and frequency-domain analysis of wider regions, i.e. text blocks in document images [3]. Their correctness is highly sensitive to the quality of the document images and to the level of the noise included in these images. Local approaches segment the characters, words or lines as connected components and analyze the feature that lists the black pixel runs [4]. Hence, this type of the analysis is convenient for low quality document images. It is insensitive to noise, too. However, it is more computer time intensive than global approaches.

This paper proposes the combination of the local and global approach. It is established by extracting all characters in text, which are classified and subjected to the statistical analysis. In this way, the modeling of document is performed in the manner like the cryptography. Each letter from text document is superseded with the cipher [5]. The cipher is extracted from the probability distribution of

G. Agre et al. (Eds.): AIMSA 2014, LNAI 8722, pp. 162–169, 2014.

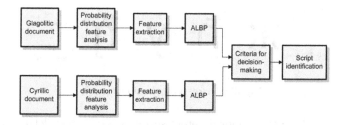

Fig. 1. The flow of the algorithm

the letter heights. Then, the feature extraction based on spatial micro pattern is carried out. It is achieved with statistical measures extracted from the gray-level co-occurrence matrix (GLCM) [6]. Furthermore, adjacent local binary pattern (ALBP) is used as classification to [7]. Taking into account the results, a linear discrimination function is determined. It is used as criteria for characterization of a certain script.

The rest of the paper is organized as follows. Section 2 addresses all aspects of the proposed algorithm. It describes the text-line definition and script modeling. Furthermore, ALBP is defined in order to extract and classify the features of the script. Section 3 defines the experiment and shows the elements of the custom oriented database. Section 4 presents and discusses the effect of the experimentation. As a consequence, the criteria for the script identification are proposed. Section 5 draws conclusions and points out further research direction.

2   The Proposed Algorithm

2.1   Procedure

Suppose that exists the same printed document written in Glagolitic and Cyrillic script. First, the probability distribution of the letter heights in each text file is computed. The script types as in ref. [5] are extracted from these distributions. As a result, the cipher is obtained. It is subjected to co-occurrence analysis. The extracted features are classified by ABLP. It is a starting point for discrimination of the scripts. At the end, criteria for the script identification is determined. Fig. 1 shows the flow of the proposed algorithm.

2.2   Probability Distribution Feature Analysis

In ref. [5] the elements of the script are mapped in the similar manner like hash function. This approach is based on their position in the text-line structure. Each text-line consists of the following vertical zones [8]: (i) upper zone, (ii) middle zone, (iii) lower zone. All letters, diacritics, ligatures or signs represent the constituents of any script. They can be classified according to aforementioned zones. Fig. 2 illustrates types of script elements according to vertical zones.

Fig. 2. Script type classification

Furthermore, there exists four different script types: (i) base letter, (ii) ascender letter, (iii) descender letter, (iv) full letter. The middle zone only is occupied by the base letter (B), which height is the smallest. Hence, it is mapped to 1. The upper and middle zones are occupied by the ascender letter (A). Its height is slightly bigger. Hence, it is mapped to 2. The lower and middle zones are occupied by the descender letter (D). It is mapped to 3. The upper, middle and lower zones are occupied by a full letter (F). Accordingly, it is mapped to 4. All letters from different scripts are mapped according to the script type set S:

$$S = \{B, A, D, F\}, \tag{1}$$

or according to height set C:

$$C = \{1, 2, 3, 4\}. \tag{2}$$

If we consider the proposed concept, then the possible probability function of the letter heights will be multi-mode, i.e. four-mode type (maximum number of script type elements).

2.3   Feature Extraction

To apply the proposed concept, the probability function of the letter heights have to be investigated. Fig. 3(a) presents a typical four-mode probability distribution, while Fig. 3(b) shows a typical letter heights distribution obtained from the real text sample.

The histogram of distributions can be divided into left and right part. The left part of the histogram represents the distributions of the punctuation elements, or similar ones. In contrast, the right part of the histogram shows the distribution of letter heights. Hence, the right part of the histogram is the only one which is important for further analysis.

It incorporates the four-mode distribution. Each region represents the different script type region. The expected probability of the script type can be extracted by summing up the probability from each of these regions. In this way, the probability of each script type represents the feature that has to be extracted.

2.4   Adjacent Local Binary Pattern (ALBP)

The local binary pattern (LBP) is an image operator which transforms an image into an array of integer labels [9]. Initially, itr works on a 3×3 pixel image block. LBP represents the magnitude relation between the center pixel and its

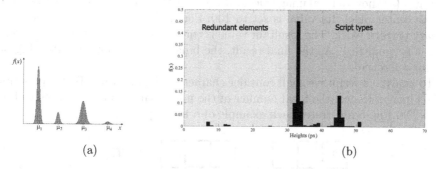

(a)   (b)

Fig. 3. The probability heights distribution: (a) Four-mode histogram, (b) Histogram from the real text sample

neighboring pixels [9]. It is obtained by thresholding the image intensity of the surrounding pixels with the strength of the center pixel. Because the neighborhood consists of 8 pixels, then the total number of different labels is $2^8 = 256$. After that, the analysis is managed by LBP histogram. The obtained binary patterns are converted into a decimal number representing a label. Then, the histogram is generated from the labels of all local image regions.

Let I be an image intensity, and $r = (x, y)^T$ be a position vector. LBP is defined as follows:

$$b_j = \begin{cases} 1, & I(r) \leq I(r + \Delta r_i) \\ 0, & \text{otherwise} \end{cases}, \qquad (3)$$

where $i = 1, ..., N_n$. N_n is the number of neighbor pixels, while Δr_i are displacement vectors from the position of center pixel r to neighbor pixels [7],[9]. For 8-connected neighborhood, N_n is equal to 8, while the displacement distance $d(\Delta r_i)$ is equal to 1 (pixel). Next, LBP $b(r)$ is converted into a decimal number. At the final stage, the LBP histogram is made by considering the decimals as labels. Fig. 4 shows an example of a micro pattern and corresponding LBP.

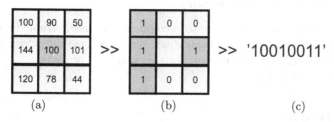

(a)   (b)   (c)

Fig. 4. LBP example of image (8-connected neighborhood): (a) Local image block, (b) Local binary pattern block, (c) Binary label

Fig. 4 (a) shows a local image block. It is thresholded according to the intensity of the center pixel. Its value is set to 100. Fig. 4 (b) shows the resultant local binary pattern block. The resultant value is equal to 0, if the intensity is smaller, else it is equal to 1. As the final result, the binary label is obtained (See Fig. 4 (c) for reference).

To analyze a script we shall consider characteristics of a text. Text represents a 1-D image. Hence, the total number of the micro patterns is $2^2 = 4$ instead of $2^8 = 256$. Fig. 5 illustrates such example of a micro pattern.

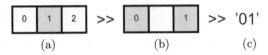

(a)   (b)   (c)

Fig. 5. LBP example of text (2-connected neighborhood): (a) Local text block, (b) Local binary pattern block, (c) Binary label

The binary pattern is converted to its decimal equivalent '01'. Then, analysis is managed by LBP histograms generated from decimal values of all micro patterns. Because the number of micro patterns is only $2^2 = 4$, the extension of LBP is mandatory.

Ref. [7] proposed a new approach called ALBP, which is called a multiple LBP. The approach starts up with the initial LBP. The, two subset LBPs are created [7]: (i) LBP(+), and (ii) LBP(×). LBP(+) considers only two horizontal and two vertical pixels. The other configuration is the LBP(×), which considers the four diagonal pixels. Furthermore, the combination of adjacent LBP(+) and LBP(×) establish ALBP [7]. LBP (+) can be linked to another LBP (+) horizontally or vertically. In contrast, LBP (×) can be linked to another LBP (×) diagonally (two diagonal directions) [7]. In our conditions, LBP(+) only matters. Due to the nature of text, the horizontal direction is only valid. Hence, ALBP is created using two adjacent LBP(+) only. In this way, a combination of two local 2-digit binary label will establish a 4-bit binary label. Hence, it represents the total number of $2^4 = 16$ micro patterns. In contrast, these labels are qualitatively different from those obtained from single LBP, because they represent the co-occurrence of LBP. This way, the results of the analysis are much more credible.

2.5   Criteria for Decision-making

The histograms are created using into account ALBP distribution. These results are the basis for establishing a linear discrimination function, which is used as criteria for the script discrimination. In this way, the procedure for the script identification is established.

3   Experiments

The experiment is created in order to estimate the quality of the proposed approach. Therefore, a custom-oriented database of 50 documents is created. These

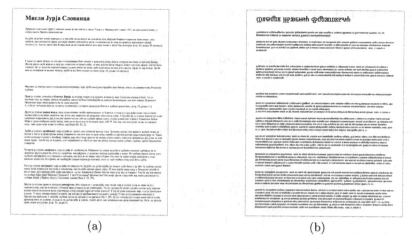

Fig. 6. Document samples from database: (a) Cyrillic, and (b) Glagolitic

documents are based on the texts form the web site [10]. The author of the initial text from the site is Juraj Slovinac (Latin: Georgius de Sclavonia, French: Georges d'Esclavonie, 1355/60 - 1416), which was the professor at College de Sorbonne around 1400. The text from the site was written in Tours, France in 1411. It represents the parts of the book "Castle of the Virginity." The text is divided into the number of separated documents. The excerpt is given in Fig. 6. All documents are scanned as bi-level images at the resolution of 300 dpi.

4   Results and Discussion

The feature extraction based on the ALBP gives 16 referent micro patterns from '0000' to '1111'. However, seven micro patterns only give a valuable results, i.e.: '0000', '0010', '0011', '1000', '1010', '1011', and '1100'. To quantify the results, the normalized micro patterns evaluated by their minimum, maximum and mean values were used. The typical values of the ALBP micro patterns for text written in Glagolitic and Cyrillic script is given in Tables 1-2, respectively.

Table 1. ALBP micro pattern normalized distribution for Glagolitic documents

Glagolitic	0000	0010	0011	1000	1010	1011	1100
Minimum	0.5787	0.1292	0.0369	0.1266	0.0000	0.0058	0.0622
Maximum	0.6093	0.1487	0.0618	0.1348	0.0169	0.0112	0.0765
Mean	0.6012	0.1408	0.0463	0.1284	0.0037	0.0078	0.0717

Because of the difference in script characteristics, the analysis shows significant diversity between both scripts. Hence, it represents the key point for

Table 2. ALBP micro pattern normalized distribution for Cyrillic documents

Cyrillic	0000	0010	0011	1000	1010	1011	1100
Minimum	0.6436	0.0972	0.0410	0.1056	0.0055	0.0000	0.0234
Maximum	0.6667	0.1485	0.0585	0.1250	0.0175	0.0028	0.0528
Mean	0.6561	0.1285	0.0475	0.1132	0.0106	0.0006	0.0435

Table 3. Comparison of ALBP micro pattern distribution ratio between Glagolitic and Cyrillic script

Glagolitic/Cyrillic	0000	0010	0011	1000	1010	1011	1100
Minimum	0.8991	1.3291	0.9013	1.1986	0.0000	0.0000	2.6603
Maximum	0.9139	1.0015	1.0567	1.0787	0.9607	4.0449	1.4490
Criteria	< 1	> 1	-	> 1	< 1	-	> 1

decision-making process of script identification. Tables 3 gives the comparison between ALBP measures needed for decision-making process.

It is very important to use only measures with distinct difference in values for the different scripts. Establishing the ratio between these measures gives their relation that can be utilized as a part of the identification criteria. Considering the results, the candidates for confident ALBP micro pattern ratios are: '0000', '1000', and '1100'. By combining logical relation, the criteria can be described with the following pseudo-code:

```
program ALBP_Script_Distinction (Output)
    If [('0000' < 1) and ('1000' > 1) and ('1100' > 1)]
        then Output = "Glagolitic text"
        else Output = "Cyrillic text"
    end
end.
```

The aforementioned criteria distinct Glagolitic and Cyrillic script in at least 99% of document samples while typical recognition rate is between 93.5% and 98.5% [2]. The proposed method is computationally non-intensive, which is a real advance because it is classified as a local approach. The algorithm is coded in Matlab. Typical ALBP processing time is as low as 0.1 sec per 2K characters (typical text page length), while the feature extraction takes up to 4-5 sec for the same document.

5   Conclusions

The paper introduced a novel method for the script characterization in OCR. The algorithm incorporated the statistical analysis of the letters in text document based on their baseline status, i.e. its height. This way, the number of analyzing variables was considerably reduced. Furthermore, the analysis of the height distribution achieved the feature extraction. Then, the extended LBP called ALBP was performed to classify the features. Because of the differences in the script characteristics, the ALBP micro patterns showed significant diversity between scripts. Hence, it was used as a cornerstone for decision-making process of the script discrimination. The proposed method was tested on documents from custom oriented database. The document was written in Glagolitic and Cyrillic scripts. The experiments gave encouraging results.

The research presented in the manuscript can be used in preprocessing stages of OCR. Further research direction will be toward ALBP analysis of different scripts, including Latin, Fraktur, Bangla, etc., and its integration into the preprocessing stage of the optical character recognition (OCR).

Acknowledgments. This work was partially supported by the Grant of the Ministry of Education, Science and Technological Development of the Republic Serbia, as a part of the project TR33037 and III43011.

References

1. Ghosh, D., Dube, T., Shivaprasad, A.P.: Script Recognition - A Review. IEEE Transaction on Pattern Analysis and Machine Intelligence 32(12), 2142–2161 (2010)
2. Joshi, G.D., Garg, S., Sivaswamy, J.: A Generalised Framework for Script Identification. International Journal of Document Analysis and Recognition (IJDAR) 10(2), 55–68 (2007)
3. Busch, A., Boles, W.W., Sridharan, S.: Texture for Script Identification. IEEE Transaction on Pattern Analysis and Machine Intelligence 27(11), 1720–1732 (2006)
4. Pal, U., Chaudhury, B.B.: Identification of Different Script Lines from Multi-Script Documents. Image and Vision Computing 20(13-14), 945–954 (2002)
5. Brodić, D., Milivojević, Z.N., Maluckov, Č.A.: Recognition of the Script in Serbian Documents using Frequency Occurrence and Co-occurrence Analysis. The Scientific World Journal 2013(896328), 1–14 (2013)
6. Haralick, R., Shanmugam, K., Dinstein, I.: Textural Features for Image Classification. IEEE Transactions on Systems, Man, and Cybernetics 3(6), 610–621 (1973)
7. Nosaka, R., Ohkawa, Y., Fukui, K.: Feature Extraction Based on Co-occurrence of Adjacent Local Binary Patterns. In: Ho, Y.-S. (ed.) PSIVT 2011, Part II. LNCS, vol. 7088, pp. 82–91. Springer, Heidelberg (2011)
8. Zramdini, A.W., Ingold, R.: Optical Font Recognition Using Typographical Features. IEEE Transaction on Pattern Analysis and Machine Intelligence 20(8), 877–882 (1998)
9. Ojala, T., Pietikainen, M., Harwood, D.: A comparative study of texture measures with classification based on featured distributions. Pattern Recognition 29(1), 51–59 (1996)
10. http://www.croatianhistory.net/etf/juraj_slovinac_misli.html

New Technology Trends Watch:
An Approach and Case Study[*]

Irina V. Efimenko[1] and Vladimir F. Khoroshevsky[2]

[1] Center for Information Intelligence Applications of Institute for Statistical Studies
and Economics of Knowledge, NRU HSE, Moscow, Russia
{iefimenko,vkhoroshevsky}@hse.ru
[2] Dorodnicyn Computing Center of RAS, Moscow, Russia
khor@ccas.ru

Abstract. A hybrid approach to automated identification and monitoring of
technology trends is presented. The hybrid approach combines methods of ontology based information extraction and statistical methods for processing
OBIE results. The key point of the approach is the so called 'black box' principle. It is related to identification of trends on the basis of heuristics stemming
from an elaborate ontology of a technology trend.

Keywords: New technology trend monitoring, statistical method, linguistic method, hybrid approach, knowledge discovery, ontology-based information extraction, time series, tag cloud, Gartner Hype Cycle.

1   Introduction

One of the hot topics in science and technology (S&T) mapping is the identification
of new technology trends (technology trend monitoring, trend hunting, trend watch).
Most of approaches in this field are based on bibliometrics, patterns and time series
analysis aimed at description of current and forecasted processes, technology road-
mapping and foresight methodologies.

In this paper we focus on automated identification of new technology trends (TT)
based on hybrid approach to natural language processing. The hybrid approach combines methods of ontology-driven information extraction (ontology based information
extraction, OBIE) and statistical methods for processing OBIE results. The key point
of the approach is the so called 'black box' principle proposed by the authors. TTs are
identified on the basis of heuristics stemming from an elaborate ontology of a technology trend.

2   Related Works and Main Challenges

The task of monitoring the TTs initiates from scientometrics and bibliometrics [1]. In the
early stages, S&T mapping tasks were usually solved manually, then semi-automatically

---

[*] This paper is based on the results of the project 'Identification of the prominent research areas
in Social Science and Humanities aimed at improving efficiency of S&T management'
(Agreement № 02.602.21.0003, Identification number: RFMEFI60214X0003) which was initiated and sponsored by the Russian Ministry of Education and Science

G. Agre et al. (Eds.): AIMSA 2014, LNAI 8722, pp. 170–177, 2014.
© Springer International Publishing Switzerland 2014

with the use of statistical methods of information extraction (IE). Statistical approach is aimed at building thematic, probabilistic thematic and dynamic thematic models, with document and / or term clusters as main results [2]. Important constraints of such approaches in case of TT monitoring are related to the use of a 'bag of words' model, lack of consideration of a document structure, and the need to preset a number of topics. Moreover, defining a TT purely as a set of words which are weighted based on probability of term attribution to a topic contradicts an intuitive understanding of a TT. However, it still facilitates the routine work. For this reason, many commercial tools are based on statistical methods. Most popular tools are focused on bibliometric analysis (Vantage-Point[1], SJR[2], etc.).

In recent works, the mentioned problems are partially solved by the use of computational linguistics and various IE methods which provide key expressions instead of key words. Thesauri and ontologies are used for information extraction, labeling and synonymy analysis [3]. Thus, the current stage of research and developments (R&D) involves combination of powerful statistical models and fairly simple methods of IE [2, 3]. Most of the works however deal with the task of trend monitoring in social networks or media [4]. The minority of them pose the problem of *technology* trend monitoring with patent analysis as the most popular approach to solve this task [5].

Despite many beneficial results, almost no solutions allow users to escape from getting too general (e.g. 'semantic technologies' for IT) or garbage (e.g. 'research' for collections of scientific papers) results which make it impossible to identify trends at the stage of weak signals. Lack of attention is paid to automated labeling and merging (for the 'same' trends).

We assume that the main problems faced by developers in the field of automated TT monitoring arise from the fact that they focus solely on bibliometrics or attempt to identify trends based on key players (persons, institutes, companies, countries) while there are no semantic models for a TT notion itself. This makes it impossible to specify semantic indicators of trends in texts of various genres (papers, patents, technology news, techno blogs, etc.).

An approach proposed in the paper involves the following methods:

1. We build an elaborate ontology of a TT and form a system of indicators of a TT presence (description of a TT) in documents of different genres. The indicators are interrelated with the domain ontology of a TT through linguistic and extra linguistic markers which build lexical and, more generally, linguistic ontology.
2. The ontology-driven processing is carried out not for the complete initial text collections, but for the preprocessed documents which include only meaningful fragments identified on the basis of TT indicators ('precompressed' documents, digests).
3. Specific OBIE models are used for each type of information sources. The results are integrated within a common knowledge representation model.

IE methods provide domain (IT, medicine, nanotechnologies, etc.) independent processing within the 'black box' framework and are used to build characteristic vectors consisting of meaningful elements. Statistical methods are used for clustering and time series analysis. As a result, analysts should receive meaningful and interpretable information on new TTs.

---

[1] https://www.thevantagepoint.com/
[2] http://www.scimagojr.com/

Thus, the paper presents a hybrid approach which is based on the use and evolution of methods proposed recently in [4, 5, 6], and extrapolates them for multilingual, multi genre domain independent text collections within OBIE framework. Ontologies drive the process of 'bag of terms' construction. Linguistic and statistical processing of texts is followed by an automated generation of OWL representation of the proposed TT ontology on the instance level. Merging OWL representations for document collections of the same and different genres is carried out with consideration of the models of time.

3   'Black Box' Principle and Hybrid Approach to New Technology Trend Monitoring

3.1   Basic Hypotheses and Assumptions

We employ Gartner's Hype Cycle model as an element of our foresight methodology. It is well known that each Hype Cycle drills down into the five key phases of a technology life cycle: Technology Trigger, Peak of Inflated Expectations, Through of Disillusionment, Slope of Enlightenment, Plateau of Productivity. We use the following basic hypotheses:
1. Specific (extra) linguistic markers should be used for each phase.
2. TT monitoring should be carried out based on a combination of text genres. However, different genres correspond to each phase as the most relevant.
3. Patent analysis is most important for the phases of Slope of Enlightenment and Plateau of Productivity.
4. Surge and loss of interest could be fixed based on processing technology news and some types of foresight and analytical reports. It is relevant for the Peak of Inflated Expectations and the Through of Disillusionment.
5. Scientific papers and R&D reports are the most relevant genres for the Technology Trigger. It seems to be the most interesting phase for the TT monitoring task.

Integration of processing results for text collections of different genres is performed via the union and / or intersection of partial results.

3.2   Black-Box Principle

Our approach is based on the so called 'black box' principle. The concept of a TT is characterized by a complex nature, low formalization level (with almost no formal and commonly accepted definitions), blurred boundaries, and high degree of domain dependency leading to the need for expert knowledge. For all that, 'Big Data' in IT and 'Genome Editing' in Healthcare should have some similar features which actually allow us to name both phenomena 'a TT'. This leads us to an idea of hunting for domain independent 'external signs' (trend indicators) while letting a TT itself stay a black box for an observer. A similar principle appears in the fact that, in a sense, the number of technological solutions is much greater than the number of tasks they are created for. Globally, technological progress in all domains and periods is aimed at satisfaction of a restricted number of needs (stability, safety, health, etc.) and at fighting for a restricted number of parameters (price, efficiency, size, reliability, etc.).

Thus, the black box idea is related to extraction of knowledge on TTs based on heuristics, or trend indicators. E.g. the fact of forming a new regulatory committee is a potential evidence for a TT (the need for regulation means that the phenomenon is newish, but already having significant impact). All indicators are grouped into the corresponding semantic fields. These fields all together build an ontology system which is used for the ontology-driven information extraction. Most of indicators are either an event or a linguistic marker. Weights are assigned to each indicator or a group of indicators. Some of them are genre-specific, others are universal. All heuristics based on indicators correspond to a level of (a) external extra linguistic context; (b) metadata for a text collection; (c) document structure; (d) linguistic markers in a document body. Digests of texts and text collections are built based on identification of indicators. The most meaningful fragments are included into the digests for further processing.

The indicators may be of an abstract nature. One of the key groups of indicators includes linguistic scales which appear in the texts. They correspond to (a) new / existing needs (customers' pains); (b) resources, values; (c) S&T issues (tasks); (d) technological parameters (the change of). Indicators of this group have proved to be extremely reliable. A test subcollection for evaluation of indicators was formed (500 scientific papers and technology news). Indicators related to linguistic scales appeared in every text (100%). In the example below, expressions which contain indicators are underlined; indicator modifiers (quantification markers) are given in italics: "For the marine industry where unpredictable dynamic loading conditions are the case, MRE isolators could *greatly* decrease the level of vibrations transmitted from the machines to the shell of the ship and the opposite, resulting to smaller fatigue loads and a *much more comfortable journey*".

Semantic 'stop fields' for specification of garbage terms are also included into the ontologies.

3.3   Technology Trends Watch Ontologies

The system of ontologies was developed to automate the processes of new TT identification. The Protégé[3] system (version 4.2) was used for specification of ontologies. Key objects include concepts which are related to all phases of a technology life cycle (*innovation, driver, regulatory measures*, etc.). A TT is understood as a complex notion consisting of a number of objects with attributes and semantic relations between them (the set of objects differs depending on the level of a TT maturity). TT indicators and the corresponding linguistic and extra linguistic markers are also specified (within the 'black box' framework). Some of them are related to a TT as a whole, others are for TT components, objects or relations. The designed system of ontologies is used at the IE stage and for generation of OWL representation of ontology instances. On the level of IE results, a TT is represented as a set of weighted terms.

---

[3] http://protege.stanford.edu/

4   Implementation of Technology Trends Watch System

Common architecture of the software system aimed at identification of new TTs is shown in Fig. 1. Two platforms were used as key tools: the Protégé ontology editor and GATE[4] environment. Tools for knowledge-based analytics (time series analysis toolkit [7], tag clouds plugin TagCrowd[5] and plugin OntoGraf[6] for Protégé) were also used.

Protégé was used to build (a system of) ontologies for (a) TTs and indicators for identification of TTs in scientific and technical texts; (b) genre models of texts being processed, and (c) processes of trends watch (TT monitoring). Besides, Protégé can be used by analysts and experts for the analysis of the final or interim results and for adding new indicators, markers, and new domain terms into ontologies based on the processing results. GATE was used in OBIE component, and allowed us to implement the 'black box' principle through linguistic rules.

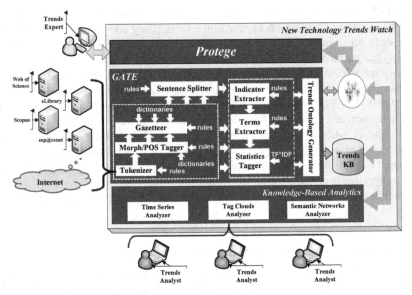

Fig. 1. Common Architecture of the TT Watch System

5   Case Study

5.1   Documents Collections

The developed system was tested using text collections in Green energy domain for the period 2002-2012. The total number of texts equals 131,477 documents in English

---

[4] https://gate.ac.uk/
[5] http://tagcrowd.com/
[6] http://protegewiki.stanford.edu/wiki/OntoGraf

and in Russian. Collections include scientific papers (130,867), blog fragments (21), texts from news sites (560), PhD theses (29).

During the research, each document was processed according to two different schemes. According to the first one, initial documents were processed with no exclusions. The second scheme involved only text fragments containing markers with consideration of their weights and other parameters. Thus, the digests consisting of meaningful fragments (which potentially included trend descriptions) were prepared for the further processing within the 'black box' framework. The results were proposed to 10 experts for comparative analysis. Decisions on enlargement of ontologies were made.

5.2   Evaluation Results

26279 single-word and 341596 multi-word terms have been extracted from the input collections according to the first processing scheme. 6636 single-word and 15693 multi-word terms have been extracted from the same collections within the second processing scheme. For frequency DF>2500, 284 one-word and 35 multi-word terms were extracted according to the first scheme, while 170 one-word and 20 multi-word terms were extracted according to the second one.

Terms extracted under the second scheme are much more relevant, according to the experts. Both relevant and general (garbage) concepts were found in the lists of single-word terms, while 99% of multi-word terms were assessed as relevant by experts. Some examples are given below (Table 1).

Table 1. Multi-word Terms, the Second Processing Scheme. DF > 1000.

Multiwords Terms	DF	TF	IDF	TF*IDF
carbon_nanotubes	1100	0.001621	4.770939	0.007733
composite_materials	4842	0.008848	3.288921	0.029101
electron_microscopy	2557	0.003771	3.927414	0.014812
tensile_strength	1091	0.001809	4.779154	0.008644
thermal_conductivity	1151	0.001755	4.725618	0.008293
thermal_stability	1102	0.001719	4.769122	0.008197
thermogravimetric_analysis	1056	0.001584	4.811761	0.007624
transmission_electron_microscopy	1143	0.001532	4.732593	0.007249
x_ray_diffraction	2177	0.002983	4.088301	0.012196

Further analysis was carried out by (a) processing synonyms and quasi-synonyms, (b) constructing time series for multi-word terms according to TF*IDF for identification of 'Bull', 'Side', and 'Bear' trends. Discussion of the results with experts shows that terms demonstrating 'Bull' trend identify an emerging TT at the Technology Trigger stage, while 'Side' and 'Bear' trends are evidence for shifting to the plateau.

Tag clouds of the most frequent terms were formed for further analysis. Results allowed us to specify new indicators of an emerging TT. 'Splitting term' is an example. Thus, results for 2002 include no terms related to biocomposite materials. In 2005, the term 'biocomposite material' appears, but it is the only 'bio'-term in all possible clouds ('the number of top words shown in a cloud' equals 100, 200, or 500). In

2012, the term 'biocomposite material' is still the only 'bio'-term for the top-100 word cloud. However, new terms such as 'bioactive', 'biodegradable', 'biomimetic', etc. appear as elements of multi-word terms in the 200 word cloud already. Clouds related to other indicators also demonstrate results considered to be relevant and interpretable by experts.

6   Conclusions and Future Work

The paper presents methods and tools for automated identification of new technology trends. The developed algorithms are based on a hybrid approach which combines methods of ontology-driven information extraction and statistical methods for processing OBIE results. Gartner's Hype Cycle model and the 'black box' principle proposed by the authors are the key elements of the methodology. Case study and analytical insights are provided. The results of experiments and conclusions made by experts confirm the applicability of the approach.

Further R&D are to be focused on:

— Extension of ontologies with elaborate domain thesauri for deeper processing of synonyms, term generalization, and for getting enhanced domain-oriented (medicine, IT, nanotechnologies, etc.) versions of the system.
— Extension of OBIE models with new semantic 'stop fields' for (a) filtering the resulting sets of terms; (b) identification of a wider context (e.g. the term 'manufacturer' can be considered either as garbage or as an indicator of a certain phase of a TT life cycle); (c) multidisciplinary TT monitoring.
— Deeper examination of relations between indicators and a TT life cycle.
— New algorithms for merging, labeling and visualizing TTs.

Development of tools for automated generation of TT ontologies based on FCA-methods [8] integrated with the hybrid approach is also assumed.

References

1. Boyack, K.W., Klavans, R., Börner, K.: Mapping the backbone of science. Scientometrics 64, 351–374 (2005)
2. Daud, A., Li, J., Zhou, L., Muhammad, F.: Knowledge discovery through directed probabilistic topic models: a survey. Frontiers of Computer Science in China 4(2), 280–301 (2010)
3. Wimalasuriya, D.C., Dou, D.: Ontology-based information extraction: an introduction and a survey of current approaches. Journal of Information Science 36(3), 306–323 (2012)
4. Preotiuc-Pietro, D., Samangooei, S., Cohn, T., Gibbins, N., Niranjan, M.: Trendminer: An Architecture for Real Time Analysis of Social Media Text. In: Workshop on Real-Time Analysis and Mining of Social Streams (RAMSS), International AAAI Conference on Weblogs and Social Media (ICWSM), Dublin (June 2012)
5. Erdi, P., Makovi, K., et al.: Prediction of emerging technologies based on analysis of the US patent citation network. Scientometrics 95(1), 225–242 (2013)

6. Li, H., Xu, F., Uszkoreit, H.: TechWatchTool: Innovation and Trend Monitoring. In: Proc. of the International Conference on Recent Advances in Natural Language Processing, RANLP 2011, Tissar, Bulgaria, pp. 660–665 (2011)
7. Khoroshevsky, V.: Discovering New Technological Trends in Texts Collections: Hybrid Models and Data Patterns Time Series Analysis. Information-measuring and Control Systems (5), 19–25 (2013) (in Russian)
8. Grissa-Touzi, A., Ben Massoud, H., Ayadi, A.: Automatic Ontology Generation for Data Mining Using FCA and Clustering. CoRR (2013), http://dblp.uni-trier.de/db/journals/corr/corr1311.html#Grissa-TouziMA13

Optimization of Polytopic System Eigenvalues by Swarm of Particles

Jacek Kabziński and Jarosław Kacerka

Institute of Automatic Control, Lodz University of Technology,
Stefanowskiego 18/22, 90-924, Lodz, Poland
{Jacek.Kabzinski,Jaroslaw.Kacerka}@p.lodz.pl

Abstract. A modified version of particle swarm optimization algorithm is proposed for minimization of maximal real part of a polytopic system eigenvalues. New initialization procedure and special projection operation are introduced to keep all particles working effectively inside a simplex of feasible positions. The algorithm is tested on several benchmarks and statistical evidences for its' high efficiency are provided.

Keywords: polytopic systems, eigenvalue optimization, particle swam optimization, constrained optimization.

1 Introduction

Optimization problems involving the eigenvalues of symmetric and nonsymmetric matrices present continuous challenge for mathematicians and engineers [1]. The most popular problem is minimization of maximum real part of system eigenvalues. The distance between the set of system eigenvalues and the imaginary axis is for example a measure of linear continuous system robustness. Instability measures of dynamical systems, that provide key information related to stabilizability and performance, are also defined by eigenvalue optimization problems [2]. Eigenvalue optimization has a wide spectrum of applications, far wider than linear system theory or design of linear electronic circuits. One may mention for example applications in physics, numerical analysis, chemistry, structural design, mechanics (vibration control), quantum mechanics, structural engineering problems, chemical engineering pertinent models and many more [1,3,4,5].

The eigenvalue optimization theory for symmetric matrices may be solved by several approaches, for example in the context of singular value optimization [5]. Unfortunately the objective function in optimization of non-symmetric matrix eigenvalues is generally non-convex, non-differentiable and possesses several local minima. Standard, smooth optimization algorithms fail to solve the problem. Some non-smooth optimization algorithms [3] are proposed in the literature and some solutions base on smooth, convex approximations may be found [4] but the problem is still opened as the proposed solutions are complicated, difficult to apply and do not have the ability to avoid dropping into local minima.

G. Agre et al. (Eds.): AIMSA 2014, LNAI 8722, pp. 178–185, 2014.

In this paper we consider linear polytopic systems i.e. the system matrix belongs to a convex hull of several matrices (called vertices) and coefficients of this convex combination are optimized parameters. Such description of a linear system is useful if we consider uncertainty caused by parameters variations, fault diagnosis or fault-tolerant control, or linear systems synthesis. Polytopic representation is one of the most general ways to describe, without any conservatism, the physical parameter uncertainty. We propose to apply particle swarm optimization (PSO) algorithm to solve the eigenvalue optimization problem for polytopic systems. Because of the special features of the formulated problem we introduce several modifications of the standard PSO algorithm to ensure efficient utilization of all particles.

1.1   Problem Formulation and Features

Let us consider a $n \times n$ matrix $A(\alpha)$ belonging to a convex-hull of N $n \times n$ matrices $A_1, A_2, \cdots, A_N$.

$$A(\alpha) = \sum_{i=1}^{N} \alpha_i A_i, \tag{1}$$

$$\alpha = [\alpha_1, \alpha_2, \cdots, \alpha_N]^T, \quad \alpha_i \geq 0, \ i = 1, 2, \cdots, N \quad \sum_{i=1}^{N} \alpha_i = 1. \tag{2}$$

Linear dynamical system

$$\frac{dx}{dt} = A(\alpha)x \tag{3}$$

is called a polytopic system (with vertices $A_1, A_2, \cdots, A_N$). The concept of polytopic systems is useful to describe system uncertainty or design flexibility. As a special case for $N = 2$ we obtain a dynamic interval system. We will investigate the following optimization problem:

$$\min_\alpha \quad \max_{1 \leq i \leq n} Re\lambda_i(A(\alpha)) \tag{4}$$

where $Re\lambda_i(A(\alpha))$ denotes real part of the $i'th$ eigenvalue of $A(\alpha)$. Because of the equality constraint in (2) we have $N - 1$ independent optimization variables and the problem may be posted as:

$$\text{Minimize } f(\alpha) = \max_{1 \leq i \leq n} Re\lambda_i(A(\alpha)) \tag{5}$$

subject to:

$$\alpha = [\alpha_1, \alpha_2, \cdots, \alpha_N]^T, \ \alpha_i \geq 0, \ i = 1, 2, \cdots, N - 1, \sum_{i=1}^{N-1} \alpha_i \leq 1, \ \alpha_N = 1 - \sum_{i=1}^{N-1} \alpha_i \tag{6}$$

So the constraints form a $N - 1$ dimensional simplex of feasible solutions $[\alpha_1, \alpha_2, \cdots, \alpha_{N-1}]^T$. It is well known that even if the vertices of the polytopic system are all stable, it is not enough for all systems defined by (1), (2) to be stable. For example both matrices $\begin{bmatrix} -0.1 & 2 \\ 0 & -0.2 \end{bmatrix}$ and $\begin{bmatrix} -2 & 1.9 \\ 1.9 & -2 \end{bmatrix}$ are stable whereas their midpoint $\begin{bmatrix} -1.05 & 1.95 \\ 0.95 & -1.1 \end{bmatrix}$ is not – eigenvalues 0.2863, -2.4363. The objective function (5) in general case of non-symmetric matrices is non-convex, non-differentiable (at any point generating multiple eigenvalues) and may possess several local minima.

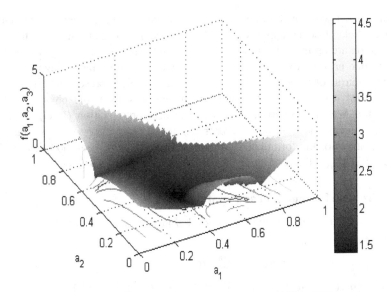

Fig. 1. Exemplary objective function for a two-dimensional problem (N=3, n=6, two local minima, global at position [0.33849, 0.15382])

The constraints (6) must be considered as 'hard' – any point outside the simplex (6) is unfeasible, even if it is arbitrary close to the boundaries. It is also quite common that the global minimum is situated on the boundary of the simplex or at the vertex. Fig.1 shows the 3D view of an exemplary objective function (5) generated by the vertices for one of the sample problems having N=3, n=6.

The proposed optimization approach must demonstrate the ability to cope with all mentioned difficulties, i.e. several local minima, hard constraints and global minimum located on the boundary of the simplex or at the vertex.

1.2   Particle Swarm Optimization

Particle swarm optimization was developed in nineties [6] as a stochastic optimization algorithm based on social simulation models. After discussion of early concepts, investigation of parameter influence and several modifications of the first version contemporary standard PSO was established. For this study we have chosen an algorithm called Unified Particle Swarm Optimization described in [7]. UPSO combines exploration and exploitation abilities of global and local variants of PSO without requiring additional function evaluations. Experimental results show its superiority against the standard PSO in terms of success rate as well as number of function evaluations [8]. Potential solutions are represented by S numbered 'particles' distributed in the M-dimensional search space at positions x_i $i = 1, 2, ..., S$. In the t'th iteration particles are evaluated by the objective function and the following values are memorized:

$p_G(t)$ - the best position for all swarm during iterations 1,2, ..., t,

$p_i(t)$ - the best position of the i'th particle, $i = 1, 2, ..., S$ during iterations $1, 2, ..., t$,

$p_{Li}(t)$ - the best position in the 'neighborhood' of the i'th particle ($i = 1, 2, ..., S$) which consists of the particles numbered $i - r, i - r + 1, \cdots, i, \cdots i + r$, during iterations $1, 2, ..., t$.

Next two position update ('velocity') components are built for $i = 1, 2, ..., S$:

$$v_{Gi}(t + 1) = \chi[v_i(t) + c_1 R_{Gi1} \circ (p_i(t) - x_i(t)) + c_2 R_{Gi2} \circ (p_{Gi}(t) - x_i(t))] \qquad (7)$$

$$v_{Li}(t + 1) = \chi[v_i(t) + c_1 R_{Li1} \circ (p_i(t) - x_i(t)) + c_2 R_{Li2} \circ (p_{Li}(t) - x_i(t))] \qquad (8)$$

where:

χ–constriction coefficient, usually 0.729 [7],

c_1, c_2– cognitive and social parameters, usually 2.05 [7],

$v_i(t)$ - is the previous 'velocity' of the i'th particle,

$R_{Gi1}, R_{Gi2}, R_{Li1}, R_{Li2}$ - are random M–dimensional vectors, which components are uniformly distributed in $[0,1]$ and $\circ$ denotes component-wise multiplication.

Then the final 'velocity' of each particle is build from 'global' and 'local' components:

$$v_i(t + 1) = u v_{Gi}(t + 1) + (1 - u) v_{Li}(t + 1) \qquad (9)$$

($u \in [0,1]$ is a 'unification factor') and positions of particles are updated:

$$x_i(t + 1) = x_i(t) + v_i(t + 1). \qquad (10)$$

The iterations are repeated until the stopping criterion is fulfilled. The algorithm parameters are: number of particles S, radius of neighborhood r, and unification factor u. In the standard UPSO algorithm it is usually assumed that the search space of feasible solution is a M–dimensional hyperrectangle defined by the bounds x_{min}, x_{max} and the velocity components are also bounded by v_{min}, v_{max}. Initialization of particle positions is usually performed on random, uniformly in a M–dimensional search space hyperrectangle and the first velocities are also chosen on random with the prescribed bounds.

2   Proposed Modifications of PSO

The main difference between the considered problem (5), (6) and the standard PSO challenge is the presence of constraints (6). The standard initialization generates starting particles position on random inside a M–dimensional hyperrectangle. It allows to use any pseudo-random number generator in a simple manner. For problem (5), (6) we have to create starting position inside a $N - 1$ – dimensional simplex. We propose the following approach:

1. Generate the particle position $\hat{\alpha} = [\alpha_1, \alpha_2, \cdots, \alpha_{N-1}]^T$, on random in the $N - 1$ – dimensional unit hypercube: $\alpha_i \epsilon [0,1]$, $i = 1, \cdots, N - 1$.

2. If $\sigma = \sum_{i=1}^{N-1} \alpha_i > 1$ replace the particle with an another one obtained by dividing $\hat{\alpha}$ by a random coefficient bigger then σ i.e.:

$$\hat{a} := \frac{\hat{a}}{\sigma(1+r)},$$ (10)

where r is random coefficient uniformly distributed in $[0, r_{max}]$.

3. Repeat steps 1 and 2 as long as all S particles are initialized.

The standard approach to cope with feasible space constraints is to apply a penalty function outside the constraints. In the considered problem (5), (6) the constraints are 'hard' and quite often the global minimum is located on a simplex face or in a vertex. A particle outside the simplex is useless – it will not contribute to the movement towards the solution, while a particle on the boundary may bring valuable information. The effectiveness of optimization will be increased if particles will possess the ability to 'slide' along a face or an edge. Motivated by the presented discussion we propose the following procedure to handle the constraints:

1. Assume that the velocity of this particle is v, the updated particle position is $\hat{a} = [\alpha_1, \alpha_2, \cdots, \alpha_{N-1}]^T$ and the previous particle position is $\hat{\beta} = [\beta_1, \beta_2, \cdots, \beta_{N-1}]^T$. The updated position may be outside the feasible simplex (6).

2. Calculate the minimum distance t_{min} from position $\hat{\beta}$ to the intersection of the half-line parallel to v starting in $\hat{\beta}$ and hyper-planes bounding the search space, i.e. $\sum_{i=1}^{N-1} \alpha_i = 1$ and $\alpha_i = 0$ for $i=1, \ldots, N-1$. Distances are normalized to the velocity of the particle, hence the intersection point is given by:

$$\widehat{a'} = \hat{\beta} + t_{min} \cdot v$$ (11)

3. Calculate the new updated particle position according to (11) and neutralize the component of v orthogonal to the hyper-plane. In case of intersection with one of the $\alpha_i = 0$ hyper-planes, the appropriate component of v is zeroed. In case of $\sum_{i=1}^{N-1} \alpha_i = 1$ the velocity is set to the component of v parallel to the hyper-plane.

4. Repeat steps 2, and 3 until particles are bounded.

3   Numerical Experiments

The proposed algorithm has been used to find the solution of randomly generated problems. For every problem of a given n and N, i.e. dimension and number of matrices respectively, elements of matrices $A_1, \ldots, A_N$ were selected as random numbers in the range [-5,5]. The N-dimensional problem space limited by (6) was sampled using a regular grid pattern covering all N-1 optimization variables at a resolution of 0.1. A value of function (5) (hereafter 'fitness function') in every sample point was calculated and a local optimization algorithm was employed to search for a local minimum starting from the sample point. The coordinates of the local minimum having the lowest fitness function value were assumed to be a global minimum and stored. Locations and fitness values of all local minima found were also stored. Global minima were inspected and problems were split into two categories according to whether the minimum is located at the search space boundary or not.

Generated problems have $N \in [3,5], n \in [2,6]$. Performed sampling revealed that the number of local minima ranged from 1 to 20 with about 80% of problems being multimodal. In about 50% of problems the global minimum was located at the search space boundary.

To examine whether the performance improvement is significant statistical tests were performed. 150 problems were generated with the aforementioned approach. The solution of every problem was searched for by:
- the proposed constrained UPSO algorithm,
- a UPSO algorithm with search space constrained by a constant penalty function.

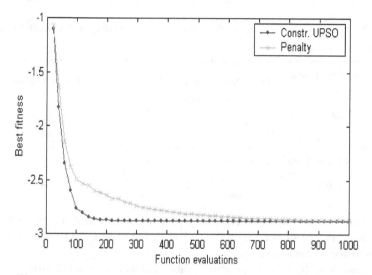

Fig. 2. Comparison of algorithm convergence for an exemplary problem

The algorithms were run 50 times for 200 iterations with the stopping criterion defined as any of the two conditions:

a) the maximum number of iterations being reached,
b) the global best fitness value found by the swarm being close to the value of global minimum for the problem with the accuracy of 10^{-3}.

A search was assumed to be successful if the condition b) was fulfilled.

Both algorithms showed a similar performance in terms of success rate: the Constrained UPSO success rate was higher by 5%, however the convergence was faster by up to 50%. The difference in the convergence of both algorithms is shown in fig. 2 for an exemplary function having $N=4$, $n=3$ and three local minima. Presented is an average best fitness for 20 executions of both algorithms.

In order to inspect the difference in algorithm performances all test problems were divided into categories according to the number of local minima, location of the global minimum - i.e. if it is located on the boundary of the search space.

Table 1. Average success rate and number of iterations depending on the number of local minima

Local minima		2	3	4	5	6	7	8	9	10	>10
Success rate	Constr. UPSO	88%	93%	83%	64%	83%	39%	62%	44%	88%	34%
	Penalty	82%	86%	87%	59%	72%	40%	71%	38%	83%	33%
Iterations	Constr. UPSO	36	31	52	65	48	116	65	68	20	93
	Penalty	77	63	93	100	92	141	78	61	34	107

Table 2. Average success rate and number of iterations depending on the location of global minimum

Location of the global minimum		Not on the boundary	On the boundary
Success rate	Constr. UPSO	68%	86%
	Penalty	67%	80%
Iterations	Constr. UPSO	68	33
	Penalty	82	83

The number of local minima (Table 1) appears to have similar influence on success rate and convergence of both algorithms. The Constrained UPSO required 41% less iterations to succeed at an average.

The location of the global minimum (Table 2) has a substantial influence on the convergence of Constrained UPSO - the number of required iterations is 60% lower if the global minimum is located on the boundary of the search space. Therefore the proposed method of constraining particles and introducing the ability of the particles to move along the edge of the feasible search space provides a great performance improvement.

4   Conclusion

Following the previous research in PSO applications in the eigenvalue problem [10] we demonstrate PSO capability to solve eigenvalue optimization problems. Proposed modifications of standard PSO allow to recommend this algorithm as an effective tool to solve eigenvalue optimization problems for polytopic systems. Although the presented discussion was limited to the case of minimization of maximum real part of system eigenvalues – objective function (5), but it is obvious that the proposed approach will work similarly for another eigenvalue optimization problems like maximization of the minimum eigenvalue (what corresponds to maximization of the lowest natural vibration frequency) or minimization of maximal eigenvalue modulus (stability investigation of discrete, linear systems). PSO application may be also generalized to eigenvalue optimization in case when a matrix depends nonlinearly on design parameters, as addressed in [9], or to the constraints defined by many nonlinear inequalities.

References

1. Lewis, A.S.: The mathematics of eigenvalue optimization. Math. Program., Ser. B 97, 155–176 (2003)
2. Chesi, G.: Measuring the Instability in Continuous-Time Linear Systems with Polytopic Uncertainty. In: 52nd IEEE Conference on Decision and Control, pp. 1131–1136 (2013)
3. Apkarian, P., Noll, D., Prot, O.: Trust region spectral bundle method for nonconvex eigenvalue optimization. SIAM. J. Optim. 19(1), 281–306 (2009)
4. Chen, X., Qi, H., Qi, L., Teo, K.L.: Smooth convex approximation to the maximum eigenvalue function. Jour. of Global Optimization 30, 253–270 (2004)
5. Blanco, A.M., Bandoni, J.A.: Eigenvalue and Singular Value Optimization. In: ENIEF 2003 - XIII Congreso sobre Métodos Numéricos y sus Aplicaciones, pp. 1256–1272 (2003)
6. Eberhart, R.C., Kennedy, J.: A new optimizer using particle swarm theory. In: Proceedings of the Sixth International Symposium on Micro Machine and Human Science, pp. 39–43. IEEE Service Center, Piscataway (1995)
7. Parsopoulos, K., Vrachatis, M.N.: Particle swarm optimization and intelligence: advances and applications, Hershey, New York (2009)
8. Parsopoulos, K.E., Vrahatis, M.N.: UPSO: A unified particle swarm optimization scheme. In: Proc. Int. Conf. Comput. Meth. Sci. Eng. (ICCMSE 2004). Lecture Series on Computer and Computational Sciences, vol. 1, pp. 868–873. VSP International Science Publishers, Zeist (2004)
9. Cao, Y.J.: Eigenvalue optimization problems via evolutionary programming. Electronics Letters 33(7), 642–643 (1997)
10. Kabziński, J.: Swarm Capability of Finding Eigenvalues. In: Ramsay, A., Agre, G. (eds.) AIMSA 2012. LNCS, vol. 7557, pp. 172–177. Springer, Heidelberg (2012)

Back-Propagation Learning of Partial Functional Differential Equation with Discrete Time Delay

Tibor Kmet[1] and Maria Kmetova[2]

[1] Constantine the Philosopher University, Department of Informatics,
Tr. A. Hlinku 1, 949 74 Nitra, Slovakia
tkmet@ukf.sk
http://www.ukf.sk
[2] Constantine the Philosopher University, Department of Mathematics,
Tr. A. Hlinku 1, 949 74 Nitra, Slovakia
mkmetova@ukf.sk

Abstract. The present paper describes the back-propagation learning of a partial functional differential equation with reaction-diffusion term. The time-dependent recurrent learning algorithm is developed for a delayed recurrent neural network with the reaction-diffusion term. The proposed simulation methods are illustrated by the back-propagation learning of continuous multilayer Hopfield neural network with a discrete time delay and reaction-diffusion term using the prey-predator system as a teacher signal. The results show that the continuous Hopfield neural networks are able to approximate the signals generated from the predator-prey system with Hopf bifurcation.

Keywords: feed-forward neural network, multilayer Hopfield neural network with a discrete time delay and diffusion-reaction term, time-dependent learning, prey-predator system with a discrete time delay and diffusion-reaction term, numerical solution.

1 Introduction

Partial functional differential equations with a discrete time delay arise in many biological, chemical, and physical systems which are characterized by both spatial and temporal variables and exhibit various spatio-temporal patterns [13]. The recurrent neural network with a time delay and diffusion term are described by a special type of such equations. One of the fundamental ideas of the recurrent neural network is its application to the dynamical memory, a desired attractor embedded into the dynamical system [1],[11]. It is well known that a neural network can be used to approximate the smooth time-invariant functions and the uniformly time-varying function [3], [9]. Besides, it has been used for the universal function approximation to solve optimal control problems with a discrete time delay forward in time to approximate the costate variable [5], [12]. The costate variables play an important role also in the back-propagation learning of infinite-dimensional dynamical systems [10], [11] in the case of time-dependent recurrent learning. Our interest here is in supervised learning of a

G. Agre et al. (Eds.): AIMSA 2014, LNAI 8722, pp. 186–193, 2014.

partial functional Hopfield neural network (PFHNN) with a discrete time delay. The supervised learning is to teach the spatio-temporal dynamics to the PFHNN by applying the back propagation algorithm [8]. In order to learn the dynamic behaviours of the partial functional differential equations based on [10], a time-dependent recurrent learning algorithm has been developed. For the neural network which determines the right hand side of PFHNN, a feed-forward neural network with one hidden layer for the state variable and one hidden layer for the delay state, a steepest descent error backpropagation rule, a hyperbolic tangent sigmoid transfer function and a linear transfer function were used.

This paper is organized as follows. In Section 2 we present a description of the back-propagation learning of infinite-dimensional dynamical systems and propose a new algorithm to calculate the gradient of cost function. Section 3 presents a short description of the prey-predator system and numerical results of the time-dependent recurrent learning using Lagrange multipliers to compute the gradients of the cost function. Conclusions are presented in Section 4.

2   Partial Functional Hopfield Neural Network Learning with a Discrete Time Delay

Let us consider supervised learning to teach the discrete time delay dynamic to the discrete time delay partial functional Hopfield neural network. We utilize the following form of multilayer continuous Hopfield neural network with a discrete time delay in the learning of complex nonlinear dynamics:

$$
\dot{x}(t) = D\frac{\partial^2 x(s,t)}{\partial s^2} + F(x(s,t), x(s,t-\tau), W)
$$

$$
= D\frac{\partial^2 x_i(s,t)}{\partial s^2} - Ax(s,t) + W^o f(W^h x(s,t) + W^{hd} x(s,t-\tau)), \quad (1)
$$

with Neumann boundary condition $\frac{\partial x_i}{\partial s}(a,t) = \frac{\partial x_i}{\partial s}(b,t) = 0$ and initial conditions $x_i(s,t) = \phi_i(s,t) \geq 0$, $a \leq s \leq b$, $t \in \langle -\tau, 0 \rangle$, $i = 1, \ldots, n$, where $x = (x_1, \ldots, x_n)$, $F = (F_1, \ldots, F_n)$, A_{nxn}, D_{nxn} are diagonal matrices, W^o_{nxn} is a weight matrix between the hidden and output layer, W^h_{nxn}, W^{hd}_{nxn} are the weight matrices between the input and hidden layer and $W = (A, W^h, W^{hd}, W^o)$. Function f is $tanh(.)$ the activation function. t denotes the time, s represents the spatial location and D is a diagonal matrix of the diffusion coefficients. For the given continuous initial condition $\phi(s,t)$, $s \in \langle a,b \rangle$, $t \in \langle -\tau, 0 \rangle$ there exists a unique solution $x(s,t)$ satisfying Eq. (1) for $s \in \langle a,b \rangle$, $t \in \langle 0,T \rangle$. The aim is to find the weight parameters W that give rise to a solution $x(s,t)$ approximately following the teacher signal $p(s,t) = (p_1(s,t), \ldots, p_n(s,t))$, where $p(s,t)$ is a solution of the following delay partial functional differential equation:

$$
\dot{p}(s,t) = D\frac{\partial^2 p(s,t)}{\partial s^2} + G(p(s,t), p(s,t-\tau)), \quad (2)
$$

with Neumann boundary condition $\frac{\partial p_i}{\partial s}(a,t) = \frac{\partial p_i}{\partial s}(b,t) = 0$ and initial conditions $p_i(s,t) = \phi_i(s,t) \geq 0$, $a \leq s \leq b$, $t \in \langle -\tau, 0 \rangle$, $i = 1, \ldots, n$. First, the cost function is defined for the weight parameters W as

$$E(W) = \int_a^b \int_0^T \frac{1}{2} \sum_{i=1}^n (x_i(s,t) - p_i(s,t))^2 dt ds. \tag{3}$$

Then the cost function (3) is minimized by the steepest descent method

$$w^{j+1} = w^j - \alpha \frac{\partial E}{\partial w}(W^j), \tag{4}$$

where $w \in W$. To compute the gradient of function (3), we use time-dependent recurrent learning (TDRL) [11]. In the TDRL algorithm, the gradients are computed by using the Lagrange multipliers $\lambda(t) = (\lambda_1(s,t), \ldots, \lambda_n(s,t))$. For a detailed explanation, see [11]. We can rewrite the cost function $E(W)$ as

$$L(W) = \int_a^b \int_0^T \sum_{i=1}^n [\frac{1}{2}((x_i(s,t) - p_i(s,t))^2 -$$
$$\lambda_i(s,t)(\dot{x}_i(s,t) - d_i \frac{\partial^2 x_i(s,t)}{\partial s^2} - F_i(x(s,t), x(s,t-\tau), W))] dt ds.$$

The partial derivatives with respect to the weight coefficients $w \in W$ are calculated as

$$\frac{\partial L}{\partial w} = \int_a^b \int_0^T \sum_{i=1}^n [(x_i(s,t) - p_i(s,t)) \frac{\partial x_i(s,t)}{\partial w} - \lambda_i(s,t) \frac{\partial \dot{x}_i(s,t)}{\partial w} +$$
$$\lambda_i(s,t) \frac{F_i(x(s,t), x(s,t-\tau), W)}{\partial w} +$$
$$\lambda_i(s,t) \sum_{j=1}^n \frac{F_i(x(s,t), x(s,t-\tau), W)}{\partial x_j(s,t)} \frac{\partial x_j(s,t)}{\partial w} +$$
$$\lambda_i(s,t) \sum_{j=1}^n \frac{F_i(x(s,t), x(s,t-\tau), W)}{\partial x_j(s,t-\tau)} \frac{\partial x_j(s,t-\tau)}{\partial w} +$$
$$\lambda_i(s,t) \frac{\partial}{\partial w} \left(d_i \frac{\partial^2 x_i(s,t)}{\partial s^2} \right) -$$
$$\frac{\lambda_i(s,t)}{\partial w} (\dot{x}_i(s,t) - d_i \frac{\partial^2 x_i(s,t)}{\partial s^2} F_i(x(s,t), x(s,t-\tau), W))] dt ds. \tag{5}$$

If $x(s,t)$ is a solution of Eq. (1) then the final term of Eq. (5) vanishes. Since $\frac{\partial x(s,t)}{\partial w} = 0$ for $t \in \langle -\tau, 0 \rangle$ the fourth term of (5) can be written by the transformation $t' = t - \tau$ as

$$\int_a^b \int_0^T \sum_{i=1}^n \lambda_i(s,t) \sum_{j=1}^n \frac{F_i(x(s,t), x(s,t-\tau), W)}{\partial x_j(s,t-\tau)} \frac{\partial x_j(s,t-\tau)}{\partial w} dt ds =$$

$$\int_0^T \int_{-\tau}^{T-\tau} \sum_{i=1}^n \lambda_i(s,t+\tau) \sum_{j=1}^n \frac{F_i(x(s,t+\tau), x(s,t), W)}{\partial x_j(s,t)} \frac{\partial x_j(s,t)}{\partial w} dt ds =$$

$$\int_0^T \int_0^T \sum_{i=1}^n \lambda_i(s,t+\tau) \sum_{j=1}^n \frac{F_i(x(s,t+\tau), x(s,t), W)}{\partial x_j(s,t)} \frac{\partial x_j(s,t)}{\partial w} \chi_{\langle 0, T-\tau \rangle} dt ds.$$

Using intergation by parts with the Neumann boundary condition $\partial \lambda_i(a,t)/\partial s = \partial \lambda_i(b,t)/\partial s = 0$ we get

$$\int_a^b \lambda_i(s,t) \frac{\partial}{\partial w} \left(d_i \frac{\partial^2 x_i(s,t)}{\partial s^2} \right) ds = \int_a^b d_i \frac{\partial^2 \lambda_i(s,t)}{\partial s^2} \frac{\partial x_i(s,t)}{\partial w} ds.$$

The derivatives $\frac{\partial L}{\partial w}$ become

$$\frac{\partial L}{\partial w} = \int_a^b \int_0^T \sum_{i=1}^n [(x_i(s,t) - p_i(s,t)) \frac{\partial x_i(s,t)}{\partial w} - \lambda_i(s,t) \frac{\partial \dot{x}_i(s,t)}{\partial w} +$$

$$\lambda_i(s,t) \frac{F_i(x(s,t), x(s,t-\tau), W)}{\partial w} + d_i \frac{\partial^2 \lambda_i(s,t)}{\partial s^2} \frac{\partial x_i(s,t)}{\partial w} +$$

$$\lambda_i(s,t) \sum_{j=1}^n \frac{F_i(x(s,t), x(s,t-\tau), W)}{\partial x_j(s,t)} \frac{\partial x_j(s,t)}{\partial w} + \tag{6}$$

$$\lambda_i(s,t+\tau) \sum_{j=1}^n \frac{F_i(x(s,t+\tau), x(s,t), W)}{\partial x_j(s,t)} \frac{\partial x_j(s,t)}{\partial w} \chi_{\langle 0, T-\tau \rangle}] dt ds.$$

The Lagrange multipliers are solutions of the following partial functional differential equations with the Neumann boundary condition $\partial \lambda_i(a,t)/\partial s = \partial \lambda_i(b,t)/\partial s = 0$ and terminal condition $\lambda(s,T) = 0$.

$$- \dot{\lambda}_i(s,t) = d_i \frac{\partial^2 \lambda_i(s,t)}{\partial s^2} + \sum_{j=1}^n \lambda_j(t) \frac{F_j(x(s,t), x(s,t-\tau), W)}{\partial x_i(s,t)} + \tag{7}$$

$$\sum_{j=1}^n \lambda_j(s,t+\tau) \frac{F_j(x(s,t+\tau), x(s,t), W)}{\partial x_i(s,t)} \chi_{\langle 0, T-\tau \rangle} + (x_i(s,t) - p_i(s,t)).$$

Since the Lagrange multipliers $\lambda(s,t)$ satisfy Eq. (7) with Neumann boundary condition $\partial \lambda_i(a,t)/\partial s = \partial \lambda_i(b,t)/\partial s = 0$ and terminal condition $\lambda(s,T) = 0$, and $\frac{\partial x(s,t)}{\partial w} = 0$ for $t \in \langle -\tau, 0 \rangle$ all the terms of Eq. (6) but the third vanish. The partial derivatives $\frac{\partial J}{\partial w}$ can be calculated by the following form:

$$\frac{\partial L}{\partial w} = \int_a^b \int_0^T \sum_{i=1}^n \lambda_i(s,t) \frac{F_i(x(s,t), x(s,t-\tau), W)}{\partial w} dt ds. \tag{8}$$

Algorithm 1. Time dependent recurrent learning algorithm to determine the weight matrix of a time delay continuous Hopfield neural network

---

Input: Choose T, τ, $p(s,t)$ - teacher signal, $maxit$, ε_E, - stopping tolerance, $\phi(s,t)$, $t \in \langle -\tau, 0 \rangle$, $s \in \langle a,b \rangle$, - initial condition.

Output: Weight matrix $\mathbb{W} = (A, W^o, W^h, W^{dh})$;

1   Set the initial weight $\mathbb{W} = (A, W^o, W^h, W^{dh})$, $i = 0$

     while $err_E \geq \epsilon_E$ **and** $i \leq maxit$ **do**

2   Compute solution $x(s,t)$ of Eq. (1) on the interval $\langle 0, T \rangle$ with initial condition $\phi(s,t), t \in \langle -\tau, 0 \rangle$, $s \in \langle a,b \rangle$

3   Compute solution $\lambda(s,t)$ of Eq. (7) on the interval $\langle T, 0 \rangle$ with terminal condition $\lambda(s,T) = 0$

4   Compute $E(W)$ by Eq. (2)

5   Compute $\frac{\partial L}{\partial W} = \int_a^b \int_0^T \sum_{i=1}^n \lambda_i(s,t) \frac{F_i(x(s,t), x(s,t-\tau), W)}{\partial W} dt ds$ by Eq. (8)

6   Compute $\alpha^* = min \; g(\alpha) = E\left(W^i - \alpha \frac{\partial J(W^i)}{\partial W} \right)$

7   Set $W^{i+1} = W^i - \alpha^* \frac{\partial L(W^i)}{\partial W}$

8   Compute $E(W^{i+1})$ by Eq. (2)

9   Set $err_E = abs(E(W^{i+1} - E(W^i))$

10 **return** $\mathbb{W}^{i+1} = (A^{i+1}, W^{o,i+1}, W^{h,i+1}, W^{dh,i+1})$

We can state the following algorithm for time dependent recurrent learning. To find the minimizer weight matrix W using the Algorithm 1 we can also use the Fletcher-Reeves, DFP and BFGS methods [7]. For the PFHNN Eq. (1) the gradients are calculated as:

$$\frac{\partial L}{\partial a_{ii}} = \int_a^b \int_0^T x_i(s,t)\lambda_i(s,t) dt ds, \quad \frac{\partial L}{\partial w_{ij}^o} = \int_a^b \int_0^T \lambda_i(s,t) f_j(s,t) dt ds$$

$$\frac{\partial L}{\partial w_{ij}^h} = \int_a^b \int_0^T \sum_{k=1}^n \lambda_k(s,t) w_{ki}^o f_i'(s,t) x_j(s,t) dt ds,$$

$$\frac{\partial L}{\partial w_{ij}^{hd}} = \int_a^b \int_0^T \sum_{k=1}^n \lambda_k(s,t) w_{ki}^o f_i'(s,t) x_j(s,t-\tau) dt ds,$$

where $f_j(s,t) = tanh\left(\sum_{k=1}^n (w_{jk}^h x_k(s,t) + w_{jk}^{hd} x_k(s,t-\tau)) \right)$.

The derivatives $\partial L / \partial w$ are computed by the discretization of the diffusion term by finite dimensional approximation [6]. The resulting semidiscrete approximation of (1) amounts to an $N \times N$ system of delay differential equations. The delay differential equations are integrated by the Euler methods using linear spline approximation described in [2].

3   Discrete Time Delay Prey-Predator Ecological Model

This section presents experimental studies of applying the time-dependent learning algorithm developed in Section 2. Let us consider the following prey-predator model [4] with the discrete time delay τ and diffusion term.

$$\dot{p}_1(s,t) = d_1 \frac{\partial^2 p_1(s,t)}{\partial s^2} + rp_1(s,t)(1 - p_1(s,t)/K) - \sigma_1 p_1(s,t) + \sigma_2 p_2(s,t),$$

$$\dot{p}_2(s,t) = d_2 \frac{\partial^2 p_2(s,t)}{\partial s^2} + zp_2(s,t)(1 - p_2(s,t)/L) + \sigma_1 p_1(s,t) - \sigma_2 p_2(s,t) - \frac{\alpha p_2(s,t)p_3(s,t)}{a + p_2(s,t)},$$

$$\dot{p}_3(s,t) = d_3 \frac{\partial^2 p_3(s,t)}{\partial s^2} + \frac{\beta \alpha p_2(s,t-\tau)p_3(s,t-\tau)}{a + p_2(s,t-\tau)} - dp_3(s,t) - \gamma p_3(s,t)^2,$$

$$(9)$$

with Neumann boundary condition

$$\frac{\partial p_i}{\partial s}(a,t) = \frac{\partial p_i}{\partial s}(b,t) = 0 \tag{10}$$

and initial conditions

$$p_i(s,t) = \phi_i(s,t) \geq 0, \ a \leq s \leq b, \ t \in \langle -\tau, 0 \rangle, \ i = 1, \ldots, 3, \tag{11}$$

where p_1, p_2 denote the prey populations and p_3 is a predator population. The

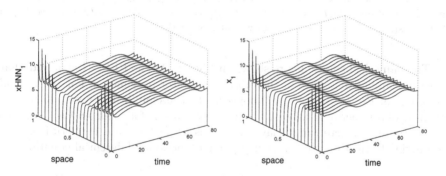

Fig. 1. Numerical solution of the prey-predator model described by Eq. (9) (right part) and numerical solution of the PFHNN Eq. (1) (left part) for $\tau = 3.7$, with teacher initial condition $\phi_1(s,t) = 7(1 + cos(2\pi s))$, $\phi_2(s,t) = 3(1 + cos(2\pi s))$, $\phi_3(s,t) = 2(1 + cos(2\pi s))$, for $t \in \langle -\tau, 0 \rangle$.

equilibria of the model were determined and the behavior of the system was investigated around the equilibria in [4]. Jana et al. [4] obtain that the time delay can cause a stable equilibrium to become unstable and even a simple Hopf bifurcation occurs when the time delay passes through its critical value. The prey-predator model (9) was numerically analyzed in [4] for the given set of parameters ($r = 0.9$, $K = 9$, $\sigma_1 = 0.2$, $\sigma_2 = 0.15$, $z = 0.8$, $L = 14$, $\alpha = 2.5$, $a = 1.2$, $\beta = 0.32$, $d = 0.3$, $\gamma = 0.1$ It was obtained that the solution of Eq. (9) for $\tau = 0.5$ converges to an equilibrium point for $\tau = 3.7$ Eq. (9) has periodic solution [4]. We can use the periodic solution $p(s,t)$ of Eq. (9) for

$\tau = 3.7$ as teacher signals to verify Alg. 1. After 1896-iterative learning the following network weight matrix was obtained for $\tau = 3.7$:

$$W^{(\tau=3.7)} = \left(A, W^o, W^h, W^{hd} \right) =$$

$$\begin{pmatrix} 0.124 & 0 & 0 & 0.6976 & 0.3013 & 0.1362 & 0.0587 & 0.0203 & 0.0128 & 0.4667 & -0.0324 & 0.1636 \\ 0 & 0.0728 & 0 & -1.2416 & 1.7255 & 1.4959 & -0.1268 & 0.2905 & 0.0064 & 0.5857 & -0.8518 & -0.2421 \\ 0 & 0 & 0.1591 & 0.3649 & -0.1866 & 1.0633 & 0.0386 & -0.0933 & -0.1187 & 0.0794 & 0.0771 & -0.3198 \end{pmatrix}.$$

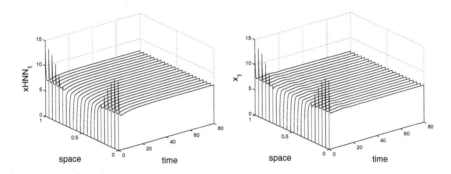

Fig. 2. Numerical solution of the prey-predator model described by Eq. (9) (right part) and numerical solution of the PFHNN Eq. (1) (left part) for $\tau = 0.5$, with the initial condition $\phi_1(s,t) = 7(1 + cos(2\pi s))$, $\phi_2(s,t) = 3(1 + cos(2\pi s))$, $\phi_3(s,t) = 2(1 + cos(2\pi s))$, for $t \in \langle -\tau, 0 \rangle$ and weight matrix $W^{(\tau=3.7)}$.

The numerical solutions are shown in Figs. 1, 2. The PFRNN achieved qualitatively similar dynamics in the original predator-prey model. For the $\tau = 0.5$ solutions of Eq. (1) and Eq. (9) converge to the equilibrium point. If $\tau = 3.7$ then we obtain periodic solutions for both systems. It follows from Figs. 1, 2 that the proposed discrete time delay continuous Hopfield neural network is able to approximate the discrete time delay partial functional differential equations.

4   Conclusion

The purpose of the paper is to develop an efficient time-dependent recurrent learning algorithm to determine the weight matrix of the discrete time delay partial functional Hopfield neural network. A signal generated from the simple predator-prey model is used as a learning example. Depending on the discrete time delay τ, the Hopf bifurcation incurred into the system for a given set of parameters of the model. The MATLAB simulations show that the proposed time-dependent learning algorithm is able to determine the weight matrix W and the partial functional Hopfiel neural network gives a good approximation of the predator-prey model.

Acknowledgment. The present paper was made possible thanks to the scientific project VEGA 1/0699/12.

References

1. Becerikli, Y., Konar, A.F., Samad, T.: Inteligent optimal control with dynamic neural networks. Neural Network 16, 251–259 (2003)
2. Farmer, J.D.: Chaotic Attractors of an infinite-dimensional dynamic system. Physica 4D, 366–393 (1982)
3. Hornik, M., Stichcombe, M., White, H.: Multilayer feed forward networks are universal approximators. Neural Networks 3, 256–366 (1989)
4. Jana, S., Chakraborty, M., Chakraborty, K., Kar, T.K.: Global stability and bifurcation of time delayed prey-predator system incorporating prey refuge. Mathematics and Computers in Simulation 85, 57–77 (2012)
5. Kmet, T., Kmetova, M.: Adaptive Critic Neural Network Solution of Optimal Control Problems with Discrete Time Delays. In: Mladenov, V., Koprinkova-Hristova, P., Palm, G., Villa, A.E.P., Appollini, B., Kasabov, N. (eds.) ICANN 2013. LNCS, vol. 8131, pp. 483–494. Springer, Heidelberg (2013)
6. Morton, K.W.: Numerical solution of Convection-Diffusion Problems. Chapman - Hall, London (1996)
7. Polak, E.: Optimization Algorithms and Consistent Approximation. Springer, New York (1997)
8. Rumelhart, D.F., Hinton, G.E., Wiliams, R.J.: Learning internal representation by error propagation. In: Rumelhart, D.E., McClelland, D.E., PDP Research Group (eds.) Parallel Distributed Processing: Foundation, vol. 1, pp. 318–362. The MIT Press, Cambridge (1987)
9. Sandberg, E.W.: Notes on uniform approximation of time-varying systems on finite time intervals. IEEE Transactions on Circuits and Systems-1: Fundamental Theory and Applications 45(8), 305–325 (1998)
10. Tokuda, I., Hirai, Y., Tokunaga, R.: Back-propagation learning of infinite-dimensional dynamical system. In: Proceedings of 1993 International Conference on Neural Network, pp. 2271–2275 (1993)
11. Tokuda, I., Tokunaga, R., Aihara, K.: Back-propagation learning of infinite-dimensional dynamical systems. Neural Networks 16, 1179–1193 (2003)
12. Werbos, P.J.: Approximate dynamic programming for real-time control and neural modelling. In: White, D.A., Sofge, D.A. (eds.) Handbook of Intelligent Control: Neural Fuzzy, and Adaptive Approaches, pp. 493–525. Van Nostrand Reinhold, New York (1992)
13. Wu, J.: Theory and Applications of Partial Functional Differential Equations. Applied Math. Sciences, vol. 119. Springer, New York (1996)

Dynamic Sound Fields Clusterization Using Neuro-Fuzzy Approach

Petia Koprinkova-Hristova and Kiril Alexiev

Inst. of Inf. and Comm. Technologies, Bulgarian Academy of Sciences
Acad. G. Bonchev str. bl.25A,
Sofia 1113, Bulgaria
{pkoprinkova,alexiev}@bas.bg

Abstract. In the presented investigation a recently proposed approach for multidimensional data clustering was applied to create a 3D "sound picture" of the data collected by a microphone array antenna. For this purpose records of acoustic pressure at each point (a microphone in the array) collected for a given period of time were used. Features for classification are extracted using overlapping receptive fields based on the model of direction selective cells in the middle temporal (MT) cortex. Next the clustering procedure using Echo state network and subtractive clustering algorithm is applied to separate these receptive fields into proper number of classes. Obtained for each time step two dimensional "sound pictures" were combined to create a 3D representation of dynamic changes in the sound pressure. We compare our results with the sonograms created by the original software of the producer of microphone array. Although our approach did not account for the distance to the noise source, it allows consideration of dynamically changing sounds.

Keywords: acoustic pressure, Echo state networks, subtractive clustering, receptive field, direction selective cells.

1   Introduction

Localization of sound sources is a task with numerous applications varying from military locators, seismic surveys, medical and machine diagnostic systems etc. For different practical applications there were created many specialized equipments and corresponding mathematical methods for signal processing aimed at accurate noise source localization. An example of such device is acoustic camera that consists of several microphones operating in tandem.

There are developed two basic approaches for processing of acoustic pressure measured by the microphone arrays: acoustic holography and beam forming. The first one reconstructs sound fields near to the camera and has established two realizations: near-field acoustic holography (NAH) [16] and statistically optimal near-field acoustic holography (SONAH) [4]. There are several strong requirements that cumber NAH implementation and limit its application to small sound sources at low frequencies. That is why SONAH was developed to alleviate some of these requirements. The

G. Agre et al. (Eds.): AIMSA 2014, LNAI 8722, pp. 194–205, 2014.

other basic approach – beam forming (BF) – was created for localization of medium and long distance sound sources. In both cases the core of the task to be solved is to divide the area observed by the acoustic camera into sub-areas in dependence on multidimensional measurement data. Hence we decided to apply a clustering approach to solve it.

In spite of numerous developments, clustering of multidimensional data sets is still a challenging task [6]. There are numerous approaches for solving it including intelligent techniques based on fuzzy logic and neural networks. In [8, 11] we proposed a new multidimensional data clustering approach that combines model of direction selective cells in the middle temporal (MT) cortex and recurrent neural networks for features extraction and fuzzy subtractive clustering for blind separation of data into clusters. Variations of the algorithm were successfully applied by now to different static and dynamic data sets: landscape classification using multi-spectral satellite image of a mountain region in Bulgaria [9, 10], clustering of dynamic data taken from an experiment that tests visual discrimination of complex dot motions [11] and classification of accumulated acoustic pressure measured by a microphone array [12].

By far the algorithm was used to create 2D picture of clusters of multidimensional data sets. The present investigation extends application of the algorithm to 3D visualization including time course of data as third dimension. The suggested approach is simplified one – it uses sound intensity only and doesn't account for the distance to the sound source. However it will allow detecting not only of static but also of moving sound sources – function that is not included in the original software version supplied by Brüel & Kjær. The obtained results were compared with the 2D sonograms created by original software. Our future intention is to incorporate it in the system we have and to extend it with ability to detect moving sound sources.

The paper is organized as follows: in next section we describe the experimental set-up, the equipment (acoustic antenna) and data collection procedure; section 3 gives short description of our algorithm with accent to its extension to 3D dynamic task; in section 4 results of dynamic data clustering and 3D visualization are presented and discussed in comparison with the sonograms created by the original software of Brüel&Kjær; the paper finishes with conclusions and directions for future work.

2   Experiment Set-up

2.1   Brüel and Kjær Microphone Array

Multidimensional data for testing of our approach was collected using the system from Brüel & Kjær for sound analysis shown on Figure 1 (a). It consists of 18 microphones array placed randomly in a wheel grid called antenna. At the center of antenna is mounted camera. All microphones are connected to a front-end panel. Both camera and front-end are connected to a computer (via USB and LAN cables correspondingly) with software for sensor information processing. The system measures acoustic pressure and visualizes "sound picture" of the observed by camera area as it is shown on Figure 1 (b).

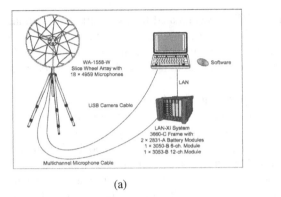 

(a)   (b)

Fig. 1. Brüel & Kjær system for sound analysis (a) and created by it "sound picture" (b)

2.2   Raw Data Collection

Our multidimensional data set consists of raw measurement data from all 18 micro-phones in antenna array. A piezo beeper WB 3509 (standard Brüel & Kjær equipment – the red box in the right low corner on the picture on Figure 1 (b)) with frequency of 2.43 kHz was used as sound source. After switching on the beeper the system collects acoustic pressure in Pa for 15.9ms – period of time predetermined by the system software – from all 18 microphones. The measurements were taken with time step $1.53*10^{-5}$ s. The collected data are periodic signals with variable amplitude and constant frequency of the noise source (the beeper). The input signal amplitude is different for each microphone due to attenuation of different beeper – microphone path loss.

3   Clusterization Algorithm

3.1   Initial Feature Extraction Procedure

In [11], following the model of human visual perception from [1, 2], we used the receptive fields of MST neurons to pre-process time series of our dot motion data. This model has been widely used to examine the emergence of complex motion pattern properties [1, 2]. The receptive fields are direction selective cells in middle temporal (MT) cortex described by the following equation [1]:

$$f_{il}(t)=\frac{1}{N}\sum_{k=1}^{N}\exp\left(\frac{-\left(\mu_i-s_k(t)\right)^2}{2\sigma^2}\right) \qquad (1)$$

Here $f_{il}(t)$ is the response of i-th MT unit to k-th input stimuli $s_k(t)$ for the l-th receptive field (area of stimuli collection) at time t; μ_i is center and σ is variance of Gaussian curve defining each filter response; N is number of inputs, i.e. stimuli received in the l-th field. In present work we divide the area of stimuli (in considered example these are microphone sensors readings) into several overlapping regions, each containing at least one stimulus (sensor) input. In [12] we accumulate receptive fields' outputs at each area and average them over all time period of measurements thus accounting for accumulated acoustic pressure. In present work we use receptive fields' outputs for a given moment in the time in order to account for time changes in the sound picture.

Division of observed by acoustic camera area into 16 overlapping square regions is shown on Figure 2 (each region is surrounded by a dashed line square with rounded edges). The small red dots with numbers represent corresponding microphone position and the big red dot in the center marks camera position. Each region contains at least one microphone (e.g. microphone 5 is the only one in upper right region). Maximal number of microphones in region is four and it is situated at the center of antenna (e.g. region containing microphones 1, 3, 8 and 7).

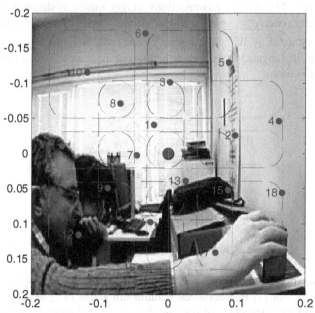

Fig. 2. Regions positions at the antenna area

Next, in order to design our receptive filed units, we divide the dynamic range of raw data (that is from -0.4 to 0.4 Pa) into 11 intervals. For each interval we define a filter with center μ_i at the center of interval and variance σ equal to one third of interval size. Thus our receptive fields overlap covering intervals from -3σ to $+3\sigma$ around their centers.

Equation (1) describes obtained at this first step feature vectors of area number l where $i=1 \div n_f$ and $n_f=11$ is number of filters in our experiment. Thus the obtained for the period of time from 0 to t_f data set of features is:

$$f_{il}(t)\big|_{t=1+t_f,\ i=1\div11,\ l=1\div16} \tag{2}$$

These features are inputs to the recurrent neural network used in the second step of feature extraction procedure.

3.2   Final Feature Extraction

At this step we exploit the equilibrium states of neurons of a special kind of recurrent neural network – Echo state network (ESN) [5, 13] – as final features extracted from multidimensional data. The basic idea was proposed for the first time in [8]. Here we'll describe it briefly.

The structure of ESN is presented on Figure 3 bellow. It consists of a randomly generated dynamic reservoir of interconnected neurons having also feedback from their own outputs. The reservoir connections weight matrix is denoted by W^{res}. All reservoir neurons receive as input a vector denoted here by u multiplied by input weight matrix denoted by W^{in}. The output of reservoir is a simple sigmoid function (usually hyperbolic tangent) that depends on current input as well as on previous state of the reservoir neurons.

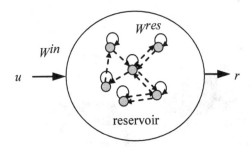

Fig. 3. Echo state network (ESN) structure

Following the proposed in [14, 15] algorithm for initial tuning of reservoir weights and conclusions from [7] that achieved equilibrium states of reservoir neurons after such tuning reflect the structure of training data set, in [8] emerged the idea that the reservoir equilibrium can serve as a feature vector.

Inputs to our second feature extractor – ESN – are initial features extracted by receptive fields, i.e. for l-th area of stimuli collection:

$$u_l(t) = [f_{1l}(t)\quad f_{2l}(t)\quad \dots f_{11l}(t)]_{t=1+t_f,\ l=1\div16} \tag{3}$$

Final features used for data clustering are equilibrium states r_e of trained ESN neurons that are calculated by presenting each vector as constant input to the ESN until all neurons outputs settle down, i.e.:

$$r_{el}(t) = \tanh\left(diag(a)W^{res}r_{el}(t) + diag(a)W^{in}u_l(t) + b\right) \tag{4}$$

Here a and b are additional vectors of parameters used to tune the reservoir according to [13, 14].

3.3 Overall Clustering Procedure

Measurement data (stimuli) are collected for given period of time from all sensors in considered area. The clustering algorithm is as follows:

- The collected data are pre-processed using first step feature extraction procedure and data set (2) is generated;
- A random ESN reservoir is generated and tuned to these data;
- The trained reservoir equilibriums are determined according to (4); then they are scaled within interval [-1, +1];
- All possible two dimensional projections between equilibrium states of every two different neurons in the reservoir i and j and for each period of time step t are generated as follows:

$$P_{ij}(t) = \left[r_{el}^i(t), \quad r_{el}^j(t)\right]_{t=1+16} \tag{5}$$

- Subtractive clustering procedure [17] is applied to all projections (5) in order to determine number and centers of data clusters. This procedure was chosen since it is reported as one of the best options in the case of unknown number of clusters [3];
- The projections with highest number of clusters are selected;

4   Results and Discussion

It was observed that acoustic pressure data is periodic with period of about 0.412 ms or approximately 28 time steps as it is shown on Figure 4. Hence we decided to investigate time changes of "sound picture" during one period as well as for all the time of measurements with 0.412 ms time step.

At the second step of described above feature extraction algorithm we used ESN reservoirs with different sizes: 10, 30 and 50 neurons. In all cases the number of inputs of ESN was determined by the number of features, i.e. 11 according to the number of receptive fields. For each new generation of reservoir number of obtained by our algorithm clusters varies but in most cases we have mainly 3 or 4 clusters.

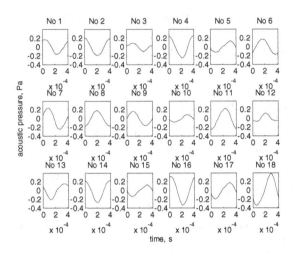

Fig. 4. Microphone signals for the first 28 time steps (approximately one period of signal)

The bigger was number of neurons in ESN reservoir, the bigger is number of possible two dimensional projections with maximal number of clusters. The number of obtained clusters however is smaller in comparison with those obtained in [12] (2D clusters case) where we had about 6 clusters.

We consider each sub-region in antenna area as an area in the picture taken from camera that has to be classified. Each cluster is covered by rectangles with different color. Figures 5, 6 and 7 present classification results obtained by using ESN reservoirs with 10, 30 and 50 neurons respectively. On each figure (a) is "unfolded" 3D picture for the first period of measurements and (b) – for all periods of measurements with time step equal to period duration (0.412 ms).

From all figures (a) we can observe "movement" of the sound wave coming from the noise source through receptive fields for the first period of time. The last picture (beginning of new period) is the same as the first one, i.e. our classification is able to reveal periodical characteristics of data. In spite of roughness of our sensing fields, the position of the beeper can be exactly estimated without usage of any information about free-space path loss formula and propagation delay.

On figures (b) we observe time changes of "sound picture during all time of measurements. The "unfolded" 3D picture reveals changes in the acoustic pressure amplitude with time. Although all pictures are from the beginning of current period, they gradually change from the beginning of the measurements to their end. The change in pictures is due to beating frequency of inexact correspondence of sampling frequency and beeper frequency. The pictures visualize 2D beating frequency propagation and can serve as an instrument for spectral analysis of the input signal.

Comparison with Figure 1 (b) is to rough but it is clear that in all pictures the area with noise source (low right corner) is recognized to be different from the other areas in the picture. Having in mind that "sound picture" form original software is accumulative for all time of collecting data and our "sound pictures" are for different time steps, the observed differences reveal dynamic nature of the sound signal.

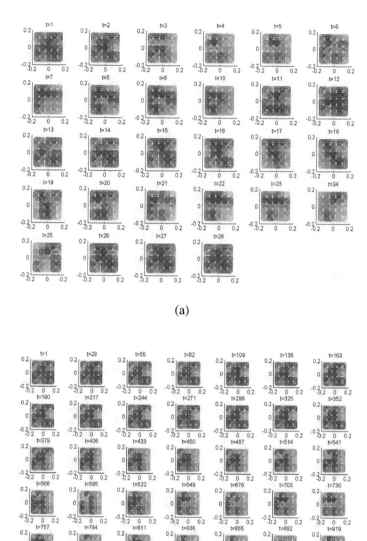

Fig. 5. Clusters obtained with 10 neurons (a) for the first period and (b) for all the time with step 0.412 ms

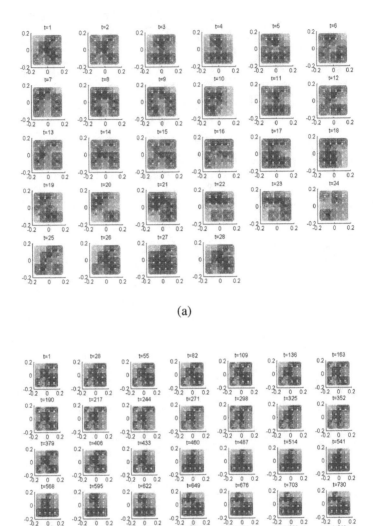

(a)

(b)

Fig. 6. Clusters obtained with 30 neurons (a) for the first period and (b) for all the time with step 0.412 ms

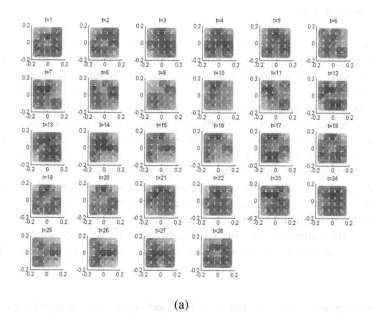

Fig. 7. Clusters obtained with 50 neurons (a) for the first period and (b) for all the time with step 0.412 ms

5　Conclusions

The presented in the paper application of currently developed algorithm for multidimensional data clustering and its extension to dynamic data representation in 3D showed promising results and pointed out to directions for future developments.

First of all, testing of our approach on yet another multidimensional data set and comparison of the results with a professional signal processing software demonstrated that our approach although not that refined is promising and gives similar results.

Possible refinement of the results could be obtained via fuzzy visualization of clusters. By the moment although we use fuzzy clustering algorithm, we classify each dot on the picture based on only its distance to the cluster centers. However subtractive clustering allows us to have also overlapping clusters with fuzzy membership of each dot. Hence we can obtain pictures with gradual change of colors that will be much closer to that of original software.

Another interesting way for further improvement is the application of described scheme for multisource input signal, localization of each source and spectral identification.

Acknowledgments. The research work reported in the paper is partly supported by the project AComIn "Advanced Computing for Innovation", grant 316087, funded by the FP7 Capacity Program (Research Potential of Convergence Regions).

References

1. Beardsley, S.A., Ward, R.L., Vaina, L.M.: A neural network model of spiral–planar motion tuning in MSTd. Vision Research 43, 577–595 (2003)
2. Grossberg, S., Pilly, P.K.: Temporal dynamics of decision-making during motion perception in the visual cortex. Technical Report BU CAS/CNS TR-2007-001 (February 2008)
3. Hammouda, K.: A comparative study of data clustering Techniques. SYDE 625: Tools of Intelligent Systems Design, Course Project (August 2000)
4. Jacobsen, F., Jaud, V.: Statistically optimized near field acoustic holography using an array of pressure-velocity probes. J. Acoust. Soc. Am. 121(3), 1550–1558 (2007)
5. Jaeger, H.: Tutorial on training recurrent neural networks, covering BPPT, RTRL, EKF and the "echo state network" approach. GMD Report 159, German National Research Center for Information Technology (2002)
6. Jain, A.K., Murty, M.N., Flynn, P.J.: Data clustering: A review. ACM Computing Surveys 31(3), 264–323 (1999)
7. Koprinkova-Hristova, P., Palm, G.: ESN intrinsic plasticity versus reservoir stability. In: Honkela, T. (ed.) ICANN 2011, Part I. LNCS, vol. 6791, pp. 69–76. Springer, Heidelberg (2011)
8. Koprinkova-Hristova, P., Tontchev, N.: Echo state networks for multi-dimensional data clustering. In: Villa, A.E.P., Duch, W., Érdi, P., Masulli, F., Palm, G. (eds.) ICANN 2012, Part I. LNCS, vol. 7552, pp. 571–578. Springer, Heidelberg (2012)

9. Koprinkova-Hristova, P., Alexiev, K., Borisova, D., Jelev, G., Atanassov, V.: Recurrent neural networks for automatic clustering of multispectral satellite images. In: Bruzzone, L. (ed.) Image and Signal Processing for Remote Sensing XIX, October 17. Proceedings of SPIE, vol. 8892, p. 88920X (2013), doi:10.1117/12, ISSN: 0277-786X, ISBN: 9780819497611

10. Koprinkova-Hristova, P., Angelova, D., Borisova, D., Jelev, G.: Clustering of spectral images using Echo state networks. In: 2013 IEEE International Symposium on Innovations in Intelligent Systems and Applications, IEEE INISTA 2013, Albena, Bulgaria, June 19-21 (2013), doi:10.1109/INISTA.2013.6577633, ISBN: 978-147990661-1

11. Koprinkova-Hristova, P., Alexiev, K.: Echo State Networks in Dynamic Data Clustering. In: Mladenov, V., Koprinkova-Hristova, P., Palm, G., Villa, A.E.P., Appollini, B., Kasabov, N. (eds.) ICANN 2013. LNCS, vol. 8131, pp. 343–350. Springer, Heidelberg (2013)

12. Koprinkova-Hristova, P., Alexiev, K.: Sound fields clusterization via neural networks. In: 2014 IEEE Int. Symposium on Innovations in Intelligent Systems and Applications, Alberobello, Itally, June 23-25 (accepted paper, 2014)

13. Lukosevicius, M., Jaeger, H.: Reservoir computing approaches to recurrent neural network training. Computer Science Review 3, 127–149 (2009)

14. Schrauwen, B., Wandermann, M., Verstraeten, D., Steil, J.J., Stroobandt, D.: Improving reservoirs using intrinsic plasticity. Neurocomputing 71, 1159–1171 (2008)

15. Steil, J.J.: Online reservoir adaptation by intrinsic plasticity for back-propagation-decoleration and echo state learning. Neural Networks 20, 353–364 (2007)

16. Williams, E.G., Maynard, J.D., Skudrzyk, E.J.: Sound source reconstructions using a microphone array. J. Acoust. Soc. Am. 68(1), 340 (1980)

17. Yager, R., Filev, D.: Generation of fuzzy rules by mountain clustering. Journal of Intelligent & Fuzzy Systems 2(3), 209–219 (1994)

Neural Classification for Interval Information

Piotr A. Kowalski[1,2] and Piotr Kulczycki[1,2]

[1] Cracow University of Technology,
Department of Automatic Control and Information Technology,
ul. Warszawska 24, PL-31-155 Cracow, Poland
{pkowal,kulczycki}@pk.edu.pl
[2] Systems Research Institute, Polish Academy of Sciences,
ul. Newelska 6, PL-01-447 Warsaw, Poland
{pakowal,kulczycki}@ibspan.waw.pl

Abstract. The subject of the presented research is to determine the complete neural procedure for classifying inaccurate information, as given in the form of an interval vector. For such a formulated task, a basic functionality Probabilistic Neural Network was extended upon the interval type of information. As a consequence, a new type of neural network has been proposed. The presented methodology was positively verified using random and benchmark data sets. In addition, a comparative analysis of existing algorithms with similar conditions was made.

Keywords: neural networks, probabilistic neural networks, data analysis, classification, interval data, imprecise information.

1 Introduction

Recently, in many applications, there has been a growth of interest in interval analysis. The basis of this concept is the assumption that the only possessed information about the tested quantity x, is the fact that it fulfils the relationship $\underline{x} \leq x \leq \overline{x}$, and, consequently, it may be identified with the interval:

$$[\underline{x}, \overline{x}]. \tag{1}$$

Interval analysis is a separate area of mathematics which has its own formal apparatus based on the axiom theory [15]. Formerly, its primary use was to provide the required accuracy within numerical calculations, as these are often affected by the control error resulting from rounding [1]. However, as a result of its continuous development, this area is becoming frequently used in engineering, econometrics, and other related fields [5]. Its main advantage is the fact that, by its nature, it is modelling the uncertainty of an examined quantity by using the simplest possible formula. In many applications, interval analysis has found to be completely sufficient, and it requires low computation effort (which allows its employment in very complex tasks). Moreover, this methodology is easy to identify and interpret, while also having a convenient formalism based on a mathematical apparatus.

G. Agre et al. (Eds.): AIMSA 2014, LNAI 8722, pp. 206–213, 2014.

The goal of the research is to reveal the complete neural procedure for classifying inaccurate information (1) as applied in cases of multi-dimensional data that are expressed in the form of the interval vector:

$$[[\underline{x}_1, \overline{x}_1], [\underline{x}_2, \overline{x}_2], \ldots [\underline{x}_n, \overline{x}_n]]^T , \tag{2}$$

where $\underline{x_k} \leqslant \overline{x_k}$ for $k = 1, 2, ..., n$, when the patterns of individual classes are determined on the basis of unambiguously defined sets of items, that is

$$\underline{x_k} = \overline{x_k} \text{ for } k = 1, 2, ..., n. \tag{3}$$

The concept of classification is based on employing the Probabilistic Neural Network approach by way of using Bayes theorem, when provided with a minimum of potential losses resulting from misclassification. For such a task, a formulated statistical kernel estimator methodology is used. This procedure is not dependent on arbitrary assumptions about character patterns. Their identification will be an integral part of the presented algorithm.

2   Kernel Density Estimator

The Statistical Kernel Density Estimators (KDE) belong to the set of non-parametric methods. They allow the designation and illustration of the characteristics of random variable distribution, without possessing the information on the membership of a particular class.

Consider a n-dimensional random variable whose distribution has density function f. Its kernel estimator $\hat{f}$ is determined on the basis of the m-element random sample:

$$x_1, x_2, \ldots, x_m \tag{4}$$

and is defined by the formula:

$$\hat{f}(x) = \frac{1}{mh^n} \sum_{i=1}^{m} K\left(\frac{x - x_i}{h}\right). \tag{5}$$

The positive coefficient h is called 'smoothing parameter'. A measurable function K, symmetric with respect to zero at this point, having weak local maximum and satisfying the condition $K(x): \mathbb{R} \to [0, \infty)$, is referred as a 'kernel'. The form of the kernel K practically does not affect the statistical quality of the estimation. In this work, we will use the one-dimensional Cauchy kernel

$$K(x) = 2/\pi(x^2 + 1)^2. \tag{6}$$

In the case of multivariate situation, this will be generalized using the concept of a kernel product.

More detailed information about the practical issues of employing KDE methods, as well as usage examples, can be found in cited references [11] and [19].

3   Neural Network for Interval Imprecise Information

The Probabilistic Neural Network (PNN), which is very often considered as being a neural realization of a set of KDE, is a special type of a Radial Neural Network. It is used mainly for regression [16], prediction [18], classification [14] [3] and identification [2] tasks, but also for non-linear time series analysis.

In this part of this paper, the generalization of PNN as used in processing interval information, will be introduced. This neural structure is based on Specht's Probabilistic Network [17], but it has a several new elements which enable us to classify interval information. Figure 1 reveals the topological scheme of a generalized probabilistic neural network. This structure, in this paper, is treated as a network implementation of the interval information classifier.

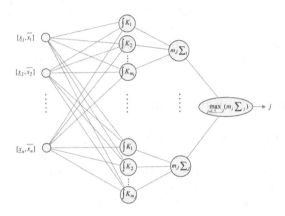

Fig. 1. The structure of a PNN extended for processing imprecise information

In this created network, there are four layers. The first is the input layer, with size equal m, wherein the inputs correspond to the dimensions of the interval element (3) under classification. The next layer is a subset of neurons representing the successive patterns of classes. Each of these consists of an appropriate number of neurons whose function is to bring about the operation of integration (9). The third layer provides a summation of neuronal signals within a pattern class, as well as a multiplication of the result value by group cardinality (8). The final single neuron, located in the output layer, determines the highest values obtained from the pattern layer and fixes the final result of this classification task.

In a situation wherein information is given by the interval $[\underline{x}, \overline{x}]$, the tested element, based on Bayes Theorem, belongs to the class for which the value:

$$\frac{m_1}{\overline{x} - \underline{x}} \int_{\underline{x}}^{\overline{x}} \hat{f}_1(x)\, dx, \ \frac{m_2}{\overline{x} - \underline{x}} \int_{\underline{x}}^{\overline{x}} \hat{f}_2(x)\, dx, \ldots, \ \frac{m_J}{\overline{x} - \underline{x}} \int_{\underline{x}}^{\overline{x}} \hat{f}_J(x)\, dx \qquad (7)$$

is the largest. This is a natural extension of Bayes' theorem into interval information type. In the above formula, the positive constants $1/(\overline{x} - \underline{x})$, can be omitted as these are negligible for the optimization problem under consideration. Therefore finally, it can be presented in the following form:

$$m_1 \int_{\underline{x}}^{\overline{x}} \hat{f}_1(x)\,dx, \ m_2 \int_{\underline{x}}^{\overline{x}} \hat{f}_2(x)\,dx, \ldots, \ m_J \int_{\underline{x}}^{\overline{x}} \hat{f}_J(x)\,dx \qquad (8)$$

Moreover, for every $\hat{f}_1, \hat{f}_2, \ldots, \hat{f}_J$ one can note:

$$\int_{\underline{x}}^{\overline{x}} \hat{f}(x)\,dx = \hat{F}(\overline{x}) - \hat{F}(\underline{x}), \qquad (9)$$

where $\hat{F}$ means the primitive of the function $\hat{f}$. For the Cauchy Kernel (6) used here, the following analytical formula can be obtained:

$$\hat{F}(x) = \frac{1}{m} \sum_{i=1}^{m} \left[\frac{(x^2 - 2xx_i + x_i^2 + h^2)\arctan(\frac{x-x_i}{h}) + h(x - x_i)}{x^2 - 2xx_i + x_i^2 + h^2} + \frac{\pi}{2} \right] \qquad (10)$$

(note that the constant $\pi/2$ could be again omitted for equal cardinality of pattern sets). In the multidimensional case, when information is represented by the interval vector (3), this can be easily extended by using the concept of a product kernel [9,13] .

4   Numerical Verification

The correctness of the presented method was verified through was conducted by the way of numerical simulation. Due to the specific conditioning of the presented method, this type of data was not found in public repositories. The following are the results for data obtained using the random number generator with normal distribution. This was done using a given vector of expected value and a covariance matrix. This, in turn, was derived from the implemented multivariate normal distribution generator based on the concept of Box-Muller [4].

The quality assessment methods presented here were obtained by generating a set of random numbers of the assumed distribution, and provide an analysis of the correctness of the results of the classification procedures used for data that was either of an interval type, or (for comparison) of an unambiguous type. In order to ensure the reproducibility of the results, for each of the pseudo-random sets, the seed value that defines it was strictly determined.

After obtaining the sequences of pseudo-random patterns representing the different classes, test data consisting of classified items was generated. These included those of an interval type, and occasionally those that were of an unambiguous type, for comparative purposes. Each class corresponded to a set of a size of 1,000 items.

4.1   Basic Synthetic Data

This section will firstly put forward the basic form of the research conducted for the classification method developed herein that is build upon the information interval and upon uniform given patterns. In the case of a one-dimensional ($n = 1$)

pattern the first class was obtained by using a pseudo-random number generator with a normal distribution $N(0,1)$ and the other as $N(2,1)$. The results of this example are presented in Table 1.

The classified elements were obtained through generation by one of the aforementioned generators with normal distribution of the first pseudo-random number, as well as the second taken from a generator with uniform distribution. This defines the location of the first as within an interval of an arbitrarily assumed length. Moreover, this represents information of interval type when there are no circumstances for the considered imprecision, although its size is known. Such an interpretation seems to be the most appropriate for the majority of practical interval analysis applications. The tables below show the results with the following size of patterns: 10, 20, 50, 100, 200, 500 and 1000. In the mentioned tables, each cell contains the results obtained from 100 tests, giving an average classification error that is defined on the basis of these 100 random samples.

Table 1. Average classification error for the basic concept of Neural Classification

interval length no. of elements	0.00	0.1	0.25	0.5	1.00	2.00	5.00
10	0.1713	0.1720	0.1720	0.1723	0.1729	0.1761	0.1944
20	0.1655	0.1669	0.1669	0.1672	0.1680	0.1713	0.1888
50	0.1602	0.1605	0.1606	0.1609	0.1617	0.1652	0.1848
100	0.1596	0.1601	0.1602	0.1604	0.1615	0.1650	0.1827
200	0.1596	0.1602	0.1604	0.1609	0.1618	0.1650	0.1840
500	0.1591	0.1595	0.1596	0.1602	0.1613	0.1647	0.1844
1000	0.1579	0.1584	0.1588	0.1591	0.1603	0.1637	0.1833

4.2   Iris Benchmark Data

In the following studies, a real data set was employed. This is derived from a well-known repository located at the Center for Machine Learning and Intelligent Systems at the University of California, Irvine. The pattern set and the reference sample testing are not distinguished. For this reason, in the study, data elements were randomly divided into subsets of elements which include both reference patterns and test samples. The results shown in Table 2 are the average of 1000 tests made into random divisions. The intervals were generated in the same manner as in previous studies.

The results that are displayed, underline the many positive features of employing the classification methods mentioned in this paper. The first is small in practice, and, while sensitivity is often mentioned as being the curse of dimensionality, yet, herein, the classification of a four-dimensional feature vector has been satisfactorily performed on the basis of patterns containing about 25 items. Additional confirmation of the effectiveness of the method proposed within this paper was obtained by comparison with the results presented in [7] for the unambiguous data. In the aforementioned article, a classification error of no less

Table 2. The results of the numerical verification for the *Iris data*

interval length	0.00×0.00	0.10×0.10	0.25×0.25	0.50×0.50	1.00×1.00	2.00×2.00
mean error	0.041	0.045	0.047	0.048	0.049	0.066

than 4.5 % was obtained. A similar result was obtained in this study for the unambiguous data (cf. second column of the Table 2). Despite reducing the accuracy of the classified information by processing it into inaccurate information, the results had not deteriorated to the length of the interval of 1.0 - which is worth especially underlining.

4.3 Comparison with Similar Algorithms Used for Classification

The purpose of the following research is to compare the quality of classification of inaccurate information with other works available in the literature which are suitable for adoption.

The first one is based on a method very broadly used today due to its certain advantages, that of support vectors machines; while the other is employed for comparing the number of elements of each pattern contained in the investigated interval.

The results were obtained using the technique of supporting vectors, according to the algorithm presented in the work [20]. As a result of this procedure, three types of decisions are generated: assignment of an interval element to the first or to the second class, or the lack thereof. The study considered the amount of misclassification, lack of decision, and, in addition, the total error which is the sum of bad decisions and those of the elements for which there is no decision. Information found inside the latter was classified by drawing lots in relations proportional to the number of patterns. Upon comparing the results in the base case, it is clear that the results obtained using the method of support vectors are worse by 5% to up to 50%.

The second, relatively simple method of classifying interval type information is the procedure for patterns counting. This consists of reckoning how many elements of the learning sample are contained in the interval which is under consideration. In each case study of this algorithm, the obtained results were distinguished to be within four situations: the amount of misclassification is equal to the cardinality of elements drawn from both patterns belonging to the tested element of the interval; that the interval elements do not contain any element that is referenced; that further total error is the sum of wrong decisions; and that the consequential errors draw a ratio of 0.5 and 0.5 for those cases where the number of elements of both patterns were the same.

The effects obtained from the use of the concept of counting, revealed themselves to be absolutely worse than those obtained using the method developed in this article. However, current methods for classifying interval data are not limited to those presented this subsection. There is also a very interesting algorithm with similar conditions described in [6]. A comparison of the proposed neural algorithm with the cited method will be the subject of further research.

5   Conclusions

In conclusion, the results presented in the previous section, through numeric verification, confirm the correctness of the developed herein neural classification method of the interval type for dealing with contained inaccurate information. The results were compared with those obtained when the element was classified as being uniquely defined, as well as with those gained through utilizing other algorithms commonly employed for classifying interval data. In all the studies, enlarging the cardinality patterns resulted in a decrease of the average value of the error classification. This, in practice, allows the gradual improvement of the quality of the classification as new data is acquired. Furthermore, with the increasing length of the interval, classification errors were seen to increase to a certain limit that is justified by the data structure.

These conclusions are worth emphasizing from the application point of view. This is because they indicate that it is possible to increase the quality of classification by way of enlarging the available information through placing this in the form of numerous patterns, and by accurately classifying the interval element. In practical matters, therefore, it becomes necessary to establish a compromise between the amount of available data and the quality of the results. In situations in which there are very large representations of classes, the neural networks size rapidly increases. For this reason, we recommend using a method of reducing the sample size. With respect to employing a generalized PNN on interval information, particularly advantageous results are gained through enlisting the reduction method described in [10].

What is more, if there are no previously distinguished classes before the learning process is undertaken, the data set should be divided into smaller groups by the way of utilising a clustering method. If the number of classes is known, the application of the simple *k-means* method is recommended. Otherwise, the algorithm that is required should determine the optimal number of groups during the process of clustering. An example of a procedure satisfying the above task is an algorithm based on the Kernel Estimators Methodology [8,12].

The issue of information classification on the basis of interval data can be illustratively interpreted when unambiguous examples of the patterns contain specific, precisely measured data, while the compartments represent limitations within the plans or estimates, or when it is difficult to perform the measurements. In particular, this neural method can be used for generating a classification where a set of unambiguous data is treated as being specific information from the past (for example, temperature or exchange rates), while the classification element represents the inaccuracies forecast as being naturally limiting.

References

1. Alefeld, G., Hercberger, J.: Introduction to Interval Computations. Academic Press, New York (1986)
2. Araghi, L.F., Khaloozade, H., Arvan, M.R.: Ship identification using probabilistic neural networks (PNN). In: Proceedings of the International MultiConference of Engineers and Computer Scientists, vol. 2, pp. 18–20 (2009)
3. Bascil, M.S., Oztekin, H.: A Study on Hepatitis Disease Diagnosis Using Probabilistic Neural Network. J. Med. Syst. 36, 1603–1606 (2012)
4. Brandt, S.: Data Analysis. Springer, Heidelberg (1999)
5. Jaulin, L., Kieffer, M., Didrit, O., Walter, E.: Applied Interval Analysis. Springer, Berlin (2001)
6. Kaytoue, M., Kuznetsov, S.O., Napoli, A., Duplessis, S.: Mining gene expression data with pattern structures in formal concept analysis. Information Science 181(10), 1989–2001 (2011)
7. Kotsiantis, S.B., Pintelas, P.E.: Logitboost of Simple Bayesian Classifier. Informatica 29, 53–59 (2005)
8. Kowalski, P.A., Lukasik, S., Charytanowicz, M., Kulczycki, P.: Data-Driven Fuzzy Modelling and Control with Kernel Density Based Clustering Technique. Polish Journal of Environmental Studies 17, 83–87 (2008)
9. Kowalski, P.A.: Bayesian Classification of Imprecise Interval-Type Information (in Polish). SRI, Polish Academy of Sciences, Ph.D. Thesis (2009)
10. Kowalski, P.A., Kulczycki, P.: Data Sample Reduction for Classification of Interval Information Using Neural Network Sensitivity Analysis. In: Dicheva, D., Dochev, D. (eds.) AIMSA 2010. LNCS (LNAI), vol. 6304, pp. 271–272. Springer, Heidelberg (2010)
11. Kulczycki, P.: Statistical Inference for Fault Detection: A Complete Algorithm Based on Kernel Estimators. Kybernetika 38(2), 141–168 (2002)
12. Kulczycki, P., Charytanowicz, M., Kowalski, P.A., Lukasik, S.: The Complete Gradient Clustering Algorithm: properties in practical applications. Journal of Applied Statistics 39(6), 1211–1224 (2012)
13. Kulczycki, P., Kowalski, P.A.: Bayes classification of imprecise information of interval type. Control and Cybernetics 40, 101–123 (2011)
14. Kusy, M., Kluska, J.: Probabilistic Neural Network Structure Reduction for Medical Data Classification. In: Rutkowski, L., Korytkowski, M., Scherer, R., Tadeusiewicz, R., Zadeh, L.A., Zurada, J.M. (eds.) ICAISC 2013, Part I. LNCS, vol. 7894, pp. 118–129. Springer, Heidelberg (2013)
15. Moore, R.E.: Interval Analysis. Prentice-Hall, Englewood Cliffs (1966)
16. Rutkowski, L.: Computational Intelligence: Methods and Techniques. Springer, Berlin (2008)
17. Specht, D.F.: Probabilistic Neural Networks. Neural Networks 3, 109–118 (1990)
18. Tran, T., Nguyen, T., Tsai, P., Kong, X.: BSPNN: boosted subspace probabilistic neural network for email security. Artif. Intell. Rev. 35, 369–382 (2011)
19. Wand, M.P., Jones, M.C.: Kernel Smoothing. Chapman and Hall, London (1995)
20. Zhao, Y., He, Q., Chen, Q.: An Interval Set Classification Based on Support Vector Machines. In: 2nd International Conference on Networking and Services, Silicon Valley, pp. 81–86 (2005)

FCA Analyst Session and Data Access Tools in FCART

A.A. Neznanov and A.A. Parinov

National Research University Higher School of Economics,
Pokrovskiy bd., 11, 109028, Moscow, Russia
{ANeznanov,AParinov}@hse.ru

Abstract. Formal Concept Analysis Research Toolbox (FCART) is an integrated environment for knowledge and data engineers with a set of research tools based on Formal Concept Analysis. FCART allows a user to load structured and unstructured data (including texts with various metadata) from heterogeneous data sources into local data storage, compose scaling queries for data snapshots, and then research classical and some innovative FCA artifacts in analytic sessions.

Keywords: Data Analysis, Formal Concept Analysis, Knowledge Extraction, Text Mining, Software.

1   Introduction

Formal Concept Analysis [1] is a mature group of mathematical models and good foundation for methods of intelligent system construction. In previous articles [2] we described the stages of development of a software system for information retrieval and knowledge discovery from various data sources (textual data, structured databases, etc.). Formal Concept Analysis Research Toolbox (FCART) was designed especially for the analysis of unstructured (textual) data. The core of the system supports knowledge discovery techniques, including those based on Formal Concept Analysis [1], clustering [2], multimodal clustering [2, 3], pattern structures [4, 5] and others. FCART was already successfully applied to analyzing data in medicine, criminalistics, and trend detection.

There are many FCA-based tools. The most well-known open source projects are ConExp [6], Conexp-clj [7], Galicia [8], Tockit [9], ToscanaJ [10], FCAStone [11], Lattice Miner [12], OpenFCA [13], Coron [14], Cubist [15]. These tools have many advantages. However, they suffer from the lack of the abilities to communicate with various data sources and poor data preprocessing. It prevents researchers from using these programs for analyzing complex big data without additional third party tools.

In this article, we describe distributed architecture of FCART, data access tools and analyst session (for data and knowledge accumulation with experiment reproducibility). The main goal of the current release is to develop architecture, which can work with really big datasets. We have tested external data access and querying on Amazon movie review dataset [16] and other collections of CSV, SQL, XML and JSON data with unstructured text fields.

G. Agre et al. (Eds.): AIMSA 2014, LNAI 8722, pp. 214–221, 2014.

2   Methodology and Technology

The DOD-DMS is a universal and extensible software platform intended for building data mining and knowledge discovery tools for various application fields. The creation of this platform was inspired by the CORDIET methodology (abbreviation of Concept Relation Discovery and Innovation Enabling Technology) [17] developed by J. Poelmans at K.U. Leuven and P. Elzinga at the Amsterdam-Amstelland police. The methodology allows one to obtain new knowledge from data in an iterative ontology-driven process. The software is based on modern methods and algorithms of data analysis, technologies for processing big data collections, data visualization, reporting, and interactive processing techniques. It implements several basic principles:

1. Iterative process of data analysis using ontology-driven queries and interactive artifacts (such as concept lattice, clusters, etc.).
2. Separation of processes of *data querying* (from various data sources), *data preprocessing* (of locally saved immutable snapshots), *data analysis* (in interactive visualizers of immutable analytic artifacts), and *results presentation* (in report editor).
3. Extendibility on three levels: customizing settings of data access components, query builders, solvers and visualizers; writing scripts (macros); developing components (add-ins).
4. Explicit definition of analytic artifacts and their types. It allows one to check the integrity of session data and provides links between artifacts for end-user.
5. Realization of integrated performance estimation tools.
6. Integrated documentation of software tools and methods of data analysis.

FCART uses all these principles, but does not have an ontology editor and does not support the full C-K cycle. Current version of FCART consists of the following components.

- Core component includes
  - multiple-document user interface of research environment with session manager and extensions manager,
  - snapshot profiles editor (SHPE),
  - snapshot query editor (SHQE),
  - query rules database (RDB),
  - session database (SDB),
  - main part of report builder.
- Local Data Storage (LDS) for preprocessed data.
- Internal solvers and visualizers.
- Additional plugins, scripts and report templates.

We use Microsoft and Embarcadero programming environments and different programming languages (C++, C#, Delphi, Python and other). Native executable (the core of the system) is compatible with Microsoft Windows 2000 and later and has not any additional dependences. Scripting is an important feature of FCART: generating and transforming artifacts, drawing, and building reports can be done by scripts. For scripting we use Delphi Web Script and Python languages.

From the analyst point of view, basic FCA workflow in FCART has four stages (see Fig. 1). On each stage, a user has the ability to import/export every artifact or add it to report.

1. Filling Local Data Storage (LDS) of FCART from various external SQL, XML or JSON-like data sources (querying external source described by External Data Query Description - EDQD). EDQD can be produced by some External Data Browser.
2. Loading a data snapshot from local storage into current analytic session (snapshot described by Snapshot Profile). Data snapshot is a data table with annotated structured and text attributes, loaded in the system by accessing LDS.
3. Transforming the snapshot to a binary context (transformation described by Scaling Query).
4. Building and visualizing formal concept lattice and other artifacts based on the binary context in a scope of analytic session.

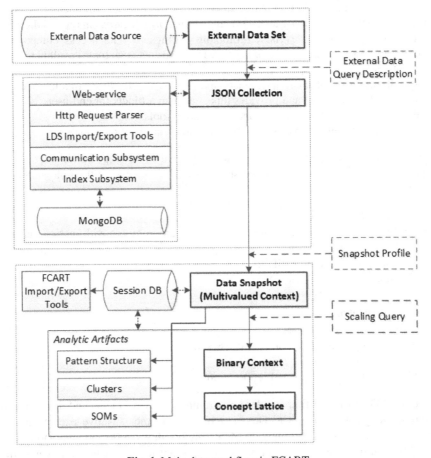

Fig. 1. Main data workflow in FCART

3   Working with Analytical Session

3.1   Analytical Session Structure

Some of FCA entities appear to be fundamental to information representation. In FCART, we use the term "analytic artifact" to denote the definition of abstract interface, describing the entity of the analytic process.

The basic artifact for FCA-based methods is that of "formal context", i.e., object-attribute representation of a subject domain. Most important artifacts also include "concept lattice" and "formal concept".

Artifact instances can be linked by "origination" relationship. For example, we can generate the concept lattice from the formal context. In this case, the formal concept will be an "origin artifact" or simply "origin" for the lattice. Another example is lattice and "association rules": the lattice is the origin of the rules. Any artifact instance is *immutable*. It means that an instance cannot be changed after creation, but can be visualized in various ways.

If we have the predefined set of artifacts in most cases we can use the term "artifact" instead of "artifact instance" without ambiguity. Collection of all artifacts in current analytic cycle forms so-called *analytical session*.

3.2   Solvers, Visualizers, and Reports

Analyst works with analytic artifacts using solvers, visualizers and reports.

Solvers. All types of artifacts are generated by solvers. Each solver requires one or many artifact instances of preassigned types as input and produces one artifact instance of preassigned type as output (Fig. 2).

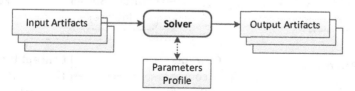

Fig. 2. Solver, it's input and output

Parameters profile of a solver allow to save/load parameters and automatically create dialogs for user input. Each parameter has Id, title, data type, hint, group and default value.

Having predefined types of artifacts and links (assigned by solvers) between immutable artifact instances we can check an integrity of data of particular analytical session. Without explicit user action a session cannot lose any artifact instances and links, and guarantees integrity of a session.

Solver without input artifacts is called *Generator*. Generator can use 1) predefined import format for loading artifact from session or external file, 2) script for automatic building artifact from scratch, 3) manual editor for changing artifact by user.

Visualizers. Artifact visualizer is a special solver that generates user-oriented visual representation of input artifact (or several artifacts) instance. From a technical point of view, visualizer produces interactive or non-interactive window with some elements of user interface. Of course, one artifact can have different kinds of visual appearance.

Each artifact type has *default visualizer*, which invoked by clicking on artifact in session. The parameters profile of visualizer defines common visual properties and drawing parameters (for example, coordinates of concepts in concept lattice). In interactive visualizer, a user has the set of tools for changing the drawing parameters "on the fly".

Usually, visualizer is the last in a chain of solvers. However, we can get a visual representation of each artifact in a session. For example, lattice browser generates a diagram of a lattice and allows a user to manipulate the diagram, but this browser itself does not generate new artifacts. We need to distinguish generation of new artifact and drawing of existing artifact for various purposes: working in the batch mode, increasing efficiency of long chains of solvers, benchmarking, etc. Of course, complex visualizer can help invoking next solvers by simplifying select of some input artifacts.

Fig. 3 illustrates all above concepts as a state of a session of simple FCA workflow from manually created binary context to visual representation of formal concept lattice and concepts indices (with default names of all artifacts).

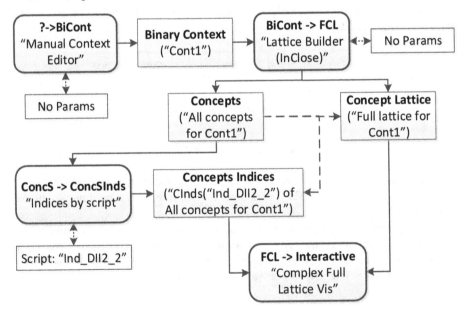

Fig. 3. Session state for simple FCA workflow

Reports. Report is a result of research. Every scientific environment must provide rich text editor of a report with additional functionality to avoid mistakes while converting and moving multiple results with metadata into an external editor. The main feature of the editor is the automatic insertion of fully decorated artifact representation in the resulting report

4   External Data Access and Preprocessing in FCART

4.1   Role of Local Data Storage

FCART Local Data Storage (LDS) plays important role in effectiveness of whole data analysis process because all data from external data storages, session data and intermediate analytic artifacts saves in LDS. All interaction between user and external data storages goes through the LDS. There are many data storages, which contain petabytes of collected data. For analyzing such Big Data analyst cannot use any of the tools mentioned above. FCART provides different scenarios for working with local and external data storages. In the previous release of FCART analyst can work with quite small external data storage because all data from external storage converts into JSON files and saves into LDS. In the current release, we have improved strategies of working with external data. Now analyst can choose between loading all data from external data storage to LDS and accessing external data by chunks using paging mechanism. FCART analyst should specify the way of accessing external data at the property page of the generator.

4.2   Abilities of Web-service Interface to LDS

All interaction between client program and LDS goes through the web-service. Client construct http-request to the web-service. The http-request to web service constructed from two parts: prefix part and command part. Prefix part contains domain name and local path (e.g. http://zeus.hse.ru/lds/). The command part describes what LDS has to do and represents some function of web-service API (See Table 1). Using described commands FCART client can query data from external data storage to the LDS.

Table 1. Main LDS web-service commands

№	Command	Description
1	db.count	get the number of databases
2	db.names	get the names of databases
3	<dbName>.cnum	get the number of collections in <dbName>
4	<dbName>.cnames	get the names of collections in <dbName>
5	<dbName>.<collectionName>.load|{<EDQDText>}	load data using EDQD to the collection
6	<dbName>.<collectionName>.elnum	get the number of elements in the Collection
7	<dbName>.<collectionName>.elids	get the field id of elements if they are present
8	<dbName>.<collectionName>.elems	get all elements from the collection
9	<dbName>.<collectionName>.find(<QueryText>)	get the elements by query <query text>
10	<dbName>.<collectionName>.insert|{<JsonText>}	insert element to the collection
11	.<dbName>.<collectionName>.iter|NUMBER: NUMBER	gets one element or slice from the collection
12	<dbName>.<collectionName>.niter|NUMBER:NUMBER	create new iterator
13	<dbName>.<collectionName>.niter|next	get next value from last created niter
14	<dbName>.<collectionName>.niter|cur	return value of niter current position

5 Conclusion and Future Work

In this paper we have introduced the version 0.9.3 of FCART client in the form of local Windows application and version 0.2 of LDS Web-service. The demo-version of FCART client can be downloaded from http://ami.hse.ru/issa/Proj_FCART.

We have tested the current release with Amazon movie reviews dataset (7.8 millions of documents). This dataset was loaded to FCART LDS and can be accessed from http://zeus2.hse.ru:8443.

The release of the version 1.0 of FCART is planned for August 2014. It will be a useful research environment and prototype for other practical applications. LDS will be appended with new universal indexer API. We already test new solvers based on concept stability [17] and similarity [18]. Biclustering techniques [2] are being actively tested; we are going to extend our platform to triadic concept analysis and noise-robust triclustering methods [3].

Acknowledgements. This work was carried out by the authors within the project "Mathematical Models, Algorithms, and Software Tools for Intelligent Analysis of Structural and Textual Data" supported by the Basic Research Program of the National Research University Higher School of Economics.

References

1. Ganter, B., Wille, R.: Formal Concept Analysis: Mathematical Foundations. Springer (1999)
2. Neznanov, A.A., Ilvovsky, D.A., Kuznetsov, S.O.: FCART: A New FCA-based System for Data Analysis and Knowledge Discovery. Contributions to the 11th International Conference on Formal Concept Analysis, pp. 31–44 (2013)
3. Mirkin, B.: Mathematical Classification and Clustering. Springer (1996)
4. Ignatov, D.I., Kuznetsov, S.O., Magizov, R.A., Zhukov, L.E.: From Triconcepts to Triclusters. In: Kuznetsov, S.O., Ślęzak, D., Hepting, D.H., Mirkin, B.G. (eds.) RSFDGrC 2011. LNCS (LNAI), vol. 6743, pp. 257–264. Springer, Heidelberg (2011)
5. Kuznetsov, S.O.: Pattern Structures for Analyzing Complex Data. In: Sakai, H., Chakraborty, M.K., Hassanien, A.E., Ślęzak, D., Zhu, W. (eds.) RSFDGrC 2009. LNCS, vol. 5908, pp. 33–44. Springer, Heidelberg (2009)
6. Yevtushenko, S.A.: System of data analysis "Concept Explorer". In: Proceedings of the 7th National Conference on Artificial Intelligence KII 2000, Russia, pp. 127–134 (2000) (in Russian)
7. Conexp-clj, http://daniel.kxpq.de/math/conexp-clj/
8. Valtchev, P., Grosser, D., Roume, C., Hacene, M.R.: GALICIA: an open platform for lattices. In: Using Conceptual Structures: Contributions to the 11th Intl. Conference on Conceptual Structures (ICCS 2003), pp. 241–254. Shaker Verlag (2003)
9. Tockit: Framework for Conceptual Knowledge Processing, http://www.tockit.org
10. Becker, P., Hereth, J., Stumme, G.: ToscanaJ: An Open Source Tool for Qualitative Data Analysis. In: Proc. Workshop FCAKDD of the 15th European Conference on Artificial Intelligence (ECAI 2002), Lyon, France (2002)
11. Priss, U.: FcaStone - FCA file format conversion and interoperability software. In: Conceptual Structures Tool Interoperability Workshop, CS-TIW (2008)
12. Lahcen, B., Kwuida, L.: Lattice Miner: A Tool for Concept Lattice Construction and Exploration. Supplementary Proceeding of International Conference on Formal Concept Analysis, ICFCA 2010 (2010)
13. Borza, P.V., Sabou, O., Sacarea, C.: OpenFCA, an open source formal concept analysis toolbox. In: Proc. of IEEE International Conference on Automation Quality and Testing Robotics (AQTR), pp. 1–5 (2010)
14. Szathmary, L.: The Coron Data Mining Platform, http://coron.loria.fr
15. Cubist Project, http://www.cubist-project.eu/
16. Web data: Amazon movie reviews, http://snap.stanford.edu/data/web-Movies.html
17. Poelmans, J., Elzinga, P., Neznanov, A., Viaene. S., Kuznetsov, S.O., Ignatov D., Dedene G.: Concept Relation Discovery and Innovation Enabling Technology (CORDIET). In: Concept Discovery in Unstructured Data. CEUR Workshop proceedings, vol. 757 (2011)
18. Kuznetsov, S.O., Obiedkov, S., Roth, C.: Reducing the Representation Complexity of Lattice-Based Taxonomies. In: Priss, U., Polovina, S., Hill, R. (eds.) ICCS 2007. LNCS (LNAI), vol. 4604, pp. 241–254. Springer, Heidelberg (2007)

Voice Control Framework for Form Based Applications

Ionut Cristian Paraschiv[1], Mihai Dascalu[1], and Stefan Trausan-Matu[1,2]

[1] University Politehnica of Bucharest, Computer Science Department, Romania
[2] Research Institute for Artificial Intelligence of the Romanian Academy
ionut.paraschiv@cti.pub.ro,
{mihai.dascalu,stefan.trausan}@cs.pub.ro

Abstract. Enabling applications with natural language processing capabilities facilitates user interaction, especially in the case of complex applications such as a mobile banking. In this paper we introduce the steps required for building such a system, starting from the presentation of different alternatives alongside their problems and benefits, and ending up with integrating them within our implemented system. However, one of the main problems with voice recognition models is that they tend to use different approximations and thresholds that aren't completely reliable; therefore, the best solution consists of combining multiple approaches. Consequently, we opted to implement two different and complementary recognition models, and to detail in the end how their integration within the framework's architecture leads to encouraging results.

Keywords: Voice recognition, Natural Language Processing, Sentence similarity, Internet banking application, Discourse analysis.

1   Introduction

As the number of mobile devices connected to the Internet increased along with their bandwidth capabilities, life tends to become easier as multiple facilities have become available: realtime navigation with GPS, applications for ordering taxies, for scheduling events, office functionalities (e.g. document viewing and editing, e-mails), etc. Most of these applications run on devices with touch screens, so they tend to have simple and user-friendly interfaces. But if we look at some applications that require completing many fields inside forms, or perform complex actions that require multiple inputs from the user, we could harder adapt them to the small screen of a mobile device.

In this case, the adaptation of complex applications like Internet banking can be achieved by extending the touch screen interface of the mobile device with a voice recognition interface. Therefore, instead of completing many fields and finding the submenu of payments, users can expresses their desires in oral form (e.g., "Go to the payments screen." or "Make a payment of 100 EURO to Mother. Confirm."). Also, receiving a sound feedback from the application would increase interactivity.

In this paper we propose a system built to integrate voice recognition models, as to obtain better and more accurate results. The final framework contains two different

G. Agre et al. (Eds.): AIMSA 2014, LNAI 8722, pp. 222–227, 2014.

models that are used separately to match a spoken input command to an application state and to perform a specific action. After describing the overall architecture and the two integrated models, the forth section is focused on experimental results that support our approach, while the last section is centered on conclusions and future improvements.

2   State of the Art

Before applying complex processing models to the input text, basic preprocessing such as tokenization, stop words removal, stemming, lemmatization and spelling correction [1] needs to be applied. In many cases, due to the lack of sound quality and environmental noise, the obtained text transcriptions is not that good; therefore, using a proper processing pipeline is essential. This section focuses on presenting the methods used to assign actions from the application space to a user's input command.

Firstly, Pucella and Chong [2] provided an initial framework suitable for assigning user commands to an action based application. The main problem with their approach is that it turned out to be unfeasible to implement, as it is very hard to assign categories to every element from an input phrase, due to the complexity of the vocabulary. Nevertheless, the main steps of processing an user command in natural language can be generalized and applied to subsequent approaches:

1. *Parsing*: use categorical grammars to test whether the user phrase is or not well formed; for each atom of the phrase, a category is associated from the lexicon and the final parsing structure of the sentence is computed; at this stage, the user interface is divided into constants and predicates;
2. *Evaluation*: evaluate the command using a Lambda-calculus model capable to match the input to a state and an associated action from the application model;
3. *Reporting*: Report the result to the application interface, such that the user will check in real time whether his command executed successfully or not.

Secondly, Li, Mclean, Bandar, O'Shea and Crockett [3] proposed a model that focuses on sentence similarity using word semantic information and the order in which they appear. Unlike Latent Semantic Analysis [4] that extracts semantic relations from a large and sparse term-document matrix, their approach concentrates on small texts and sentences rather than big documents, which is better suited for the problem at hand. More specifically, for every two phrases from the input set, the model composes two *semantic word vectors* alongside the two corresponding *word order vectors* or feature vectors that reflect words occurring in the sentence pair. The semantic vector is extended with the general meaning of the whole phrase/corpus and the model computes a semantic similarity along with an order similarity, later on to combine them as a weighted average in order to obtain a general sentence similarity measure. Within our implemented model we opted to experiment different semantic similarity functions between two individual words [5].

3   Model Description

This section describes the voice recognition model implemented in a real application. The main focus in the platform was to build a model that can be further developed with voice recognition algorithms, as to enhance its accuracy. The current version integrates two voice recognition models: one specifically built for this problem and one derived from the previous sentence similarity model.

In general, a form-based application must have an interface with clear and intuitive text labels, which can be easily selected by the user, reducing in the end the cognitive load. The same principles are also applicable when it comes to voice recognition. The input commands must be as simple as possible and the user interface must expose a clear set of instructions available for execution. In this context, each action from a screen can be associated with a main verb and a main object that the verb refers to. Along with a verb (or a list of verbs) and a matching object, a predefined screen action must have as result a unique screen action identifier that signals to the client application what action must be performed. A set of predefined actions for the user interface can be observed in Fig. 1.

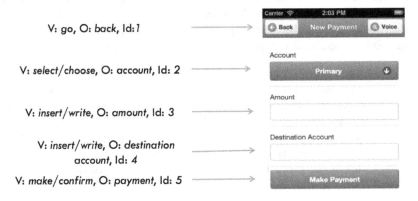

Fig. 1. Assigning verbs (V) and objects (O) to elements from the user interface

If, for example, the model received the text "*Go back*", it would match the first command in the current screen and it would return to the interface with identifier "*1*". The interface, after receiving an identifier, should use an internal algorithm aligned to the input types. If the received identifier is a button, the model will programmatically execute the action assigned to that specific button; if it is an input type, the model will listen for another voice input to complete the text, and in case of a select input identifier, a voice command to select the given value is required. In this manner, the application logic implementation is separated from the voice recognition interface.

In terms of implementation, the mobile client is responsible for invoking the Nuance Speech to Text module [6] and sending the text to the voice recognition model. The client will receive from the model an action identifier that must be interpreted as described previously, or it will receive a parse error that must be outputted to the device's loudspeaker.

The model uses the input text received from the client and finds for the current screen (sent from the mobile client) the matched action using a predefined list of verbs and objects. To find the phrase's verb, the underlying algorithm comprises of the following steps:

1. Build the phrase's dependency graph [7];
2. Build a recursive function on this graph in order to determine the central verb – induced action;
3. Lemmatize the verb;
4. Compare the verb to all of the verbs that exist in the current screen (using Wu-Palmer, Lin and path length [5]);
5. Iterate through the previous list and try to match using similarity measures the objects with any other words that appear in the dependency sub-graph of the current phrase's verb;
6. Retrieve the most probable action and execute if overall reliability is above a pre-imposed threshold.

However, it is very hard to define a sufficient number of predefined actions to consider all the possible input commands. Another problem arises from the fact that the same action can be expressed in more complex ways than a simple verb and object: for example "Go back" and "Application, change the page where I was before". The solution is therefore to integrate additional voice recognition models and facilities. The current recognition system receives simple input commands, which can be seen as small sentences, and thus some phrases associated with each command can be predefined (later on extended in an automatic manner).

Our optimization considers extracting the contextual possible sentences for the current application screen along with their command ids. For each screen and possible command, a large number of instructions can be defined to better describe that specific action. When the engine receives a command, it extracts all the possible sentences for the current page, it finds the most probable sentence over a predefined threshold, and returns its associated action identifier.

We have presented in the third section two different methods that return an action id starting from a given command. Overall, the methods are complementary one to another as their main focuses are different (identification of verb and corresponding object, respectively pattern matching on given instructions), but they enlarge the horizon of searching feasible actions to be performed. Because the verb/object model does not return a score, but a true/false response, a boolean response was enforced for the second model, as well; the final combination of partial results uses the following logic:

- If the models return the same action id, the specific action it is returned;
- If only one of the models found an action id, the available action is returned;
- If the models return two different action ids, the input command is considered unclear and the user is requested to repeat the command.

By executing the two models in parallel a faster response was obtained, a critical criterion in such a system, as the user should receive feedback from the application almost immediately. Moreover, the overall accuracy of the system was increased.

4   Results

In order to better understand how the current model can be applied to a real life situation, we opted to focus on a concrete scenario tightly connected to the introduction – a mobile banking application. For example, if a user wants to make a small payment, he/she must issue a command like 'Make a new payment', followed by 'Amount' and 'Select account'; in the end, the user would say 'Make the payment' in order to complete the transaction. Such a scenario was given to the initial testers of the application to notice how many of them would manage to control the application using simple voice commands.

The first tests were made on a small set of people (10 users) with no previous knowledge of the predefined actions and screens, that were asked to just open the application and follow the scenario. We observed that by using a simple interface with suggestive labels, they got used to the voice command interpreter quite easily as a natural representation of desired action. Also, users were guided as the interface outputted all the possible actions when first entering a screen. All initial testers reacted positively to the natural language extension of the application, and moreover many of them said that they would use it in their real life when performing other concomitant complex actions such as driving their car or riding a bike. Only a few users started by inputting complex commands such as 'Make a payment of 1000 to the account RO30XX1234', that are not supported in the current version, but, after seeing that the interface responds to simple commands, they got used to them quickly and started to effectively exploit the application.

Overall, only 3 of the 10 alpha testers managed to fully complete the test scenario. This is not a very good result, but the conclusions taken from the experiment will certainly help us improve the model. The precision of the system was around 70% for the input commands issued by the previous users. Many errors came from speech-to-text recognition, mostly because of environmental noise and bad pronunciations due to a non-native speaker. In conclusion, the positive reactions of the first testers approved the future applicability of our model. The key is to clearly define the user interface along with its possible actions, and to generate as many expressions as possible, task that can be automated as well.

5   Conclusions and Future Work

Enabling complex form based applications with natural language processing capabilities can increase user interactivity while performing other complex actions like driving or bicycling. Our method simplifies the interaction with an application with many complex forms, but can be used as well to support people with disabilities (mainly visually impaired) to interact with such systems.

From a different point of view, the most important aspects to consider when developing a framework used directly by users are its responsiveness, accuracy and simplicity. The current architecture, the proposed evaluation models and the received user feedbacks support the previous evaluation dimensions. Moreover, the extensibility towards other languages is limited only by the required integration of additional dependency parsing libraries and lexicalized ontologies.

The next step in extending the system will be to integrate it with a complex discourse analysis model such as the cohesion graph implemented in *ReaderBench* [8, 9] developed within our research group, and try to add more interactivity to the system such that the user will better interact with the mobile device. Moreover, we will try to improve the complexity of the input phrases as to include more than one simple command.

Acknowledgements. The undergone research was proposed and partially supported by the Romanian company Advahoo Business Software Solutions that desired to extend a mobile banking application with a voice recognition model and by the Sectoral Operational Programme Human Resources Development 2007-2013 of the Ministry of European Funds through the Financial Agreement POSDRU/159/1.5/S/134398.

References

1. Jurafsky, D., Martin, J.H.: An introduction to Natural Language Processing. Computational linguistics, and speech recognition. Pearson Prentice Hall, London (2009)
2. Pucella, R., Chong, S.: A Framework for Creating Natural Language User Interfaces for Action-Based Applications. In: 3rd Int. AMAST Workshop on Algebraic Methods in Language Processing, TWLT Report 21, pp. 83–98 (2003)
3. Li, Y., Mclean, D., Bandar, Z.A., O'Shea, J.D., Crockett, K.: Sentence similarity based on semantic nets and corpus statistics. IEEE Transactions on Knowledge and Data Engineering 18(8), 1138–1150 (2006)
4. Landauer, T.K., Dumais, S.T.: A solution to Plato's problem: the Latent Semantic Analysis theory of acquisition, induction and representation of knowledge. Psychological Review 104(2), 211–240 (1997)
5. Budanitsky, A., Hirst, G.: Evaluating WordNet-based Measures of Lexical Semantic Relatedness. Computational Linguistics 32(1), 13–47 (2006)
6. Nuance Communications: Dragon Software Developer Kits, http://www.nuance.com/for-developers/dragon/index.htm
7. Toutanova, K., Klein, D., Manning, C.D., Singer, Y.: Feature-Rich Part-of-Speech Tagging with a Cyclic Dependency Network. In: HLT-NAACL 2003, pp. 252–259. ACL, Edmonton (2003)
8. Dascălu, M.: Analyzing Discourse and Text Complexity for Learning and Collaborating. SCI, vol. 534. Springer, Heidelberg (2014)
9. Dascalu, M., Dessus, P., Bianco, M., Trausan-Matu, S., Nardy, A.: Mining texts, learners productions and strategies with ReaderBench. In: Peña-Ayala, A. (ed.) Educational Data Mining. SCI, vol. 524, pp. 345–377. Springer, Heidelberg (2014)

Towards Management of OWL-S Effects by Means of a DL Action Formalism Combined with OWL Contexts

Domenico Redavid, Stefano Ferilli, and Floriana Esposito

Dipartimento di Informatica - Università di Bari Aldo Moro, Bari 70126, Italy
{redavid,ferilli,esposito}@di.uniba.it

Abstract. The implementation of effective Semantic Web Services (SWS) platforms allowing the composition and, in general, the orchestration of services presents several problems. Some of them are intrinsic within the formalisms adopted to describe SWS, especially when trying to combine the dynamic aspect of SWS effects and the static nature of their ontological representation in Description Logic (DL). This paper proposes a mapping of OWL-S with a DL action formalism in order to evaluate executability and projection by means of the notion of Contexts.

1 Introduction

OWL-S[1] is an OWL ontology that enables semantic description of Web services. One of its purposes is the automation of four use cases, namely Discovery, Selection, Composition and Invocation. In OWL-S, preconditions and effects are represented with logical formulas. These formulas cannot always be translated into OWL DL without losing some of their original semantics. The reasons are manifold. Firstly, the OWL-S specification enables to describe such logical formulas with different languages. Secondly, incompatibility between those languages and Description Logic (DL) makes not feasible the translation between candidate languages into DL themselves. Last, but not least, even if we restrict the representation to formalisms which are compatible with DL, the dynamicity of effects is not expressible natively in DL. In particular, the language at the state of the art that offers a greatest level of compatibility with DLs is the Semantic Web Rule Language (SWRL) [1] in its decidable fragment [2]. The adoption of this formalism for encoding preconditions and effects (partially) solves the first two compatibility issues mentioned above. However, the problem of managing effects without breaking DL semantics remains open. The issues are of ontological nature because DLs are monotonic languages. Although some DLs extensions could be used to deal with the dynamic aspects mentioned above, they are not included in the current OWL specifications. Moreover, DL monotonicity rules out any retraction primitives, i.e., there is no way of removing an axiom from a knowledge base, making it hard to deal with knowledge that changes over time. The consequences vary from wrong results for queries to inconsistencies in the knowledge base. This happens because the effect of a service is intended as an *alteration* of the state of the world. It is quite obvious that the new information should replace the old one, rather than just be added to the knowledge base.

---

[1] OWL for Services. http://www.w3.org/Submission/OWL-S/

G. Agre et al. (Eds.): AIMSA 2014, LNAI 8722, pp. 228–235, 2014.

2   OWL-S and SWRL Characteristics

OWL-S enables semantic descriptions of Web services using the *Service Model* ontology, which defines the OWL-S process model. Each process is based on the IOPR (Inputs, Outputs, Preconditions, and Results) model. *Inputs* represent the information required for the execution of the process. *Outputs* represent the information the process returns to the requester[2]. *Preconditions* are conditions imposed on *Inputs* that have to hold in order to invoke the process in a correct manner. Since an OWL-S process may have several results with corresponding outputs, the *Results* provide a mean to specify this situation. Each result can be associated to a result condition, called *inCondition*, which specifies when that particular result can occur. It is assumed that such conditions are mutually exclusive, so that only one result can be obtained for each possible situation. When an *inCondition* is satisfied, there are properties associated with this event that specify the corresponding output and, possibly, the *Effects* produced by the execution of the process. The OWL-S conditions (*Preconditions*, *inConditions* and *Effects*) are represented as logical formulas. Since OWL-DL offers limited support to formulate constructs like property compositions without becoming undecidable, a more powerful language is required for the representation of OWL-S conditions. One of the proposed languages is Semantic Web Rule Language (SWRL) [1]. Although SWRL is undecidable, a solution has been proposed in [2] where decidability is achieved by restricting the application of SWRL rules only to the individuals explicitly introduced in the *ABox*. This kind of SWRL rules, called DL-safe, makes this language the best candidate to describe OWL-S conditions [3]. Let us now briefly mention the characteristic of SWRL that are relevant to our scope. SWRL extends the set of OWL axioms to include *Horn-like* rules in the form of implications between an antecedent (body) and consequent (head), both consist of zero or more conjunctive atoms having one of the following forms:

- $C(x)$, with C an OWL class, $P(x, y)$, with P an OWL property,
- *sameAs*(x, y) or *differentFrom*(x, y), equivalent to the respective OWL properties,
- $builtIn(r, z_1, \ldots, z_n)$, functions over primitive datatypes.

where x, y are variables, OWL individuals or OWL data values, and r is a built-in relation between $z_1, \ldots, z_n$ (e.g., $builtIn(greaterThan, z_1, z_2)$). The intended meaning can be read as: whenever the conditions specified in the antecedent hold, then the conditions specified in the consequent hold also. A rule with conjunctive consequent can be transformed into multiple rules by means of Lloyd-Topor transformations. Each rule has an atomic consequent.

3   Action Formalism for OWL-S

The OWL-S composition methodology based on SWRL DL-safe rules presented in [3] explain the procedure to encode the OWL-S services process model by means of the following (abstract) SWRL rule:

---

[2] Inputs, Outputs and Local variables (entities used within the process) are SWRL variables and their types are defined in the domain ontology.

$$Preconditions \wedge inCondition \rightarrow \{output\} \wedge Effect$$

If the service has more *Results*, multiple rules having different *inCondition, output* and/or *Effect* are used. In order to evaluate executability and projection we propose to map this abstract rule with the action formalism proposed in [4] based on $\mathcal{ALCQIO}$, an OWL-DL fragment. In detail, given:

- N_X and N_I disjoint and countably infinite sets of variables and individual names;
- an acyclic TBox $\mathcal{T}$;
- a set of primitive literals for $\mathcal{T}$ corresponding to the ABox assertions $A(a)$, $\neg A(a)$, $r(a,b)$, $\neg r(a,b)$, with A primitive concept in $\mathcal{T}$, r a role name, $a,b \in N_I$.

An atomic action is defined as $\alpha = (pre; post)$ where:

- pre is a finite set of ABox assertions, the preconditions;
- $post$ is a finite set of conditional postconditions of the form φ/ψ, where φ is an ABox assertion and ψ is a primitive literal for $\mathcal{T}$.

An operator for $\mathcal{T}$ is a parametrised atomic action for $\mathcal{T}$, i.e., an action in which some variables from N_X may occur in place of individual names. The postconditions of the form $\top(t)/\psi$ are called unconditional and denoted just by ψ. For the sake of simplicity, we consider an OWL-S atomic service with an arbitrary number of preconditions and results, with a single effect[3]. This implies that there are as many *inConditions* as *Effects*. Supposing that *inConditions* and *Effects* are formed by single SWRL atoms, the proposed mapping with the DL action formalism is:

$$pre : \{pre_1(at_1), \dots, pre_1(at_l), pre_2(at_1), \dots, pre_2(at_m), pre_t(at_1), \dots, pre_t(at_n)\}$$
$$post: \{inCond_a/Effect_a, \dots, inCond_z/Effect_z\}$$

with $pre_t(at_n)$ the n-th atom of the t-th OWL-S precondition, and $inCond_z/Effect_z$ the z-th conditional post-condition. The traceability amongst the atoms in the action pre and the service preconditions is guaranteed by the corresponding SWRL rule. The formalism in [4] allows to encode only simple inCondition and Effect of the form specified above ($A(a)$, $\neg A(a)$, $r(a,b)$, $\neg r(a,b)$).

4   A Mapping Example

To describe the proposed mapping, we use a simplified version of the OWL-S Atomic Service *ExpressCongoBuy*[4], based on a subset of inputs (renamed here to improve the readability of service conditions), and with simplified *inConditions* and *Effects* so that the limitation imposed by the definition of action in [4] are respected. The service features are described in Figure 1,a); the SWRL rules representing the service are depicted in Figure 1,b); finally, the resulting service described in terms of actions is described in Figure 1,c). When OWL-S atomic services present a single result, the inCondition can be omitted. In this case the service effect will be an unconditional postcondition.

---

[3] Outputs are not logical conditions and can be omitted because do not generate KB variations. Generally, built-in atoms are not a single OWL class or property; they are left as future work.

[4] http://www.ai.sri.com/daml/services/owl-s/1.2/CongoProcess.owl

---

a) ExpressCongoBuy Service Characteristics

INPUTS:

	Type
x : $ECBCreditCardType$	$CreditCardType$
y : $ECBBookISBN$	$profileHierarchy : Book$
z : $ECBSignInInfo$	$SignInData$
t : $ECBCreditCardNumber$	$xsd : decimal$

OUTPUT:

o : $ECBOutput$	$ECBOutputType\,[ECBOT]$

LOCAL variables (SWRL variables):

u : $ECBAcctID$	$AcctID\,[AID]$
v : $ECBCreditCard\,[ECBCC]$	$CreditCard$

PRECONDITIONS:

$\quad$ precond$_1$: $validity(?v, Valid) \wedge cardNumber(?v, ?t)\,[V(?v, Valid) \wedge CN(?v, ?t)]$

$\quad$ precond$_2$: $hasAcctID(?z, ?u)\,[hasAID(?z, ?u)]$

RESULT 1:

$\quad$ inCondition : $OutOfStockBook(?y)\,[OSBook(?y)]$

$\quad\quad$ Effect : $FailureNotification(?o)\,[FN(?o)]$

RESULT 2:

$\quad$ inCondition : $InStockBook(?y)\,[ISBook(?y)]$

$\quad\quad$ Effect : $OrderShippedAcknowledgement(?o)\,[OSA(?o)]$

* The namespace is the base one for the service. $ExpressCongoBuy$ is shortened to ECB for readability. Further shortenings are in [...].

b) The service described with SWRL rules

1 : $[inputs] \wedge hasAID(?z, ?u) \wedge CN(?v, ?t) \wedge V(?v, Valid) \wedge safe(?y, ?o)$
$\quad \wedge OSBook(?y) \rightarrow ECBOT(?o) \wedge FN(?o)$

2 : $[inputs] \wedge hasAID(?z, ?u) \wedge CN(?v, ?t) \wedge V(?v, Valid) \wedge safe(?y, ?o)$
$\quad \wedge ISBook(?y) \rightarrow ECBOT(?o) \wedge OSA(?o)$

*$safe$ is an always true property needed to guard the SWRL safety condition.

c) The resulting service described in terms of actions

pre : $\{\, CN(?v, ?t), ECBCC(?v, Valid), hasAID(?z, ?u)\,\}$

post : $\{\, holds(?y, ?o), OSBook(?y)\,/\,FN(?o), ISBook(?y)\,/\,OSA(?o)\,\}$

Fig. 1. ExpressCongoBuy mapping with Action formalism

The proposed mapping allows to exploit the theoretical results about projection and executability working in DL. In detail, executability and projection together grant that a composite action, i.e., a service or sequence of services, can be executed to completion, thus obtaining the original goal. However, this is subject on each action in the sequence being *consistent* with the knowledge base as it evolves. When this is not the case, in order to obtain the goal, either a new plan must be computed, or a solution to inconsistent actions must be designed, so that the knowledge base can be modified to be consistent with the actions. Let us introduce an example of inconsistency due to the *Effects* of rules being applied, based on the service in Figure 1. The service is invoked twice, with ?o matched to an individual, i_o. The first invocation generates *OrderShippedAcknowledgement(i_o)* assertion, while the second generates *FailureNotification(i_o)*. Both assertions end up in the knowledge base. If these two classes are disjoint in the ontology defining them, the two type of assertions would force i_o to belong to a class and its complement, therefore making the knowledge base inconsistent. This is an example of the kind of inconsistencies arising from lack of retraction primitives that this paper aims at addressing.

5 Contexts for Incompatibility Management

In this work, the notion of *contexts* is used to justify a mechanism able to handle *actions* which are inconsistent with respect to a knowledge base. Recalling the definition 2 in Milicic et al. [4], given $\mathcal{T}$ an acyclic TBox, $\alpha = (pre, post)$ an atomic action for $\mathcal{T}$, and $\mathcal{I}, \mathcal{I}'$ models of $\mathcal{T}$ respecting the unique name assumption (UNA) and sharing the same domain and interpretation of all individual names. We say that α may transform

$\mathcal{I}$ to $\mathcal{I}'$ ($\mathcal{I} \Rightarrow_{\alpha}^{\mathcal{T}} \mathcal{I}'$) iff, for each primitive concept A and role name r, a change of the interpretation is defined as follows:

$$A^{\mathcal{I}'} := (A^{\mathcal{I}} \sqcup A^{\mathcal{I}'}_{A(a)}) \setminus A^{\mathcal{I}'}_{\neg A(a)}, r^{\mathcal{I}'} := (r^{\mathcal{I}} \sqcup r^{\mathcal{I}'}_{r(a,b)}) \setminus r^{\mathcal{I}'}_{\neg r(a,b)}$$

where:

$$
\begin{aligned}
A^{\mathcal{I}'}_{A(a)} &= \{a^{\mathcal{I}} \mid \varphi/A(a) \in post \wedge \mathcal{I} \models \varphi\} \\
A^{\mathcal{I}'}_{\neg A(a)} &= \{a^{\mathcal{I}} \mid \varphi/\neg A(a) \in post \wedge \mathcal{I} \models \varphi\} \\
r^{\mathcal{I}'}_{r(a,b)} &= \{(a^{\mathcal{I}}, b^{\mathcal{I}}) \mid \varphi/r(a,b) \in post \wedge \mathcal{I} \models \varphi\} \\
r^{\mathcal{I}'}_{\neg r(a,b)} &= \{((a^{\mathcal{I}}, b^{\mathcal{I}}) \mid \varphi/\neg r(a,b) \in post \wedge I \models \varphi\}
\end{aligned}
$$

The composite action $\alpha_1, \ldots \alpha_k$ may transform $\mathcal{I}$ to $\mathcal{I}'$ ($\mathcal{I} \Rightarrow_{\alpha_1,\ldots,\alpha_k}^{\mathcal{T}} \mathcal{I}'$) iff there are models $\mathcal{I}_0, \ldots, \mathcal{I}_k$ of $\mathcal{T}$ with $\mathcal{I} = \mathcal{I}_0, \mathcal{I}' = \mathcal{I}_k$, and $\mathcal{I}_{i-1} \Rightarrow_{\alpha}^{\mathcal{T}} \mathcal{I}_i$ for $1 \le i \le k$.

This definition does not cover the situation in which a postcondition generates an inconsistency due to the lack of retraction primitives, as in the example presented in Sect. 4. OWL contexts are introduced as a possible solution to this kind of problems. In particular, by identifying contexts with the portions of ABox it is possible to manage this kind of inconsistencies without modifying the definition of interpretation change. This is done by defining *context relations* that enable a dynamic partitioning of the knowledge base, simulating the retraction of conflicting assertions. The proposed solution is inspired to the work presented in [5], even though the work proposed in [6], if suitably adapted for the contexts, could be a valuable alternative. For our aims, the *content* of a context is a set of complete OWL axioms.

Definition 1 (Content of an OWL context). *Given a context C_{TX}, a signature $S = C \cup R \cup I$ where C are* concept names, *R are* role names, *and I are* individuals, *the content of C_{TX} consists of the union of:*

- *a TBox $\mathcal{T}_C$ whose concepts and nominals are included in $C \cup I$;*
- *a RBox $\mathcal{R}_C$ whose roles are included in R;*
- *an ABox $\mathcal{A}_C$ whose individuals are included in I.*

$\mathcal{T}_C, \mathcal{R}_C$ and $\mathcal{A}_C$ are expressed in OWL.

The *parameters* representation adopted is the one outlined in [5]:

Definition 2 (Parameter or Contextual Relation). *Given a context C_{TX}, a parameter P and a value X for P, the relation $P(C_{TX}, X)$ associates C_{TX} with X and is represented with the triple (C'_{TX}, P', X') where C'_{TX} and P' are URIs assigned to C_{TX} and P respectively, and X' is a RDF node identifying a resource, another context or a datatype value.*

Two *parameters* or *contextual relations* (CR) are defined as:

Definition 3 (EXTENDS Contextual Relationship). *Given contexts C_{TX1} and C_{TX2}, the expression C_{TX2} EXTENDS C_{TX1} is interpreted as:*
the content of C_{TX2} includes the content of C_{TX1}, and therefore transitively the content of any context D_i that C_{TX1} is declared to EXTEND.

Definition 4 (INCOMPATIBLE Contextual Relationship). *Given contexts C_{TX1} and C_{TX2}, the expression C_{TX2} INCOMPATIBLE C_{TX1} is interpreted as: called $\mathcal{K}$ the knowledge base containing the content of C_{TX1} and C_{TX2}, $\mathcal{K}$ is inconsistent. The INCOMPATIBLE relation is symmetric: C_{TX2} INCOMPATIBLE C_{TX1} is equivalent to C_{TX1} INCOMPATIBLE C_{TX2}.*

These definitions allow to describe a situation where two contexts C_{TX2} and C_{TX3} extend a third context C'_{TX1} in two different, incompatible ways, i.e., C_{TX3} contains $\neg A(x)$ and C_{TX2} contains $A(x)$ (where A is defined in $\mathcal{T}_{C_{TX3}}$ and x is a named individual belonging to $\mathcal{A}_{C_{TX1}}$ and $\mathcal{A}_{C_{TX2}}$, but not necessarily to $\mathcal{A}_{C_{TX3}}$). C_{TX2} and C_{TX3} are incompatible, since any knowledge base containing both $\neg A(x)$ and $A(x)$ will be inconsistent. With reference to the definition for change of interpretation given at the beginning of this section, C_{TX2} and C_{TX3} correspond to $A^{\mathcal{I}'}_{A(a)}$ and $A^{\mathcal{I}'}_{\neg A(a)}$, and therefore implement the operator for the change of interpretation. This implies that the change needed to fulfil the goal of the action sequence (i.e. the right context and therefore the right change of interpretation) is chosen as the actual state of the knowledge base. It is straightforward that a means to know when to apply contextualization is needed. A possible strategy can be summarized as follows. We keep track of the current model and use a heuristic function $t(e)$ (where e is a type or role assertion) to estimate whether the change produced by e affects a constrained part of the model, therefore possibly producing inconsistencies, or an unconstrained part, in which case the change can be applied safely.

6 Executability and Projection Evaluation by Means of Contexts

This section presents the algorithm for verifying the validity of a sequence of services. Suppose that we have a set of sequences of SWRL rules that achieve an input goal. Then the procedure can be summarized as follows:

- For each sequence S_i, every SWRL rule is mapped with an action, as described in Sect. 3, resulting in the sequence of actions AS_i;
- *Executability* and *projection* are then used to determine whether the sequence of actions $AS_i = \{\alpha_1, \ldots, \alpha_n\}$ is executable and produces the required effect.

Assuming that the sequence AS_i is executable in $\mathcal{A}$ w.r.t. $\mathcal{T}$, the *projection* verifies that the assertion φ is a consequence of applying $\alpha_1, \ldots, \alpha_n$ in $\mathcal{A}$ w.r.t. $\mathcal{T}$ iff for all the models $\mathcal{I}$ of $\mathcal{A}$ and $\mathcal{T}$ and for all $\mathcal{I}'$ with $\mathcal{I} \Rightarrow^{\mathcal{T}}_{\alpha_1, \ldots, \alpha_n} \mathcal{I}'$, $\mathcal{I}' \models \varphi$.
Consider a sequence S containing an action α_i, e.g., the action described in Figure 1. For this action, the postcondition φ_i/ψ_i is $OSBook(?y) \, / \, FN(?o)$.
When the projectability verification reaches α_i, it is evaluated according to the heuristic $t(e)$ delineated in Sect. 5. If t does not detect the need for a contextualization of the knowledge base, i.e., no potential inconsistency is detected, the projection can continue to $i+1$. When, instead, a possible inconsistency is detected by t, e.g., because $OSA(?o)$ has been asserted previously by an action α_j with $j < i$, the proposed contextualization will be applied, creating alternate ABoxes for the conflicting assertions, while the TBox will be left untouched. With reference to the example given in sect. 4, the Knowledge base at $i - 1$: $\mathcal{K}_{i-1} = \mathcal{T} \sqcup C_{i-1}$ produce the following contexts and context relations:

- Content of context $C_{TXi-1} = \mathcal{A} \sqcup \{OSA(?o)\}$
- Content of context $C'_{TXi-1} = C_{TXi-1} \setminus \{OSA(?o)\}$
- Content of context $C'_{TXi} = \{OSA(?o)\}$
- Content of context $C''_{TXi} = \{FN(?o)\}$
- C'_{TXi} EXTENDS C'_{TXi-1}
- C''_{TXi} EXTENDS C'_{TXi-1}
- C'_{TXi} INCOMPATIBLE C''_{TXi}

If the projection requires one or more contextualizations, it is necessary to verify the executability in the new contexts. Since C''_{TXi} is the new KB for the sequence $\alpha_i, \ldots, \alpha_n$, we need to verify whether the preconditions belonging to $\alpha_{i+1}, \ldots, \alpha_n$ actions hold in C''_{TXi}. This is equivalent to split in two parts the actions sequence: the executability of the new sequence $S'' = \{\alpha_{i+1}, \ldots, \alpha_n\}$ must be verified against $\mathcal{A}'$ w.r.t. $\mathcal{T}$, where $\mathcal{A}' = C''_{TXi}$. If this new sequence is not executable, the original sequence S will not be considered valid, since it cannot be automatically executed. In fact, its execution would either result in an inconsistent knowledge base (if executed without contextualization) or some preconditions would not be verified when the services mapped by S' are executed, resulting in unreliable behavior of the system.

7   Related Work

In [7], the notion of contexts is used for the operation of SWS composition. The adopted meaning of context, i.e. any information that can be used to characterize the situation of an entity, comes from context-aware computing [8]. The dynamic description logic adopted in [9] uses static and dynamic context information in order to compose Web services adapting user, provider and broker contexts. A dynamic DL knowledge base has, in addition to TBox and ABox, an ActionBox that contains assertions about actions. The reasoning tasks about actions proposed, i.e. executability and projection, are based on [4]. However, all the approaches above do not provide a concrete mapping with OWL-S services, and the management of non-monotonic problematic situations that can be caused by service *Effects* is missing as well. A representative work on OWL contexts is [10]. It proposes a modification of OWL semantics in order to contextualize ontologies, and formalizes a particular type of rules, called bridge-rules, allowing the creation of explicit mappings for the contexts management. Furthermore, it handles, among the others, the management of localized inconsistency and its propagation in this type of contexts. The OWL contexts we proposed in this paper do not require modifications to OWL semantics, nor the use of rules for their management.

8   Conclusions

In this paper we analyzed the current limitations that prevent the implementation of an OWL-S based service management platform. We observed that even removing or mitigating the incompatibility between the languages for specifying domain ontologies and those for annotating services that operate in such domains, the problem of static ontology against the dynamic services remains. We proposed the adoption of a theoretical framework that accomodates dynamicity inside DL [4]. However, such framework requires the implementation of non standard operations on the interpretations of Knowledge base. In particular, these operations allow for overcoming the absence of retraction primitives and avoiding inconsistencies due to DLs monotonicity property. We describe a possible solution based on the notion of contexts for an OWL knowledge base. As future work, we plan the integration of SWRL built-in atoms and design of a concrete implementation of heuristic function.

Acknowledgments. This work fulfills the objectives of the PON 02_00563_3489339 project "Puglia@Service - Internet-based Service Engineering enabling Smart Territory structural development" funded by the Italian Ministry of University and Research (MIUR).

References

1. Horrocks, I., Patel-Schneider, P.F., Bechhofer, S., Tsarkov, D.: OWL rules: A proposal and prototype implementation. J. of Web Semantics 3, 23–40 (2005)
2. Motik, B., Sattler, U., Studer, R.: Query answering for owl-dl with rules. Journal of Web Semantics: Science, Services and Agents on the World Wide Web 3, 41–60 (2005)
3. Redavid, D., Ferilli, S., Esposito, F.: Towards dynamic orchestration of semantic web services. T. Computational Collective Intelligence 10, 16–30 (2013)
4. Milicic, M.: Planning in action formalisms based on dls: First results. In: Calvanese, D., Franconi, E., Haarslev, V., Lembo, D., Motik, B., Turhan, A.Y., Tessaris, S. (eds.) Description Logics. CEUR Workshop Proceedings, vol. 250. CEUR-WS.org (2007)
5. Stoermer, H., Bouquet, P., Palmisano, I., Redavid, D.: A context-based architecture for RDF knowledge bases: Approach, implementation and preliminary results. In: Marchiori, M., Pan, J.Z., de Sainte Marie, C. (eds.) RR 2007. LNCS, vol. 4524, pp. 209–218. Springer, Heidelberg (2007)
6. Rosati, R.: On the complexity of dealing with inconsistency in description logic ontologies. In: Walsh, T. (ed.) Proceedings of the 22nd International Joint Conference on Artificial Intelligence, IJCAI 2011, Barcelona, Catalonia, Spain, July 16-22, pp. 1057–1062. IJCAI/AAAI (2011)
7. Niu, W., Shi, Z., Chang, L.: A context model for service composition based on dynamic description logic. In: Shi, Z., Mercier-Laurent, E., Leake, D. (eds.) Intelligent Information Processing IV. IFIP, vol. 288, pp. 7–16. Springer, Boston (2008)
8. Schilit, B.N., Adams, N., Want, R.: Context-aware computing applications. In: Proceedings of the Workshop on Mobile Computing Systems and Applications, pp. 85–90. IEEE Computer Society (1994)
9. Chang, L., Shi, Z., Qiu, L., Lin, F.: Dynamic description logic: Embracing actions into description logic. In: Calvanese, D., Franconi, E., Haarslev, V., Lembo, D., Motik, B., Turhan, A.Y., Tessaris, S. (eds.) Description Logics. CEUR Workshop Proceedings, vol. 250. CEUR-WS.org (2007)
10. Serafini, L., Borgida, A., Tamilin, A.: Aspects of distributed and modular ontology reasoning. In: Kaelbling, L.P., Saffiotti, A. (eds.) IJCAI, pp. 570–575. Professional Book Center (2005)

Computational Experience with Pseudoinversion-Based Training of Neural Networks Using Random Projection Matrices

Luca Rubini[1], Rossella Cancelliere[1], Patrick Gallinari[2],
Andrea Grosso[1], and Antonino Raiti[1]

[1] Università di Torino,
Department of Computer Science Turin, Italy
{luca.rubini,rossella.cancelliere,andrea.grosso}@unito.it,
253081@studenti.unito.it
[2] Laboratory of Computer Sciences, LIP6,
Université Pierre et Marie Curie
Paris, France
patrick.gallinari@lip6.fr

Abstract. Recently some novel strategies have been proposed for neural network training that set randomly the weights from input to hidden layer, while weights from hidden to output layer are analytically determined by Moore-Penrose generalised inverse; such non-iterative strategies are appealing since they allow fast learning. Aim of this study is to investigate the performance variability when random projections are used for convenient setting of the input weights: we compare them with state of the art setting i.e. weights randomly chosen according to a continuous uniform distribution. We compare the solutions obtained by different methods testing this approach on some UCI datasets for both regression and classification tasks; this results in a significant performance improvement with respect to conventional method.

Keywords: random projections, weights setting, pseudoinverse matrix.

1 Introduction

Methods based on gradient descent (and among them the large family of techniques based on backpropagation [1]) have largely been used for training of one of the most common neural architecture, the single hidden layer feedforward neural network (SLFN). The start-up of these techniques assigns random values to the weights connecting input, hidden and output nodes; such values are then iteratively modified according to the error gradient steepest descent direction. The main criticisms about gradient descent-based learning are concerned with high computational cost because of slow convergence and zigzagging behavior showed by such methods, and relevant risk of converging to poor local minima on the landscape of the error function [2].

G. Agre et al. (Eds.): AIMSA 2014, LNAI 8722, pp. 236–245, 2014.

The reduction of computational efforts in training is of great interest and may become imperative for learning the kind of complicated high-level relations required e.g. in vision [3,4], natural language processing [5,6], and other typical artificial intelligence tasks.

A wave of interest has recently grown around some non-iterative procedures based on the evaluation of generalized pseudoinverse matrices. The idea of using these appealing techniques, usually employed to train radial basis function neural networks [7], also for different neural architectures was suggested e.g. in [8]. The work by Huang et al. [9] gave rise to a great interest in neural network community, originating many application-oriented studies in the last years devoted to the use of these single-pass techniques, easy to implement and computationally fast; some are described e.g. in [10,11,12,13]. A yearly conference is currently being held on the subject, the International Conference on Extreme Learning Machines (ELM), and the method is currently dealt with in some journal special issue, e.g. Soft Computing [14] and the International Journal of Uncertainty, Fuzziness and Knowledge-Based Systems [15].

In the pseudoinverse framework input weights and hidden neurons biases are selected randomly, usually according to a uniform distribution in the interval $[-1,1]$, and no longer modified, while output weights are analytically determined by a single computation of the Moore-Penrose (MP) generalized inverse. Since incremental adjustment of weights is completely avoided these techniques turn out to be very fast when compared to classical gradient descent approaches; the problem of the possible convergence to poor local minima is handled by repeatedly applying the method with a number of random initializations (multistart), thereby obtaining a sampling "at large" of the landscape of the error function.

This paper proposes an improvement to the state-of-the-art and focuses in initializing input weights and hidden neurons biases with "special" random structures — specifically, random projection matrices. The theoretical rationale for this approach can be found in many studies, showing random projections as a powerful method for dimensionality treatment [16,17,18] thanks to their property to be *almost orthogonal* projections. This feature makes them a potentially useful tool in order to improve performace when dealing with input data relevant features. This argument will be deepened in section 3.

The paper is organized as follows. We recall main ideas on SLFN learning by pseudoinversion in section 2; in section 3 we present fundamentals ideas on random projection and finally in section 4 we report results comparing weights setting.

2   Training by Pseudoinversion

In this section we introduce notation and we recall basic idea concerning the use of generalized inverse for neural training.

Fig. 1 shows a standard SLFN with P input neurons, M hidden neurons and Q output neurons, non-linear activation functions ϕ in the hidden layer and linear activation functions in the output layer.

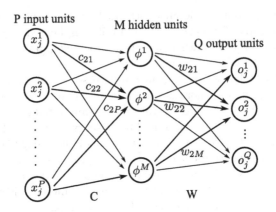

P input units   M hidden units

Q output units

Fig. 1. A Single Layer Feedforward Neural Network

Considering a dataset of N distinct training samples of (input, output) pairs $(\mathbf{x}_j, \mathbf{t}_j)$, where $\mathbf{x}_j \in \mathbb{R}^P$ and $\mathbf{t}_j \in \mathbb{R}^Q$, the learning process for a SLFN aims at producing the matrix of desired outputs $T \in \mathbb{R}^{N \times Q}$ when the matrix of all input instances $X \in \mathbb{R}^{N \times P}$ is presented as input.

As stated in the introduction, in the state of the art pseudoinverse approach input weights c_{ij} (and hidden neurons biases) are randomly sampled from a uniform distribution in a fixed interval and no longer modified.

After having fixed input weights C, the use of linear output units allows to determine output weights w_{ij} as the solution of the linear system $HW = T$, where $H \in \mathbb{R}^{N \times M}$ is the hidden layer output matrix of the neural network, $H = \Phi(XC)$.

Since H is a rectangular matrix, the least square solution W^* that minimises the cost functional $E_D = ||HW - T||_2^2$, as shown e.g. in [19, 20] is:

$$W^* = H^+T. \tag{1}$$

H^+ is the Moore-Penrose generalised inverse (or pseudoinverse) of matrix H.

Direct use of expression (1) is not anyway the best choice because most learning problems are ill-posed; regularisation methods have to be used [21, 22] to turn the original problem into a well-posed one, i.e. roughly speaking into a problem insensitive to small changes in initial conditions. Among them, Tikhonov regularisation is one of the most common [23, 24]: it minimises the error functional

$$E \equiv E_D + E_R = ||HW - T||_2^2 + \lambda ||W||_2^2. \tag{2}$$

With regularisation we introduce a penalty term that not only improves on stability, but also contains model complexity avoiding overfitting, as largely discussed in [25]. Applications to different neural network models are discussed for instance in [26, 27, 28].

If we consider the singular value decomposition (SVD) of H

$$H = U\Sigma V^T, \tag{3}$$

the regularised solution $\hat{W}$ that minimises the error functional (2) has the form (see e.g. [29]):

$$\hat{W} = VDU^TT. \tag{4}$$

$U \in \mathbb{R}^{N \times N}$ and $V \in \mathbb{R}^{M \times M}$ are orthogonal matrices and $D \in \mathbb{R}^{M \times N}$ is a rectangular diagonal matrix whose elements, built using the singular values σ_i of matrix Σ, are:

$$D_i = \frac{\sigma_i}{\sigma_i^2 + \lambda}. \tag{5}$$

Therefore in our work we always utilise regularised pseudoinversion. Input weights setting is discussed in next section.

3   Basic Ideas on Random Projections

If $X_{N \times P}$ is the original set of N P-dimensional observations,

$$X_{N \times K}^{RP} = X_{N \times P} C_{P \times K} \tag{6}$$

is the projection of the data onto the new K-dimensional space.

Strictly speaking, a linear mapping such as (6) is not a projection because C is generally not orthogonal and it can cause significant distortions in the data set. However, and unfortunately, orthogonalizing C is computationally expensive. Instead, we can rely on a result presented by Hecht-Nielsen [30]: in a high-dimensional space, there exists a much larger number of *almost orthogonal* than strictly orthogonal directions. Besides, Bingham and Mannila [31] performed an extensive experimentation which allows them to claim that vectors having random directions might be sufficiently close to orthogonality and equivalently that C^TC would approximate an identity matrix. They estimate the mean squared difference between C^TC and the identity matrix is about $1/K$ per element.

This key idea is confirmed also by the Johnson-Lindenstrauss lemma [32]: if a set of points in a vector space is randomly projected onto a selected space of suitable dimension, then the original distances between the points are approximately preserved in the new space, with only minimal distortions. For a simple proof of this result, see [33]. This property appears to be really appealing because suggests the possibility to preserve the topological structure of the initial input space while allowing the creation of a new optimal data representation in the hidden layer space, able to easy the classification/diagnosis task and to increase performance.

Therefore we can use random projections to project the original P-dimensional data into a K-dimensional space, using a random entries matrix $C_{K \times P}$ whose columns have unit norm.

Besides, random projection is very simple from a computational standpoint: the process of forming the random matrix C and projecting the data matrix X into K dimensions has complexity of order $O(PKN)$; moreover, if the data matrix X is sparse with about G nonzero entries per column, the complexity is of order $O(GKN)$.

Table 1. UCI datasets characteristics

Dataset	Type	N. Instances	N. Attributes	N. Classes
Abalone	Regression	4177	8	-
Cpu	Regression	209	6	-
Delta Ailerons	Regression	7129	5	-
Housing	Regression	506	13	-
Iris	Classification	150	4	3
Wine	Classification	178	13	3
Diabetes	Classification	768	8	2
Landsat	Classification	4435	36	7

Actually, a large variety of zero mean, unit variance distributions of elements c_{ij} result in a mapping that still satisfies the Johnson-Lindenstrauss lemma: among them, entries of C can be randomly sampled from a gaussian distribution. Another appealing possibility is using sparse random projections which have only a small fraction of nonzero elements. For example, Achlioptas [34] shows that generating random entries c_{ij} by

$$c_{ij} = \sqrt{3} \cdot \begin{cases} +1 & \text{with probability } 1/6 \\ 0 & \text{with probability } 2/3 \\ -1 & \text{with probability } 1/6 \end{cases} \tag{7}$$

one obtains a valid random projection with (expected) density 33%.

A difficulty arises because random projections are mainly used for linearly separable tasks although many real world problems are not linearly separable. Neural networks feature among the tools available to deal with the latter class of problems, so we propose to join these techniques using random projections matrices for the setting of input weights while the subsequent processing by hidden nodes nonlinear activation function will account for the non-linearity of the problem.

4   Experimental Investigation

In this section we report results of some numerical experiments performed on the eight benchmark datasets from the UCI repository [35] listed in Table 1, and investigate neural networks with the architecture shown in Fig. 1 and sigmoidal hidden neuron activation functions. The number of input and output neurons is determined by dataset features.

For the sake of comparison input weights are selected according to i) the conventional strategy, where c_{ij} is sampled from a uniform random distribution in the interval $[-1, 1]$, that in the following will be referred to as Unif. or ii) using random projection matrices with elements c_{ij} gaussian distributed, with mean value 0 and variance 1 (in the following referred to as Gauss.), or iii) using sparse random projection matrices with 33% average density. All simulations are carried out in Matlab 7.10 environment.

4.1 Regularisation Parameter Calibration

To determine the regularisation parameter value for the three cases, for each dataset we gradually increase the number of hidden nodes by unit steps in an unregularised framework (eq. (2), $\lambda = 0$); for each selected hidden layer size, average RMSE (for regression tasks), or average misclassification rate (for classification tasks) were computed over 100 different initial trials for each input weight setting, i.e. uniform and gaussian.

All datasets show, after an initial steep decrease, a fast error growth as a function of the hidden layer size, opposite to the monotonically decreasing training error.

This effect is typically caused by overfitting, arising when a large amount of free parameters is available to reproduce almost exactly training data.

The best performance is associated to an interval of hidden neurons, that we name critical dimension, in which we decided to look for, according to a cross validation scheme, the value of λ resulting in the best score: its determination concludes the calibration phase.

4.2 Computational Results

Comparison of the relative strengths of the approaches studied in this work is assessed by evaluation of the mean test error resulting from 100 trials for each fixed size of SLFN in the regularised framework: the test performance is reported in Table 2.

We underline that the regularised test error features a monotonic decrease as a function of hidden neurons number, proving that regularisation is necessary to provide overfitting control, and to allow optimal exploitation of the superior potential of larger architectures.

In the "Error" columns we report the average value (of 100 trials) and standard deviation of the RMSE for regression datasets; for classification datasets, we report average value (of 100 trials) and standard deviation for the percentage misclassification error. On each row, the lowest average error figure is highlighted in bold whenever we can prove a statistically significant dominance of the random-projection initialization over the random-uniform initialization, assessed with at least a confidence level of 95%in the Student's test.

We also report the number of hidden neurons N_H and the value of λ emerged from the calibration phase.

As far as the testing performance is concerned, we can claim a substantial dominance of the random projections based approach over the classical uniform initialization.

We then compared the test performance of networks with initialization based on random projections and trained by pseudoinversion against the test performances of networks trained with a classical backpropagation method. The comparison is shown in Table 3; for each dataset, the "PINV" columns report the error statistics for the winner observed in Table 2. Statistics for backpropagation are taken from `tunedit.org`, except for the Wine dataset, for which we got

Table 2. Random projections vs. random-uniform setting. For Delta Ailerons, the average errors and standard deviations are multiplied by 10^{-4}.

Dataset	Unif. Error Avg	StD	N_H	λ	Gauss. Error Avg	StD	N_H	λ	Sparse Error Avg	StD	N_H	λ
Abalone	2.165	0.004	128	$3\cdot10^{-2}$	2.169	0.009	129	$3\cdot10^{-1}$	**2.162**	0.006	118	$3\cdot10^{-2}$
Mach. Cpu	57.35	1.7	98	$4\cdot10^{-2}$	**56.85**	2.8	61	$8\cdot10^{-1}$	57.86	1.6	89	$5\cdot10^{-1}$
Delta Ail.(10^{-4})	1.636	$2\cdot10^{-3}$	244	$3\cdot10^{-3}$	**1.630**	$4\cdot10^{-3}$	272	$3\cdot10^{-2}$	1.636	$2\cdot10^{-3}$	225	$3\cdot10^{-3}$
Housing	3.61	0.21	130	$8\cdot10^{-3}$	3.58	0.19	200	$5\cdot10^{-2}$	3.64	0.18	180	$7\cdot10^{-2}$
Iris	1.00	1.1	102	$3\cdot10^{-4}$	1.88	1.1	120	$3\cdot10^{-2}$	1.08	1.0	266	$3\cdot10^{-3}$
Diabetes	20.312	0.8	266	$3\cdot10^{-3}$	20.430	1.0	173	$3\cdot10^{-2}$	**20.086**	1.0	192	$3\cdot10^{-3}$
Landsat	10.438	0.32	579	$3\cdot10^{-3}$	**9.848**	0.30	600	$3\cdot10^{-2}$	10.394	0.32	600	$3\cdot10^{-3}$
Wine	2.2542	1.5246	60	$3\cdot10^{-2}$	2.0847	1.6313	70	$2\cdot10^{-1}$	2.5593	1.5704	80	$8\cdot10^{-2}$

Table 3. Random projections based training (pseudoinversion) vs. backpropagation

Dataset	PINV Avg	StDev	Backprop. Avg	(N_{tests})	StDev
Abalone	**2.162**	0.006	2.3044	(35)	0.1908
Mach. Cpu	**56.85**	2.8	28.6673	(5)	27.3535
Delta Ail.	**1.630 · 10^{-4}**	$4\cdot10^{-7}$	$2\cdot10^{-3}$	(10)	0.0
Housing	**3.58**	0.19	4.5492	(35)	0.9517
Iris	**1.00**	1.1	1.73	(10)	0.85
Diabetes	**20.086**	1.0	26.52	(31)	2.38
Landsat	**9.848**	0.30	13.03	(5)	0.63
Wine	**2.0847**	1.6313	3.77	(10)	0

better results than tunedit's ones by running the backpropagation method on our own under WEKA. For all datasets in the table we can claim dominance of the pseudoinversion based approach with a 99% confidence level.

In our experiments, the running times of all the pseudoinversion-based approaches are substantially equivalent, hence we base the comparison only on the average error. As far as the comparison with backpropagation is concerned, pseudoinversion based methods save a relevant amount of time, being up to 10 times faster than backpropagation. For example, 10 runs of pseudoinversion-based training on the Wine dataset require 0.078 seconds on average whereas backpropagation requires on average 0.721 seconds (times on a laptop with Pentium CPU, 2 GHz clock, 4 GB RAM); other tests gave roughly similar results.

5   Conclusions

We considered pseudoinversion-based techniques for training of neural networks feeding them by random projections (gaussian and sparse) matrices of input weights and biases instead of the classical uniform-random initialization. We believe that the computational results presented in this paper assess initialization by random projection matrices as a useful tool for improving performances

In future research we will consider hybridizing the pseudoinversion-based training technique with basic descent techniques. The rationale behind this is that pseudoinversion-based techniques mostly rely on a pure random sampling of input weights and biases, whereas it could make sense trying to profit also from some local exploration of the error landscape.

Acknowledgment. The activity has been partially carried on in the context of the Visiting Professor Program of the Gruppo Nazionale per il Calcolo Scientifico (GNCS) of the Italian Istituto Nazionale di Alta Matematica (INdAM).

References

1. Rumellhart, D.E., Hinton, G.E., Williams, R.J.: Learning internal representations by error propagation. In: Parallel Distrib. Process.: Exploration in the Microstructure of Cognition, vol. 1, pp. 318–362. MIT Press, Cambridge (1986)
2. LeCun, Y.A., Bottou, L., Orr, G.B., Müller, K.-R.: Efficient backProp. In: Orr, G.B., Müller, K.-R. (eds.) NIPS-WS 1996. LNCS, vol. 1524, pp. 9–50. Springer, Heidelberg (1998)
3. Larochelle, H., Erhan, D., Courville, A., Bergstra, J., Bengio, Y.: An empirical evaluation of deep architectures on problems with many factors of variation. In: 24th ICML (2007)
4. Vincent, P., Larochelle, H., Bengio, Y., Manzagol, P.-A.: Extracting and composing robust features with denoising autoencoders. In: 25th ICML (2008)
5. Collobert, R., Weston, J.: A unified architecture for language processing: Deep neural networks with multitask learning. In: 25th ICML (2008)
6. Mnih, A., Hinton, G.E.: A scalable hierarchical distributed language model. In: 23rd NIPS, pp. 1081–1088 (2009)
7. Poggio, T., Girosi, F.: Networks for approximation and learning. IEEE 78(9), 1481–1497 (1990)
8. Cancelliere, R.: A High Parallel Procedure to Initialize the Output Weights of a Radial Basis Function or BP Neural Network. In: Sørevik, T., Manne, F., Moe, R., Gebremedhin, A.H. (eds.) PARA 2000. LNCS, vol. 1947, pp. 384–390. Springer, Heidelberg (2001)
9. Huang, G.-B., Zhu, Q.-Y., Siew, C.-K.: Extreme Learning Machine: Theory and applications. Neurocomputing 70, 489–501 (2006)
10. Halawa, K.: A method to improve the performance of multilayer perceptron by utilizing various activation functions in the last hidden layer and the least squares method. Neural Processing Letters 34, 293–303 (2011)
11. Nguyen, T.D., Pham, H.T.B., Dang, V.H.: An efficient Pseudo Inverse matrix-based solution for secure auditing. In: IEEE International Conference on Computing and Communication Technologies, Research, Innovation, and Vision for the Future (2010)

12. Kohno, K., Kawamoto, M., Inouye, Y.: A Matrix Pseudoinversion Lemma and Its Application to Block-Based Adaptive Blind Deconvolution for MIMO Systems. IEEE Transactions on Circuits and Systems I: Regular Papers 57(7), 1449–1462 (2010)

13. Ajorloo, H., Manzuri-Shalmani, M.T., Lakdashti, A.: Restoration of damaged slices in images using matrix pseudo inversion. In: 22nd International Symposium on Computer and Information Sciences (2007)

14. Wang, X.-Z., Wang, D., Huang, G.-B.: Special Issue on Extreme Learning Machines. Editorial. Soft Comput. 16(9), 1461–1463 (2012)

15. Wang, X.: Special Issue on Extreme Learning Machine with Uncertainty. Editorial. Int. J. Unc. Fuzz. Knowl. Based Syst. 21(supp. 02), v–vi (2013)

16. Arriaga, R.I., Vempala, S.: An algorithmic theory of learning: robust concepts and random projection. In: 40th Annual Symp. on Foundations of Computer Science, pp. 616–623. IEEE Computer Society Press (1999)

17. Vempala, S.: Random projection: a new approach to VLSI layout. In: 39th Annual Symp. on Foundations of Computer Science. IEEE Computer Society Press (1998)

18. Indyk, P., Motwani, R.: Approximate nearest neighbors: towards removing the curse of dimensionality. In: 30th Symp. on Theory of Computing, pp. 604–613. ACM (1998)

19. Penrose, R.: On best approximate solution of linear matrix equations. Proceedings of the Cambridge Philosophical Society 52, 17–19 (1956)

20. Bishop, C.M.: Pattern Recognition and Machine Learning. Springer, Berlin (2006)

21. Badeva, V., Morosov, V.: Problemes incorrectements posès, thèorie et applications (in French). Masson, Paris (1991)

22. Cancelliere, R., De Luca, R., Gai, M., Gallinari, P., Artières, T.: Pseudoinversion for neural training: tuning the regularisation parameter. Technical report n. 149/13, Dep. of Computer Science, University of Turin (2013)

23. Tikhonov, A.N., Arsenin, V.Y.: Solutions of Ill-Posed Problems. Winston, Washington, DC (1977)

24. Tikhonov, A.N.: Solution of incorrectly formulated problems and the regularization method. Soviet Mathematics 4, 1035–1038 (1963)

25. Gallinari, P., Cibas, T.: Practical complexity control in multilayer perceptrons. Signal Processing 74, 29–46 (1999)

26. Poggio, T., Girosi, F.: Regularization algorithms that are equivalent to multilayer networks. Science 247, 978–982 (1990)

27. Girosi, F., Jones, M., Poggio, T.: Regularization theory and neural networks architectures. Neural Computation 7(2), 219–269 (1995)

28. Haykin, S.: Neural Networks, a comprehensive foundation. Prentice Hall, U.S.A. (1999)

29. Fuhry, M., Reichel, L.: A new Tikhonov regularization method. Numerical Algorithms 59, 433–445 (2012)

30. Hecht-Nielsen, R.: Context vectors: general purpose approximate meaning representations self-organized from raw data. In: Zurada, J.M., Marks II, R.J., Robinson, C.J. (eds.) Computational Intelligence: Imitating Life, pp. 43–56. IEEE Press (1994)

31. Bingham, E., Mannila, H.: Random projection in dimensionality reduction: Applications to image and text data. In: Conference on Knowledge Discovery and Data Mining, KDD 2001, San Francisco, CA, USA (2001)

32. Johnson, W.B., Lindenstrauss, J.: Extensions of Lipshitz mapping into Hilbert space. In: Conference in Modern Analysis and Probability. Contemporary Mathematics, vol. 26, pp. 189–206. Amer. Math. Soc. (1984)
33. Dasgupta, S., Gupta, A.: An elementary proof of the Johnson-Lindenstrauss lemma. Technical report TR-99-006, International Computer Science Institute, Berkeley, California, USA (1999)
34. Achlioptas, D.: Database-friendly random projections. In: ACM Symp. on the Principles of Database Systems, pp. 274–281 (2001)
35. Asuncion, A., Newman, D.J.: UCI Machine Learning Repository, University of California, Irvine, School of Information and Computer Sciences (2007), http://www.ics.uci.edu/~mlearn/MLRepository.html

Test Case Prioritization for NUnit
Based Test Plans in Agile Environment

Sohail Sarwar[1], Yasir Mahmood [1], Zia Ul Qayyum[2], and Imran Shafi[3]

[1] Department of Computing Iqra University Islamabad, Pakistan
[2] National University of Computing and Emerging Sciences (FAST) Islamabad, Pakistan
[3] Abasyn University Islamabad, Pakistan
{sohail.sarwar,yasir.mahmood}@seecs.edu.pk
zia.qayyum@nu.edu.pk, imran.shafi@gmail.com

Abstract. Test Case prioritization having a key role to play in prioritizing test scenarios from a pile of scenarios, to best of our knowledge, has not been employed in Agile environment for prioritizing test cases in Automated Test Plans. Considering automated testing in agile environment esp scrum, a prioritized test plan containing high priority test cases is emanated using Genetic Algorithms. This prioritization is courtesy to base factors such as operational profile, test scenario criticality, and faults uncovered by each test case; used to weight test scenarios. Proposed technique exhibits great performance by ameliorating the rate of fault detection by dynamically prioritizing NUnit based test scenarios.

Keywords: Test Case Prioritization, Genetic Algorithms, Agile Testing, Regression Testing, Automated Test Plans.

1 Introduction

Software testing being an ongoing process in development and maintenance of software has a crucial role to play in success of any software project. This statement can further be complemented by studies showing 50% of effort in software production and maintenance involves testing [1]. Furthermore, software bugs cost only US economy $ 59.5 billion where one third of this cost can be saved with effective testing [2].

Our focal point, the phenomenon of testing, is equally vital during software maintenance i.e. corrective as well as perfective. In other words, each time a bug/defect is fixed or new functional unit is incorporated in software, testing process is triggered to avoid any anomalous behavior of software. This process of testing is termed as regression testing. Effectiveness of regression testing is true for all software development models i.e. traditional models (waterfall, evolutionary etc.) as well as agile models (Xtreme Programming, Scrum etc.). However, we will take into account regression testing in agile environment for practicing scrum model in our organization. Soon after new build is produced during maintenance phase, we perform very brief testing exercise involving execution of manual as well as automated test plans acronym ATPs (such as smoke test, performance test and test scripts for User Interface testing). However,

G. Agre et al. (Eds.): AIMSA 2014, LNAI 8722, pp. 246–253, 2014.

execution of all manual and automated test plans is a tiresome activity consuming a lot of time, cost and may not be much effective in uncovering critical defects at earliest. We need to be conscious while selecting fewer but important tests (specifically "test cases") from this plethora of test plans. Test coverage, faults unleashed previously by certain test case (criticality of test case), and frequency of usage (operational profile) are the aspects which distinguish a vital test case from ordinary ones. These highly important test cases may be congregated to devise a high priority test plan while exploiting the phenomenon of "Test Case Prioritization" [3, 4, 5].

Test case prioritization techniques attempt to sort the given test scenarios such a way that uncovering of high priority defects, smooth functioning of critical & prioritized features is maximized. This has been carried out in various ways [3,4, 5, 6, 7] where Greedy Algorithms, classification techniques for weighing the test cases, neural networks, requirements, coverage and cost criterion have been exploited to rightly prioritize the test cases. However, we propose a simple technique that will weight a test case based on three factors (1) test case execution frequency (2) test scenario criticality and (3) faults uncovered by each test case. These weighted test cases will undergo different GA cycles in order to get a set of prioritized test cases entailing in generation of a prioritized test plan. Prioritized test plan will be automatically updated due to dynamic maneuvering of "test case weights" (based on faults uncovered in each run) so that only high priority test cases are retained. Proposed approach is purely focusing on the ATPs written using C #.NET that are executed through NUnit. Moreover, we will discuss things in the perspective of "scrum", the very development model followed in our organization. The results of proposed approach are presented using a slightly modified form of APFD metric [3] where tests with higher priority weights appear to uncover maximum number of faults.

The research efforts asserted in [8,11,12] are evident of how important test case prioritization is, however these techniques have some problems as given below:

- These techniques order the test cases for the first time but may not be dynamic enough to cater multiple executions and prioritizations each time test plans are executed.

- The factors used in computing the weight(s) of a test cases, sometimes are inefficient to comprehend at desired level.

- None of the techniques presented so far considers development model while prioritizing the test cases.

The paper is organized in the fashion given below:

Section 2 discusses proposed architecture based on our research, theoretical concepts and explains them at lowest level of abstraction. Section 3 discusses implementation strategy. The results of experiments with varying parameters are presented in section 4. Conclusion and future work are furnished in Section 5.

2   System Architecture

The approach is presented keeping in view the above concerns as well as the culture and practices in software development organization. We start off from the software development model currently employed i.e. scrum. Such models with slight variations are widely applied due to their flexible nature to comprehend highly demanding environment. Use of such model defines active involvement of testing and role of test engineers. Contrary to traditional models, testing starts much earlier than it used to be. Each tester receives two "Builds" of product he is testing and there is a possibility of receiving a "Priority Build" (carrying quick fixes or prioritized requirements) for testing. He has to run all the test plans as a part of regression testing in addition to routine testing of new requirements/enhancements and bug fixes. Now execution of all manual test plans will be a cumbersome task requiring approx 30-35 man hours and concentration of resources. One solution is to devise Automated Test Plans (ATPs) and other is prioritizing them in manual test plans (as all we cannot transform all manual test cases to automated test plans. However, we need certain prioritization mechanism for ATPs as well as all ATPs may not be equally important. So prioritization mechanism objective is to minimize the man-hour effort through reduction of prioritized test cased. The constraints are the part of every software development practice and man-hours and/or efforts are subject change.

$$\text{Minimize} \quad (\sum_{i}^{m} c_i x_i) \tag{1}$$

Where ci represents the constraints in software testing environment and Xi is priorities that may help to reduce m man hours.

These ATPs are scripted using C#.NET and are executed using NUnit. Each ATP refers to a test case in the test plan where each test case refers to a single user requirement. Normally, every test case in a test plan has following parameters:
Test Case ID, Description, Priority, Criticality, Fault occurrence frequency, Module Rank Number of defects associated with each test case, magnitude of test plan.

However, only three most important parameters have been selected for prioritizing the test cases and promoting them to a prioritized test plan using priority weight for each test case. These parameters are 1) Operational Profile of test case (2) test case criticality and (3) Number of faults uncovered by each test case. We briefly describe these parameters and their usage in the proposed approach as priority weight for each test will be computed using these parameters.

Operational Profile (OP): refers to how many times a specific function is used out of the total functional executions. Here it refers to the execution of single test case ration all the test cases in specific test plan. This profile will be valued on a scale of 1 – 10, where rank of each test case will be in accordance to its percent use.

Test case criticality (TC): is the measure of some parts that may be more critical than others in perspective of catastrophic failure and its management.

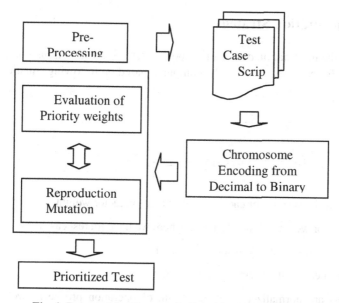

Fig. 1. Proposed Architecture for Prioritized Test Plan Generation

More critical components need stronger justification. A software criticality allows appropriate effort to be directed at each component of the software. It uses domain experts to classify high and low criticality. This parameter again is measured on a scale of 1 – 10, where 0 refers to least critical and 10 represents most critical component. It is worth mentioning that a highly critical test case may not have high operational profile and vice versa.

Number of faults uncovered (D): refers to the number of defects which have been detected by certain test case. This parameter will have a key role in evaluating a test case and its placement in prioritized test plan so during whole process of prioritizing test cases this parameter will be maintained and constantly updated as new faults are uncovered by certain test case. Hence this will play a role in dynamically updating the prioritized test plans by inducting new test cases and evicting older ones based in following phenomenon:

The faults uncovered by each test case are retained against each test for future reference so that weight and priority of such test cases over others is maintained. Consequently, these test cases based on their fault count weight can be moved to high priority test plan. A threshold of minimum weight will be required for each test case for moving it to "prioritized" test plan. The decision for each test case will be made on the basis of weight each test case has with respect to weight of test case with highest value. For example Test case 'A' having weight 10 has the highest weight so far among available test cases. Test case 'B' having weight 4.5 will be moved to prioritized test plan if and only if it meets the minimum threshold i.e. 40% of the weight of Test case 'A'. Each time test plan is executed, new faults are uncovered, weights of test cases are updated and test having weight lower than specified threshold will be evicted from prioritized test plan.

3   Implementation Strategy

In this section, the information details of proposed architecture are discussed.
Priority weight for each test case will be computed based on following equation given the parameter ranks

$$(P_w)_i = \frac{(F_w)i}{n+m} \tag{2}$$

$$(F_w)_i = D * (\textstyle\sum_{OP=1}^{m} OP + \sum_{TC=1}^{n} TC) \tag{3}$$

Where

- P_W is priority weight for each test case based on factor weight

- F_W is factor weight of each test case where we have ith test cases

- OP is operational profile where m=10 and

- TC is test criticality having n= 10.

Priority weights are normalized with maximum of operation profile as well as test criticality. Factor weight is weighted accumulation of operation profile and test criticality. Weights the updated number of defects found before regression testing. Once we have priority weight(s) associated with each test case, we will place together the Test case ID, faults detected by a test case (D) and priority weight (PW) together.

We applied Genetic Algorithms [13] on these tuples to get an optimized solution of prioritized test cases. Stepwise description of applying the GA is given below.

Each locus is represented by tuple.
Chromosome length = 8
Initial Population = 400 chromosomes
Fitness function = PW
Selection
2 chromosomes with maximum PW be selected
L1: For selected chromosomes
CrossOver
Mutation
Remove Duplicates
Check Optimization or move to L1.
Place new offspring in population and end the cycle.
If the end condition is satisfied, stop, and return the best solution in population.

During GA process PW has been used to measure the optimization of a solution. However, each time a prioritized test plan is updated new test cases are added to it and few of them are evicted. This eviction is made on the basis of phenomenon explained in the beginning of this section.

4   Results and Evaluations

In order to assess the effectiveness of proposed technique, total numbers of faults detected by a specific test case are used as a metric. JAVA API for Genetic Algorithm (JAGA), a free & open source API for evolutionary algorithmic computation allowing creating GA applications has been used for GA portion given the required parameters.

We chose about 500 scenarios, based on their high operational profile, criticality and faults detection in previous builds, are subjected them to undergo the process mentioned in section 3. The prioritized test plan carried the tests in such a way that ones with high PW appeared on top in order higher to lower.

4.1   Prioritized Vs Non-Prioritized

The benefits of prioritizing test cases are evident from defect count placing the tests in a earlier based on priority weight. Moreover, non-prioritization reveals ordinary behavior where no effort has been made to catalyze the process of uncovering faults within a given time span.

The proposed approach, compared to factor oriented prioritization [3] (red as shown in figure 5) and time aware approaches [12] (blue as shown in figure 5). The better performance has been achieved for factor weight prioritization because of its intrinsic ability, the inclusion of test case criticality and use of factor weight of each test case.

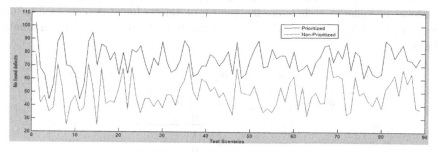

Fig. 2. Prioritized vs Non-Prioritized

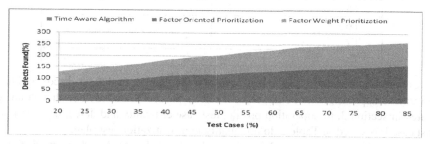

Fig. 3. Comparison of Proposed Technique with Existing techniques

Table 1 depicts the numbers of defects found with respect to priority weight. The table has been sorted according to priority weight. The numbers of defects with higher priority weight are more than those having lesser priority weight. The Test case T125 has priority weight 38 so it detects 12 defects. The test case T122 has lower priority weight so it detects lower number of defects in comparison to T125, where as it has detected more number of defects than T126 and so on.

Table 1. Defects uncovered using proposed Prioritization Approach

Test Case ID	Defects Detected	Priority Weight
T125	12	38
T122	8	31
T126	7	26
T128	7	24
T123	5	21
T124	2	13
T121	1	10
T127	0	5

The prioritized and non-prioritized test case procedures have been compared with respect to defect finding. The number of test cases chosen, in each of the technique is same but since our prioritized approach uses prior knowledge of defects found (i.e. D multiplier in equation 3) therefore the number of defects found in regression testing, are more than non-prioritized, while random selection of test cases find lesser number of test cases

Table 2. Defects uncovered using Non - Prioritization Approach

Test Case ID	Defects Detected
T1	0
T2	1
T3	0
T4	2
T5	1
T6	0
T7	1
T8	0

Comparing the table 1 and 2, it is clear that the number of defects discovered using prioritized approach is far better, since it selects the scenarios based on the parameters stated in section 3, the operation profile, test case criticality and defect detection. The test case IDs stated in Table 2 do not quality in prioritization test plan.

Keeping in view the data above, it can be stated that proposed approach for test prioritization yields better results than conventional way of testing.

5   Conclusion

Regression testing technique has been proposed specifically for agile environment where scrum is employed. This technique exploits three important parameters to calculate the priority weight for each test scenario of ATPs to place it in prioritized test plan dynamically. GA when applied on given dataset, acquired from real time test plans, provides effective sequence of test cases emanating better fault detection at early stage.

References

1. Harrold, M.: Testing a roadmap. In: International Conference on Software Engineering, Limerick, Ireland, pp. 61–72 (2000)
2. Srikanth, H., Williams, L., Osborne, J.: System Test Case Prioritization of New and Regression Test Cases. In: Proceedings of 4th International Symposium on Empirical Software Engineering (ISESE), pp. 62–71. IEEE Computer Society (2005)
3. Raju, S.: Factor Oriented Test Case Prioritization Technique in Regression Testing using Genetic Algorithm. European Journal of Scientific Research 74(3), 389–402 (2012) ISSN 1450-216X
4. Sujata, Kumar, M., Kumar, V.: Requirement based Test Case Prioritization using Genetic Algorithm. IJCST 1(2) (December 2010)
5. Kumar, A., Gupta, S., Reparia, H., Singh, H.: An approach for test case prioritization based upon varying requirements. International Journal of Computer Science, Engineering and Applications (IJCSEA) 2(3) (2012)
6. Siripong, Jirapun: Test Case Prioritization Techniques. Jouranl of Theoratical and Applied Information Technology (2012)
7. Elbaum, S., Malishevsky, A.G., Rothermel, G.: Incorporating varying test costs and fault severities into test case prioritization. In: Proceedings of the 23rd International Conference on Software Engineering, pp. 329–338 (May 2001)
8. Zhang, X., Nie, C., Xu, B., Qu, B.: Test Case Prioritization based on Varying Testing Requirement Priorities and Test Case Costs. In: Seventh International Conference on Quality Software (QSIC 2007) (2007)
9. Lehmann, E., Wegener, J.: Test case design by means of the CTE XL. In: Proc. of the 8th European International Conf. on Software Testing, Analysis & Review, EuroSTAR 2000 (2000)
10. Rothermel, G., Untch, R., Chu, C., Harrold, M.: Test Case Prioritization. IEEE Transactions on Software Engineering 27, 929–948 (2001)
11. Kaur, A., Goyal, S.: A Genetic Algorithm for Fault based Regression Test Case Prioritization. International Journal of Computer Applications 32(8) (2011) 975–8887
12. You, L., Lu, Y.: A genetic algorithm for the time – aware regression testing reduction problem. In: ICNC, pp. 596–599. IEEE (2012)
13. http://www.jaga.org

Pattern Structure Projections
for Learning Discourse Structures

Fedor Strok[2], Boris Galitsky[1], Dmitry Ilvovsky[2], and Sergei Kuznetsov[2]

[1] Knowledge Trail Incorporated
bgalitsky@hotmail.com
[2] School of Applied Mathematics and Information Science, National Research University
Higher School of Economics, Bol. Trekhsvyatitelskii 3, Moscow, Russia
{dilvovsky,skuznetsov}@hse.ru, fdr.strok@gmail.com

Abstract. We consider a graph representation for a paragraph of text. It widely uses linguistic theories of discourse to extend the set of edges between vertices corresponding to words. Parse thickets is a set of syntactic parse trees augmented by a number of inter-sentence coreference links and links based on Speech Act and Rhetoric Structures Theories. Similarity of parse thickets is defined by means of intersection operation taking common parts of the thickets. Several approaches to computing intersection of parse thickets are proposed and compared. Projections as approximation means are considered.

Keywords: Parse Thickets, Pattern Structures, Text Clustering.

1 Introduction

Ranking of search results is one of the essential topics in search engineering. The problem of relevance is reduced to the scoring system for search results ranking. In industrial search ranking is not only based on relevance, but also on location, time, and expected revenue from search results, and other parameters.

The alternative to displaying search results in a sequence is clustering search results [5,6]. The advantage is that similar search results are combined together, so a user has to navigate through clusters instead of individual search results to find what she means or looks for. One of the most promising clustering techniques is conceptual clustering, where clusters are given by pairs of the form (set of search results, common features of search results), see [1,2,3,4]. These pairs, called concepts, are naturally ordered, this order making an algebraic lattice, called concept lattice [7]. For various search tasks, e.g. in social search, when search results are delivered from different users by different pathways, it is hard to rank them in a sequence, so the lattice diagram is helpful for a receiver to quickly grasp the types and topics of answers obtained in a (social) search session.

An obvious disadvantage of concept lattices is that its construction starts from a binary object-attribute data table, which requires binarization as a preprocessing step. The binarization can blow up the representation complexity and can result in a loss of

G. Agre et al. (Eds.): AIMSA 2014, LNAI 8722, pp. 254–260, 2014.

information. In this paper we leverage the structural description of text paragraphs and mechanism of computing similarity between paragraphs on syntactic trees and semantic relations between them explored in our previous studies [14]. We combine this technique with the conceptual clustering idea to provide structural description of texts instead of binary attributes and to build a pattern structure [9] based on these descriptions. We also introduce a linguistic approximation of this representation in the form of projection, which helps us to solve the scalability issue.

2   Parse Thickets

Parse thicket [12] is defined as a set of parse trees for each sentence augmented with a number of arcs, reflecting inter-sentence relations. We used a few sources of relations:

- Coreferences [15]
- Rhetoric structure theory (RST) [11],
- Communicative Actions (CA, particular case of Speech Acts theory) [8].

We used a vocabulary of Communicative actions to

1. find their subjects,
2. add respective arcs to the parse thicket,
3. index combination of phrases as subjects of communicative actions.

For RST, we introduce explicit indexing rules which will be applied to each paragraph and

1. attempt to extract an RST relation,
2. build corresponding fragment of the parse thicket, and
3. index respective combination of formed phrases (noun, verb, prepositional), including words from different sentences.

The process of construction parse thickets is described in previous works.

3   Pattern Structures and Projections

Pattern structures gives a formal means for representing similarity of an arbitrary set of objects in terms of intersection operation defined on the set of objects descriptions [9].

Let G be a set of objects, $(D, \sqcap)$ - intersection (meet) semilattice on a set of descriptions, $\delta: G \to D$ - a mapping. Triple $(G, (D, \sqcap), \delta)$ is called a pattern structure if $\delta(G) := \{\delta(g) | g \in G\}$ generates complete semilattice $(D_\delta, \sqcap)$ for $(D, \sqcap)$. For a pattern structure $(G, (D, \sqcap) \delta)$ the following operations are defined:

$$A^\blacksquare := \sqcap_{g \in A} \delta(g) \text{ for } A \subseteq G$$
$$d^\blacksquare := \{g \in G | d \sqsubseteq \delta(g)\} \text{ for elements of semilattice } d \in D$$

A pair $< A, d >$ is a pattern concept if $A^{\blacksquare} = d, d^{\blacksquare} = A$.

For the case of graphs the operation of intersection is defined by taking the set of maximal common subgraphs [13]. Pattern concepts are naturally ordered, this order making a lattice. This lattice can be exponentially large wrt. the input size and the operation of intersection can also be intractable, like, e.g. in the case of graph descriptions. Projections are means for approximating pattern structures to attain scalability.

Formally, projection of a pattern structure is a function ψ: D→D which is monotone $(x \sqsubseteq y \Rightarrow \psi(x) \sqsubseteq \psi(y))$, contracting $(\psi(x) \sqsubseteq x)$ and idempotent $(\psi(\psi(x)) = \psi(x))$. In NLP applications where graphs stay for representing text structure and semantics, text projections can consist of meaningful pieces of text, like e.g. noun and verb phrases.

4   Pattern Structures on Parse Thickets

4.1   Linguistic Projections

In our study, the set of objects is given by a set of texts, their descriptions are represented by parse thickets. Since parse thicket is a graph with special properties, it is natural to define the intersection operation as a set of maximal common subgraphs. Since the problem of subgraph isomorphism is NP-complete, we use projections to decrease the computation complexity.

We define projection of a parse thicket as a set of maximal syntactic and extended (including RST, coreferential and CA arcs) groups (phrases) [12]. In terms of structure this is a set of maximal subtrees of a parse thicket.

The intersection operation is defined separately for each type of groups. Intersection on projections imply pair-wise intersection of groups of each type and filtering them by subsumption. Such projections allow us to do the computations faster (since we come to processing trees) and to save all connections in the paragraph.

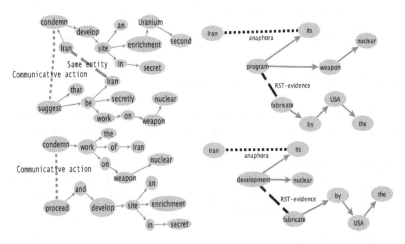

Fig. 1. Generalization based on CA (on the left) and RST-evidence (on the right)

4.2   Implementation

The approach described here is implemented as an OpenNLP contribution. It relies on the following systems:

- OpenNLP/Stanford NLP parser;
- Stanford NLP Coreference (Recas).

It includes the following components of Apache OpenNLP.similarity project provided by authors:

1. Rhetoric parser
2. Parse thicket builder and generalizer [12].
3. Pattern structure builder (AddIntent [16] is used for constructing the lattice)
4. A number of applications based on the above components, including search (request handler for SOLR), speech recognition, content generation and others.

The textual input is subject to a conventional text processing flow such as sentence splitting, tokenizing, stemming, part-of-speech assignment, building of parse trees and coreferences assignment for each sentence. Either OpenNLP or StanfordNLP implements this flow, and the parse thicket is built based on the algorithm presented in this paper. The coreferences and RST component strongly rely on Stanford NLP's rule-based approach to finding correlated mentions based on the multi-pass sieves.

The graph-based approach to generalization relies on finding maximal cliques for an edge product of the graphs for PTs being generalized. As it was noticed earlier the main difference with the traditional edge product is that instead of requiring the same label for two edges to be combined, we require non-empty generalization results for these edges. Hence although the parse trees are relatively simple graphs, parse thicket graphs reach the limit of real-time processing by graph algorithms.

This framework allows seamless integration into other open source systems available in Java for search, information retrieval and machine learning. Moreover, pattern structure construction can be embedded into Hadoop framework in the domains where offline performance is essential. Code and libraries described here are also available at http://code.google.com/p/relevance-based-on-parse-trees and http://svn.apache.org/repos/asf/opennlp/sandbox/opennlp-similarity/.

4.3   Hierarchical Text Clustering with Pattern Structures

An outline of the algorithm for text clustering can be described as follows:

1. Consider the set of texts (search results) T.
2. For each result $t_i \in T$ we build a parse thicket $p_i \in P$.
3. We use the intersection operation of parse thickets to build a pattern structure $(T,(P,\Pi),\delta)$ for all texts with AddIntent [16] or CbO [10].

If we use projections the algorithm is modified in the following way:

1. Consider the set of texts (search results) T.
2. For each result $t_i \in T$ we build a parse thicket projection $\psi(p_i) \in \psi(P)$.
3. We use generalization of projections as lattice operation and apply standard algorithm (AddIntent or CbO) to build a pattern structure $(T, (P_\psi, \Pi_\psi), \psi \circ \delta)$

4.4   Example of Constructing Pattern Structures on Parse Thickets

Let us consider 3 news:

1. *At least 9 people were killed and 43 others wounded in shootings and bomb attacks, including four car bombings, in central and western Iraq on Thursday, the police said. A car bomb parked near the entrance of the local government compound in Anbar's provincial capital of Ramadi, some 110 km west of Baghdad, detonated in the morning near a convoy of vehicles carrying the provincial governor Qassim al-Fahdawi, a provincial police source told Xinhua on condition of anonymity.*
2. *Officials say a car bomb in northeast Baghdad killed four people, while another bombing at a market in the central part of the capital killed at least two and wounded many more. Security officials also say at least two policemen were killed by a suicide car bomb attack in the northern city of Mosul. No group has claimed responsibility for the attacks, which occurred in both Sunni and Shi'ite neighborhoods.*
3. *A car bombing in Damascus has killed at least nine security forces, with aid groups urging the evacuation of civilians trapped in the embattled Syrian town of Qusayr. The Syrian Observatory for Human Rights said on Sunday the explosion, in the east of the capital, appeared to have been carried out by the extremist Al-Nusra Front, which is allied to al-Qaeda, although there was no immediate confirmation. In Lebanon, security sources said two rockets fired from Syria landed in a border area, and Israeli war planes could be heard flying low over several parts of the country.*

The Figure 4 illustrates pattern structure, obtained for these texts. We use the set of all possible phrases (noun, pronoun, verb, etc.) as a projection of a parse thicket. In intersection operations we strictly match parts of speech tags, but allow leaf vertices (corresponding to words) to be labeled by wildcards (*).

At the first level we get concepts, corresponding to the documents. The descriptions will be set of parse thicket phrases. The top-level concept contains phrases, which are common for all texts. In this case, all 3 texts are about car bombing near the capitals that is represented by phrases: [DT-a NN-car NN-bombing], [DT-the NN-capital], [VBN-killed], [JJS-least CD-* NN-*].

For the level of pair intersections the most interesting is the concept for texts 1 and 2. Since they describe the same event, specific place is common for them: [NN-* NN-* IN-in NNP-baghdad]. Both texts use same verbs [NN-* NN-bomb NN-attack], [NNS-attacks], and information about preys: [VBD-wounded], [VBD-were VBN-killed], as well as the details: [CD-* NNS-people], [CD-four NNS-*]

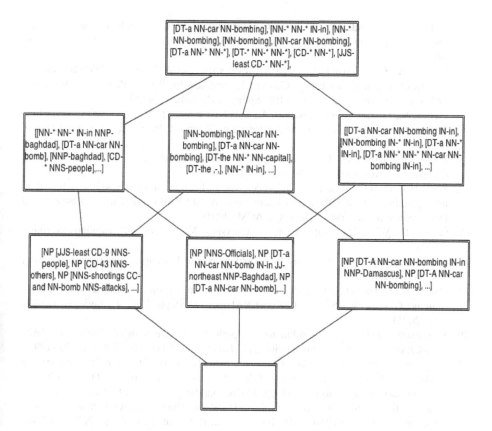

Fig. 2. Pattern structure for news texts

5   Conclusions

We studied the idea of representing short texts as graphs (parse thickets) in combination with the pattern structures approach. For this reason we introduced the intersection operation for parse thickets and used it for constructing text similarities as pattern concepts. To reduce the computation complexity projections of parse thickets were defined in the most natural way. An illustrative example of a pattern structure on parse thickets was considered. We extended our framework, which is based on OpenNLP, with the tools for computing with pattern structures.

In further studies we plan to implement a tool for visualizing pattern structures and applying the presented technique in a few tasks, such as short text categorization, clustering of search results, etc.

References

1. Carpineto, C., Romano, G.: A Lattice Conceptual Clustering System and Its Application to Browsing Retrieval. Machine Learning 24(2), 95–122 (1996)

2. Cole II, R.J., Eklund, P.W., Stumme, G.: Document Retrieval for E-Mail Search and Discovery Using Formal Concept Analysis. Applied Artificial Intelligence 17(3), 257–280 (2003)
3. Koester, B.: Conceptual Knowledge Retrieval with FooCA: Improving Web Search Engine Results with Contexts and Concept Hierarchies. In: Perner, P. (ed.) ICDM 2006. LNCS (LNAI), vol. 4065, pp. 176–190. Springer, Heidelberg (2006)
4. Messai, N., Devignes, M.-D., Napoli, A., Smaïl-Tabbone, M.: Many-Valued Concept Lattices for Conceptual Clustering and Information Retrieval. In: ECAI 2008, pp. 127–131 (2008)
5. Zamir, O., Etzioni, O.: Grouper: a dynamic clustering interface to Web search results. Computer Networks 31(11), 1361–1374 (1999)
6. Zeng, H.J., He, Q.C., Chen, Z., Ma, W.Y., Ma, J.: Learning to cluster web search results. In: Proceedings of the 27th Annual International ACM SIGIR Conference on Research and Development in Information Retrieval. ACM (2004)
7. Ganter, B., Wille, R.: Formal Concept Analysis – Mathematical Foundations. Springer (1999)
8. Searle, J.: Speech acts: An essay in the philosophy of language. Cambridge University, Cambridge (1969)
9. Ganter, B., Kuznetsov, S.O.: Pattern Structures and Their Projections. In: Delugach, H.S., Stumme, G. (eds.) ICCS 2001. LNCS (LNAI), vol. 2120, pp. 129–142. Springer, Heidelberg (2001)
10. Kuznetsov, S.O.: A fast algorithm for computing all intersections of objects in a finite semi-lattice. Automatic Documentation and Mathematical Linguistics 27(5), 11–21 (1993)
11. Mann, W.C., Matthiessen, C.M., Thompson, S.A.: Rhetorical Structure Theory and Text Analysis. In: Mann, W.C., Thompson, S.A. (eds.) Discourse Description: Diverse Linguistic Analyses of a Fund-raising Text, pp. 39–78. John Benjamins, Amsterdam (1992)
12. Galitsky, B.A., Kuznetsov, S.O., Usikov, D.: Parse Thicket Representation for Multi-sentence Search. In: Pfeiffer, H.D., Ignatov, D.I., Poelmans, J., Gadiraju, N. (eds.) ICCS 2013. LNCS, vol. 7735, pp. 153–172. Springer, Heidelberg (2013)
13. Ganter, B., Grigoriev, P.A., Kuznetsov, S.O., Samokhin, M.V.: Concept-Based Data Mining with Scaled Labeled Graphs. In: Wolff, K.E., Pfeiffer, H.D., Delugach, H.S. (eds.) ICCS 2004. LNCS (LNAI), vol. 3127, pp. 94–108. Springer, Heidelberg (2004)
14. Galitsky, B., Ilvovsky, D., Kuznetsov, S.O., Strok, F.: Matching sets of parse trees for answering multi-sentence questions. In: Proceedings of the Recent Advances in Natural Language Processing, RANLP 2013, pp. 285–294. INCOMA Ltd., Shoumen (2013)
15. Lee, H., Chang, A., Peirsman, Y., Chambers, N., Surdeanu, M., Jurafsky, D.: Deterministic coreference resolution based on entity-centric, precision-ranked rules. Computational Linguistics 39(4) (2013)
16. van der Merwe, D., Obiedkov, S., Kourie, D.: AddIntent: A new incremental algorithm for constructing concept lattices. In: Eklund, P. (ed.) ICFCA 2004. LNCS (LNAI), vol. 2961, pp. 372–385. Springer, Heidelberg (2004)

Estimation Method
for Path Planning Parameter
Based on a Modified QPSO Algorithm

Myongchol Tokgo* and Renfu Li**

Department of Aerospace Engineering, Huazhong University of Science
and Technology, Hubei, Wuhan, 430074, China
Kim Chaek University of Technology, Pyongyang, 999093, DPRK
renfu.li@hust.edu.cn, dugumingze@163.com

Abstract. This paper presents a modified natural selection based quantum behaved particle swarm optimization (SelQPSO) algorithm for the path planning of mobile robot vehicles. To ensure the global searching and the high efficiency of the QPSO's searching process, the particle swarms are sorted by fitness and the group of the particles with worst fitness are replaced by the group with best fitness in each iteration of the whole procedure. The effectiveness and feasibility of this algorithm are demonstrated by the results from numerical experiments on well-known benchmark functions. Then, this algorithm is employed to estimate the basic parameters of the mobile robot path planning in the barrier free environment. The convergency of the estimation method versus particle numbers and iteration times is studied with variation of particle dimension. A unary linear regression equation taking the particle number, maximum generation and particle dimension as variables is formulated. The results from experiments for optimal path planning of a mobile robot in complex environment justifies the estimation method.

Keywords: Quantum behaved particle swarm optimization (QPSO) algorithm, Path planning, Natural selection, Regression equation.

1 Introduction

Quantum behaved particle swarm optimization (QPSO) is one type of particle swarm optimization based on the principle of quantum mechanics, making use of properties the δ- potential well model and the quantum motion of particle swarm. The particles meet the state of aggregation quite differently in the

---

* Myongchol Tokgo,is currently working toward the Ph.D. degree in the Department of aerospace engineering, Huazhong University of Science and Technology. His research interests include mobile robot trajectory planning,swarm intelligence algorithm,trajectory control.
** Renfu(corresponding author), professor and head of the department of aerospace engineering, Huazhong University of Science and Technology. His main research interests include flight vehicle design, flight control and control optimization.

G. Agre et al. (Eds.): AIMSA 2014, LNAI 8722, pp. 261–269, 2014.
© Springer International Publishing Switzerland 2014

quantum space than in normal space and can prevail the whole feasible solution space. The QPSO then have better global searching performance than that of the standard PSO [1,2,3]. In order to improve the global searching convergence and accuracy of the QPSO, many methods have been proposed such as random distribution institution change method, the compression-expansion factor control approach and hybrid search method [4,5,6]. The introduction of interference factor into the current search particle position and use of Gaussian potential well G-QPSO algorithm should prevent premature convergence [7,8]. At present, the studies of QPSO for mobile robots have become a hot topic, breeding strategy based HQPSO, phase angle-encoded based θ - QPSO for the three-dimensional trajectory planning[9,10] and the combination of QPSO and Quadrinomial and quintic polynomials for online trajectory planning scheme [11].

The global search capability and speed of the optimization procedure have great importance in mobile robot path planning. As such, this paper propose a modified natural-selection-based QPSO algorithm and an estimation method for the basic parameter of path planning. The results from the tests on several benchmark functions and the experiments of the mobile robot path planning in complex environment have shown the good performance of the QPSO algorithm and the feasibility and reliability of estimation method.

The paper is structured as follows: Section 2 is the development of the modified QPSO algorithm based on natural selection. Numerical results on some benchmark functions are described in section 3; In section 4, basic parameter estimation method for mobile robot path planning is formulated, experiments of mobile robot path planning in the complex environment are conducted; some conclusions are drawn in section 5.

2   A Modified QPSO Method Based on Natural Selection

The standard QPSO model are as follows[3];

$$X_{i,j}(t+1) = P_{i,j}(t) \pm \alpha \cdot |P_{i,j}(t) - X_{i,j}(t)| \cdot \ln(1/u). \tag{1}$$

$$P_{i,j} = \varphi \cdot P_{i,j}(t) + (1 - \varphi) \cdot G_j(t), (1 \leq i \leq N, 1 \leq j \leq M) \tag{2}$$

$$C(t) = (C_1(t), C_2(t), ..., C_D(t)) = (\frac{1}{N}\sum_{i=1}^{N} P_{i,1}(t), \frac{1}{N}\sum_{i=1}^{N} P_{i,2}(t), ..., \frac{1}{N}\sum_{i=1}^{N} P_{i,D}(t)). \tag{3}$$

where $\varphi = c_1 r_1/(c_1 r_1 + c_2 r_2)$, N is the particle number,the maximum generation (iteration) represented M, P is the local point of each particle, u is random number distributed uniformly on (0,1),the position vector of particle i in D-dimension space are represented as $X_i = (x_{i,1}(t), x_{i,2}(t), ..., x_{i,D}(t))$, c_1 and c_2 are acceleration coefficients; the parameters r_1 and r_2 are two random numbers uniformly distributed in (0,1), i.e. $r_1, r_2 \sim U(0, 1)$; the previous position vector of each individual particle is denoted as $P_i = (P_{i,1}(t), P_{i,2}(t), ..., P_{i,D}(t))$;

$G = (G_1(t), G_2(t), ..., G_D(t))$ is the optimal position vector;the mean best position (mbest) of the all particles is defined as $C = C_1(t), C_2(t)..., C_D(t)$,$\alpha$ is called contraction-expansion coefficient,the α is a control parameter without population size and particle number, generally take $\alpha < 1.781$.

In the standard QPSO model, the convergence process of quantum groups can not always get the same direction. The quantum group often collectively behaves towards the direction of the best individual evolution, not necessarily depending on all individuals. Next, we propose an improved QPSO algorithm based on the natural selection method of quantum-behaved particle swarm evolution: first, sort the whole particle swarm fitness in each selected generation process; then replace the half of the state of worst fitness with the group of the best fitness, while retaining the original each individual historical memory by the optimal value. As such, the particle's position vector can be written as

$$\begin{cases} X_i(t) = (X_1(t), X_2(t), \cdots, X_s(t), X_{s+1}(t) \cdots, X_{N-1}(t), X_N(t)), (QPSO), \\ X_i(t) = (\tilde{X}_1(t), \tilde{X}_2(t), \cdots, \tilde{X}_s(t), \tilde{X}_1(t), \tilde{X}_2(t), \cdots, \tilde{X}_s(t), \cdots), (SelQPSO). \end{cases}$$

$$(4)$$

where $(\tilde{X}_1(t), \tilde{X}_2(t), \cdots, \tilde{X}_s(t))$ are the position vector of the particles according to the fitness value,$s = N/Tg$,Tg represents the number of segmentation for the fitness value, generally setting as 2. If the number of segmentation is 2, half of the best position of the particles replaces half of the worst position, but small particle dimension will encounter premature convergence, also do not guarantee the global search. All particles in iteration process of the QPSO is attracted to the global optimal position. So the best and worst position information combination of particles is of significance in each iteration process. The QPSO algorithm based on natural selection method is called Natural Selection Quantum-behaved Particle Swarm Optimization (SelQPSO), which is described as following,

Initialize the population;
Do
find out the mbest of the swarm

```
for t=1:Iteration
  alpha=1-(1-0.5)*t/M;mbest=sum(pbest)/N;
  for i=1:population_size
    for j=1:Dimention
      fi1=rand;fi2=rand;fai=c1*fi1+c2*fi2;
      p(i,j)=(c1*fi1*pbest(i,j)+c2*fi2*gbest(j))/fai;
      b(i,j)=alpha*abs(mbest(j)-x(i,j));v=-log(rand);
      x(i,j)=p(i,j)+(-1)*ceil(0.5+rand)*b(i,j)*v;
    end for
    if Fitness(x(i,:))<Fitness(pbest(i,:))
      pbest(i,:)=x(i,:)
    end
    if Fitness(Pbest(i,:))<Fitness(gbest)
      gbest=pbest(i,:)
```

```
      end
    fx(i)=fitness(x(i,:));Fitness=fitness(gbest);
    end for
   [sortf,sortx]=sort(fx);exIndex=round((N-1)/Tg);
   x(sortx((N-exIndex+1):N))=x(sortx(1:exIndex));
 end for
```

until the termination criterion is met

The number of segmentation should choose to protect the global searching capability and improve the precision of the direction. In this article the number of segmentation 2 and 4 are taken, which are called SelQPSO1 and SelQPSO2 algorithm, respectively.

To evaluate the performance of the SelQPSO1 and SelQPSO2, four well-known benchmark functions, such as the Sphere, the Rastrigrin, the Greiwank and Ackley functions, are used. The description of the four benchmark functions is in Table 1.

Table 1. Expressions of benchmark functions

	Functions	Initial Range
F_1	$\sum_{i=1}^{n} x_i^2$	$(-100 \leq x_i \leq 100)$
F_2	$\sum_{i=1}^{n}(x_i^2 - 10 \cdot \cos(2\pi x_i) - 10)$	$(-5.12 \leq x_i \leq 5.12)$
F_3	$\sum_{i=1}^{n} x_i/400 - \prod_{i=1}^{n} \cos(x_i/\sqrt{i}) + 1$	$(-600 \leq x_i \leq 600)$
F_4	$20 + e - 20\exp(-0.2\sqrt{\sum_{i=1}^{n} x_i/^2}) - \exp(\sum_{i=1}^{n}\cos(2\pi x_i))$	$(-100 \leq x_i \leq 100)$

The population sizes are 20, 40 and 80. The maximum generations use 1000, 1500 and 2000, corresponding to the dimensions of 10, 20 and 30 for five functions, respectively. The acceleration coefficients are set to be $c_1 = 2$ and $c_2 = 2.1$, contraction-expansion coefficient α varies from 1.0 to 0.5 linearly. The mean values of best fitness values for 50 runs of each function as shown in Table 2 and Table 3.

Table 2. Mean best of Sphere and Rastrigrin function

N	D	M	Sphere function			Rastrigrin function		
			QPSO	SelPSO1	SelPSO2	QPSO	SelPSO1	SelPSO2
20	10	1000	4.01E-40	2.39E-36	1.28E-35	4.333	4.074	4.055
	20	1500	2.58E-21	2.31E-21	8.05E-21	15.986	17.026	14.278
	30	2000	2.08E-13	8.49E-14	2.32E-13	32.031	33.657	30.726
40	10	1000	2.73E-67	7.16E-67	1.07E-66	2.271	2.495	2.078
	20	1500	4.85E-39	1.15E-37	1.04E-39	10.270	11.657	9.691
	30	2000	2.04E-27	1.96E-27	4.66E-27	21.203	21.079	21.035
80	10	1000	7.61E-96	3.08E-197	1.49E-188	2.456	1.758	1.954
	20	1500	1.61E-59	7.30E-62	3.04E-62	7.391	7.731	8.811
	30	2000	2.07E-45	2.79E-46	1.69E-44	16.439	15.645	16.614

For the Sphere function, the average best fitness of SelQPSO1 and SelQPSO2 is better than that by QPSO when population size is 80, but it does not show the better performance than QPSO when population size is 20 and 40.On the Rastrigin function, it is shown that the SelQPSO2 generates better results than QPSO and SelQPSO1 do when the number of particles is 20 and 40, but it is not better than SelQPSO1 when the number of particle is 80. In the Greiwank function, the performance of SelQPSO2 is better than QPSO and SelQPSO1 for all number of particles. For the Ackley function, the average fitness of SelQPSO2 is better than that of QPSO and SelQPSO1 with respect to all population sizes.

From the observation of the numerical results, one can find that the SelQPSO2 works best on Greiwank function and Ackley function. However, it does not perform better on Rastrigin function and Sphere function than SelQPSO1 when the number of particle is 80.

Table 3. Mean best of Greiwank and Ackley function

N	D	M	Greiwank function			Ackley function		
			QPSO	SelPSO1	SelPSO2	QPSO	SelPSO1	SelPSO2
20	10	1000	0.082	0.0886	0.0721	10.318	10.329	10.268
	20	1500	0.0218	0.0241	0.0208	10.918	10.793	10.702
	30	2000	0.0111	0.00856	0.00971	11.527	11.373	11.342
40	10	1000	0.0617	0.0584	0.04381	10.183	10.198	10.183
	20	1500	0.0181	0.0216	0.0145	10.373	10.442	10.223
	30	2000	0.0089	0.0115	0.00654	10.831	10.867	10.828
80	10	1000	0.0426	0.0424	0.0408	10.175	10.174	10.174
	20	1500	0.0154	0.016	0.0153	10.244	10.251	10.206
	30	2000	0.00697	0.0073	0.0063	10.526	10.454	10.458

So, the middle position information of the particles between the best and worst weights more than the worst information. Therefore, when the last $1/4$ of particles in terms of sorting of fitness value replace the particles with best fitness values, one can guarantee the precision and possibility of global search of the method.

3   Basic Parameter Estimation Method of Mobile Robot Path Planning

In this section we will verify the proposed SelQPSO method by the results from numerical simulation and field experiments of a mobile robot in barrier-free environment. The mobile robot path planning based on QPSO makes use of the polar coordinates [12]. Each pair of outcomes is drawn from the QPSO and SelQPSO running under the same environment and condition. The size of experiment field sets as $10m \times 10m$ with the starting point at $(0,0)$ and the target point at $(10,10)$. The particle dimension is chosen as $5 \leq D \leq 20$, the acceleration coefficients are set as $c_1 = 2$ and $c_2 = 2.1$, the contraction-expansion coefficient   varies from 1.0 to 0.5 linearly. The initial distribution set as the average distribution.

Figure 1 shows the particle number and maximum generation obtained by QPSO and SelQPSO running for 50 times. It finds that the standard errors of the optimal value and the average fitness value are both less than 0.05m. In Fig. 1 the Max and Min represent the maximum value and minimum value of the feasible particle numbers and maximum generations. One can easily see from the results in Fig.1 that the change of particle number and the maximum generation follows the exponent function law. Therefore, basic parameters of the path planning can be expressed unary linear regression equation. Using MATLAB software regress command, the regression results are as shown in Table 4.

Table 4. Regression analysis results

Algorithm	parameter	limit	$\hat{b}_0$	$\hat{b}_1$	R^2	F	$\hat{\sigma}^2$
QPSO	N	min	-1.1045	1.8445	1	1105.6	0.00
		max	-0.1215	1.5220	0.9683	427.48	0.0148
	M	min	-0.8112	2.2853	0.9720	485.15	0.0293
		max	-2.8384	3.0055	0.9706	462.07	0.0532
SelQPSO	N	min	-1.5670	1.9716	0.9832	821.08	0.0129
		max	-0.6880	1.7098	1	1670.2	0.00
	M	min	-0.6348	2.1495	0.9862	998.98	0.0126
		max	-2.1246	2.6552	1	1892.3	0.00

Through regression analysis of QPSO, the results of minimum value look better than those maximum values. A unary linear regression equation for the basic parameters by QPSO can be written as follows;

$$\begin{cases} N = round(\exp(-1.1045) \cdot D^{1.8445}) = round(0.3315 \cdot D^{1.8445}), \\ M = round(\exp(-0.8112) \cdot D^{2.2853}) = round(0.4445 \cdot D^{2.2853}). \end{cases} \quad (5)$$

The regression analysis of SelQPSO gives that the results of maximum value look better than minimum value. A unary linear regression equation of the basic parameters by SelQPSO then can be written as follows;

$$\begin{cases} N = round(\exp(-0.688) \cdot D^{1.7098}) = round(0.5026 \cdot D^{1.7098}), \\ M = round(\exp(-2.1246) \cdot D^{2.6552}) = round(0.1195 \cdot D^{2.6552}). \end{cases} \quad (6)$$

Observations from Fig. 1 show that the Eq.(5) and Eq.(6) can be sued to estimate variation of the particle number and the maximum generation with the particle dimension.

Next, we will verify the performances of the basic parameters estimation method by the mobile robot path planning experiments in a complex environment containing some obstacles. The mobile robot size is 0.4m × 0.3m, the path starting point at $(0,0)$ and target point at $(10,10)$, particle dimensions $D = 12, 16, 20$. Using Eq.(5) and Eq.(6) to calculate the particle number and the maximum generation. The algorithms are programmed in MATLAB R2009a

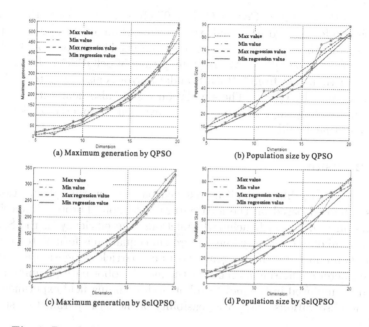

(a) Maximum generation by QPSO

(b) Population size by QPSO

(c) Maximum generation by SelQPSO

(d) Population size by SelQPSO

Fig. 1. Population size and Maximum generation analysis results

Table 5. Comparison of the length of path planning

	Algorithm	Maximum,m	Minimum,m	Mean best, m	St.Dev, m	Mean Time, s
D=12	QPSO	14.963	14.551	14.673	0.241	35.023
	SelQPSO	14.954	14.534	14.696	0.212	26.672
D=16	QPSO	14.416	14.181	14.286	0.191	146.169
	SelQPSO	14.412	14.165	14.228	0.172	112.811
D=20	QPSO	15.043	14.653	14.756	0.136	443.257
	SelQPSO	15.036	14.645	14.756	0.099	342.459

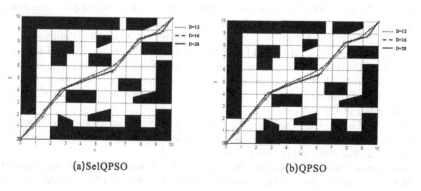

(a)SelQPSO

(b)QPSO

Fig. 2. Best path planning by SelQPSO and QPSO

and ran on a PC with Intel Core i3 CPU, 3.07 GHz. Table 5 shows the best path found using different algorithm and particle dimension after running 50 times.

As shown in Table 5, the length of the path searched by the proposed SelQPSO is better than that by QPSO. Mean running time of SelQPSO is 1.271 to 1.362 times shorter than that from QPSO. Standard error is also 1.072 to 1.373 times smaller than that by QPSO. The simulation results also demonstrate that the robot path planning time and the global convergence property by SelQPSO is shorter than that by QPSO.

4   Conclusion

In this paper, a modified quantum-behaved particle swarm optimization algorithm is proposed based on the idea natural selection. Through numerical experiments on well-known benchmark functions, the efficiency of the SelQPSO method are validated. A unary linear regression equation is formulated by taking the particle number, maximum generation and particle dimension as variables. Basic parameter estimation method for the mobile robot path planning is developed and is employed in the mobile robot path planning field experiments. The results and observations justify the feasibility of the basic parameter estimation method.

References

1. Clerc, M., Kennedy, J.: The Particle Swarm: Explosion, Stability and Convergence in a Multi-dimensional Complex Space. IEEE Transaction on Evolutionary Computation 16(1), 58–73 (2002)
2. Sun, J., Lai, C.H., Xu, W.-B., Chai, Z.: A Novel and More Efficient Search Strategy of Quantum-Behaved Particle Swarm Optimization. In: Beliczynski, B., Dzielinski, A., Iwanowski, M., Ribeiro, B. (eds.) ICANNGA 2007. LNCS, vol. 4431, Part I, pp. 394–403. Springer, Heidelberg (2007)
3. Sun, J., Feng, B., Xu, W.: Particle swarm optimization with particles having quantum behavior. In: IEEE Congress on Evolutionary Computation, pp. 325–331 (2004)
4. Sun, J., Xiaojun, W.: Convergence analysis and improvements of quantum-behaved particle swarm optimization. Information Sciences 193, 81–83 (2012)
5. Xu, W., Sun, J.: Adaptive Parameter Selection of Quantum-Behaved Particle Swarm Optimization on Global Level. In: Huang, D.-S., Zhang, X.-P., Huang, G.-B. (eds.) ICIC 2005. LNCS, vol. 3644, pp. 420–428. Springer, Heidelberg (2005)
6. Pat, A., Hota, A.R.: An Improved Quantum-behaved Particle Swarm Optimization Using Fitness-weighted Preferential Recombination. In: NBIC 2010, pp. 15–17 (December 2010)
7. Tian, N., Lai, C.-H., Pericleous, K., Sun, J., Xu, W.: Contraction-Expansion Coefficient Learning in Quantum-Behaved Particle Swarm Optimization. In: DCABES 2011, pp. 303–308 (2011), doi:10.1109,32
8. dos Santos Coelho, L., Nedjah, N., de Macedo Mourelle, L.: Gaussian Quantum-Behaved Particle Swarm Optimization Applied to Fuzzy PID Controller Design. In: Nedjah, N., dos Santos Coelho, L., de Macedo Mourelle, L. (eds.) Quantum Inspired Intelligent Systems. SCI, vol. 121, pp. 1–15. Springer, Heidelberg (2008)

9. Lu, K., Fang, K., Xie, G.: A Hybrid Quantum-behaved Particle Swarm Optimization Algorithm for Clustering Analysis. In: FSKD 2008, pp. 569–574 (2008), doi:10.1109.369

10. Fu, Y., Ding, M.: Phase Angle-Encoded and Quantum-Behaved Particle Swarm Optimization Applied to Three-Dimensional Route Planning for UAV. IEEE Transactions on Systems, Man, and Cybernetics, PART A: Systems and Humnas 42(2), 511–526 (2012)

11. Guo, J., Wang, J., Cui, G.: Online Path Planning for UAV Navigation Based on Quantum Particle Swarm Optimization. In: Wu, Y. (ed.) Advanced Technology in Teaching - Proceedings of the 2009 3rd International Conference on Teaching and Computational Science (WTCS 2009). AISC, vol. 116, pp. 291–302. Springer, Heidelberg (2012)

12. Li, R., Dokgo, M., Hu, L.: Path Planning based on PSO algorithm convergence and parameters analysis. Journal of HuaZhong University of Science and Technology (Natural Science Edition) 41(sup. I), 271–275 (2013)

On Modeling Formalisms
for Automated Planning*

Jindřich Vodrážka and Roman Barták

Charles University in Prague, Faculty of Mathematics and Physics
Malostranské nám. 2/25, 118 00 Praha 1, Czech Republic
{vodrazka,bartak}@ktiml.mff.cuni.cz

Abstract. Knowledge engineering for automated planning is still in its childhood and there has been little work done on how to model planning problems. The prevailing approach in the academic community is using the PDDL language that originated in planning competitions. In contrast, real applications require more modeling flexibility and different modeling languages were designed in order to allow efficient planning. This paper focuses on the role of a domain modeling formalism as an interface between a domain modeler and a planner.

Keywords: knowledge modeling, automated planning.

1 Introduction

We can imagine problem solving as a journey from problem specification to problem solution. The approach taken by automated planning is a model-based one - we first design a model that formally describes our knowledge about a given area of interest (domain) and later we can exploit this knowledge to solve various instances of problems in the domain using a general purpose planner. The journey in this case can be divided to two phases:

1. *knowledge modeling* - performed by human experts. All relevant information about the problem is put together in order to create a domain model. The model is usually described within some knowledge-modeling formalism.
2. *planning* - performed by an automated planner. The planner is working only with the domain model specified in the first phase and with the description of one particular problem instance.

The knowledge modeling formalism is the dividing point between the two phases.

In this paper we will show that the position of this dividing point can define a tradeoff between simplicity and usability of the modeling formalism on one side and efficiency of the resulting model on the other side. Finally we will introduce a new planning formalism designed to balance this tradeoff.

---

* This work is supported by the Czech Science Foundation under the project No. P103/10/1287.

G. Agre et al. (Eds.): AIMSA 2014, LNAI 8722, pp. 270–277, 2014.

2   Planning Domain Example

In order to illustrate our point of view we will be refering to the Petrobras planning domain [6] which was one of the domains of the Challenge track of the International Competition on Knowledge Engineering for Planning and Schedulling, ICKEPS 2012, as part of the ICAPS 2012 conference.

In the Petrobras domain we have a fleet of ships, a list of locations (ports, waiting areas, and platforms), and cargo. Each ship has limited cargo capacity and a fuel tank. The ships consume fuel while navigating between the locations.

The main goal is to deliver cargo from ports to platforms. There are six operations, *navigate, dock, undock, load, unload* and *refuel,* that each ship can do. Only refueling can run in parallel with loading or unloading, while any other pair of operations for a single ship cannot overlap in time. The ship must be docked before loading, unloading, and refueling and it must be undocked before navigating. We are given the initial locations and fuel levels of ships and the initial location, weight, and destination of each cargo item. The task is to plan operations for ships in such a way that all cargo items are delivered.

Class hierarchy. When modeling a planning domain it is natural to describe classes of involved objects. The base class hierarchy can be described independently on the formalism used. We will be refering to the following class hierarchy for the Petrobras domain:

- Ship - a class for the transport ships
- Cargo - a class for the items of cargo
- Location - a generic class for points of interest. Distances between pairs of locations are part of the planning problem specification.
    - WaitingArea - areas designated for idle ships
    - LogisticLoc - locations where a ship can be loaded/unloaded
        * Platform - location with docking capacity for a single ship
        * Port - location with docking capacity for more than one ship. Refueling is possible only at ports and selected platforms.

3   Domain Designer Perspective

Design of the planning domain model can be done in a modeling formalism that is completely independent of the planning machinery. This is the case of the Planning Domain Definition Language (PDDL) [3].

PDDL model. We have already described the class hierarchy earlier. Now we will give the description of selected object properties with *predicates* and *fluents.* Both predicates and fluents represent properties that are subject to change during the planning process. *Predicates* in PDDL are used for boolean properties whereas *fluents* can take wider range of values (e.g. numeric fluents).

In PDDL we use instances of *predicates* and *fluents* to describe a state of the world and we use *actions* to model possible transitions between those states.

Predicates: In PDDL we can represent information about the location of each ship and item of cargo, using first-order-logic *predicates* with typed arguments (e.g. (ship-at ?s - Ship ?l - Location)).

Fluents: For each ship we can express its current fuel level and free cargo capacity as *numeric fluents* (e.g. (fuel-level ?s - Ship) - number).

Actions: Legal changes that can turn one state to another state are described by *actions*. An example of a PDDL code for action load-cargo follows[1]:

```
(:action load-cargo
    :parameters (?s - Ship ?c - Cargo ?loc - Location)
    :precondition (and
                (at ?s ?loc)
                (cargo-at ?c ?loc)
                (>= (free-cargo-capacity ?s) (cargo-weight ?c))
                (isDocked ?s ?loc))
    :effect (and
                (not (cargo-at ?c ?loc))
                (cargo-at ?c ?s)
                (decrease (free-cargo-capacity ?s) (cargo-weight ?c))))
```

Each predicate, fluent, and action declaration in PDDL actually represents a schema with variables. If we assign constants to these variables we can obtain many different predicates, fluents, and actions. This process is called *grounding*.

The set of grounded predicates, that are true at the moment, together with the values of all fluents are used to describe a world state.

A grounded action is *applicable* to a given state iff all *preconditions* are satisfied. The state changes according to the *effects* of the action (i.e. some predicates are made true/false and values of some fluents are changed).

The main idea of PDDL is to describe domain physics only. However, if we consider actions one by one, the physics alone may not suffice. For example a sequential plan can contain a docking action followed immediately by an undocking action, which is a valid but not reasonable sub-plan. Additional work has to be done in order to encode action ordering rules that would be useful for a planner. Therefore we conclude that the PDDL interface is closer to the modeler.

4   Planner Perspective

The New Domain Definition Language (NDDL) is an example of a modeling formalism that is strongly influenced by the target planner. NDDL is a part of the EUROPA2 planning system [1].

---

[1] PDDL uses Lisp-like syntax.

NDDL model. In addition to the base class hierarchy we need to define special classes that represent domain *attributes*. Each ship in the Petrobras domain can be described by two numeric attributes (`fuelLevel`, `cargoCapacity`) and one multi-valued state variable `shipState`, which can take values such as: `Navigating(loc1,loc2)` (ship is navigating from `loc1` to `loc2`) or `Loading(loc3,c1)` (ship is loading `c1` at `loc3`). All possible values are described with predicates.

```
class shipState extends Timeline {
    predicate Navigating { Location from; Location to; }
    predicate Loading { LogisticLoc dock; Cargo crate; }
    ... }
```

Tokens: Each predicate defined in this way can be used in a *token* which is a triple (O, P, I) where:

O - is an object i.e. instance of some class (e.g. `shipState`)
P - is a predicate of the class (e.g. `Navigating(X,Y)`)
I - is a time interval $[s, t]$ where $s < t \leq H$ for some fixed value H (planning horizon)

Timelines: A sequence of *tokens* $T = (t_1, \ldots, t_k)$ with non-overlapping intervals:

$$\forall i \neq j : (I(t_i) = [a, b] \wedge I(t_j) = [c, d]) \Rightarrow (b \leq c \vee d \leq a)$$

is called a *timeline*, and it describes the history of a state variable. We can indicate this fact by the code `class X extends Timeline`. If the intervals defined in T cover all the time from 0 to H then T is a *completely specified* timeline. If there are some uncovered intervals the timeline is *partially specified*.

The planner starts with a set of *partially specified* timelines and its objective is filling in the gaps with matching tokens. Any set of partially specified timelines represents a *partial solution* of the original problem. In a *complete solution* all the timelines are *completely specified* in a way that satisfies all *constraints* defined in the domain model.

Constraints: To decide whether a token can extend a given partial solution, the NDDL model has to define temporal constraints (based on the Allen's interval algebra) among tokens. Figure 1 shows a diagram of token relations for action `LoadCargo`. Solid boxes represent tokens that have to exist in the solution in order to allow the addition of tokens represented by dashed boxes.

We have used only a small subset of NDDL features in our example to illustrate the position of the interface between a domain modeler and a planner. The language describes domain knowledge with timelines, tokens, and constraints – the structures used by the planner in the planning process. Valid sequences of tokens on a timeline can be deduced from the constraints described in the domain model. However, this requires an additional effort on the side of a domain modeler who has to design the constraints. This intuition leads us to conclusion that the NDDL interface is closer to the planner in this case.

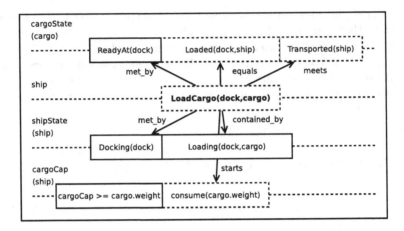

Fig. 1. Token relations in LoadCargo

5   Comparison

We have seen how two different modeling formalisms can be used to model one planning domain. Now we will analyze the action LoadCargo in more detail. Let us suppose that our knowledge can be summarized as follows.

There are three conditions: -

– both **ship** and **cargo** have to be at the same **location**,
– **ship** has to be docked,
– there has to be enough free cargo capacity on the **ship**.

If these conditions are satisfied, the action can be executed:

– **cargo** will be loaded on the **ship**,
– available cargo capacity on the **ship** will be decreased by the **cargo** weight.

Now we will review the modeling decisions taken when using PDDL and NDDL and we will discuss the differences.

PDDL: The code for the corresponding action can be found in Section 3. Once we declare the predicates and the fluents that are sufficient to encode all relevant domain knowledge we can use the PDDL syntax to describe action preconditions and effects in a way that is very close to the description at the beginning of this section.

NDDL: By using NDDL we need to take a different point of view. We suggest one possible sequence of modeling decisions:

1. Choose relevant timelines:
   – cargoState(cargo) - a timeline for the state of the loaded cargo,
   – shipState(ship) - a timeline for the state of the target ship,

- `cargoCap(ship)` - a timeline for free cargo capacity of the target ship,
- `ship` - a timeline for the target ship.
2. Allocate the token representing the action: `LoadCargo(dock,cargo)`.
3. Allocate tokens representing conditions and effects using temporal constraints. These tokens represent either conditions or effects as indicated in Figure 1.

In both cases we first need to declare some domain-specific notions (e.g. predicates and fluents in PDDL and timelines with predicates in NDDL). The difference between the two approaches is in the way that these notions are used.

It is possible to model actions in the PDDL without encoding any explicit knowledge about their possible ordering (e.g. dock and undock situation described earlier). The missing information can be difficult to obtain but there is a clear notion of action stated as a set of conditions and effects.

On the other side the NDDL model can provide the planner with some additional information but the notion of action, which is specified as a set of temporal constraints on tokens, can blur the semantics of the domain model.

6   New Knowledge Modeling Interface

We assume that a domain modeler can treat the planner as a black box. This is in accord with the *physics-only* principle employed in the PDDL. In contrast with this principle we want the modeler to give as much domain-independent information as possible.

Classes: In our formalism we distinguish between two types of classes:

1. *enumerative* classes can be used to represent discrete objects such as ships,
2. *numeric* classes allow description of quantities such as fuel tank capacity.

In addition to the class hierarchy from the Section 2, we need to specify at least one numeric class to be used for numeric values in the domain.

State variables: Domain properties that are subject to change are described as state variables of the following form:

$$s(a_1, \ldots, a_n) : r$$

where s is a name of an n-ary *state variable*, a_i represents classes of its parameters, and r is defined as a set of classes, which implicitly defines the range of values for the state variable (e.g. `cargoLocation(Cargo):{Ship,Location}` - an item of cargo can be either stored at some location or loaded on a ship).

The state of the world is described as a vector of values for all state variables defined in the planning problem instance.

Domain rules: Decisions made during the planning process often require some kind of computation (e.g. the fuel consumption depends on the trip distance). We suggest to describe each such computation as an n-ary function:

$$f : D_1 \times \ldots \times D_n \to D_{n+1}$$

called a *domain rule*, where D_i is a set of constant symbols compatible with some class C_i. Classes of both *numeric* and *enumerative* types are allowed.

Operators: Possible changes of the world state are described with operators. An operator is specified as a list of expressions. There are two types of expressions (we will refer to the example operator code below):

- *conditional expressions* are used to describe conditions among different state variable values and their parameters (e.g. line 4). The value of state variable cargoWeight(C) is constrained by X. The variable X is defined at line 5 where it represents the original value of the state variable cargoCap(S).
- *transitional expressions* define both the condition and the change (e.g. line 6). The cargo item C is transfered from D to S.

```
1 loadCargo(D - LogisticLoc, C - Cargo, S - Ship)
2    shipLoc(S) = D
3    shipState(S) = docked
4    cargoWeight(C) <= X
5    cargoCap(S): X --> (X - cargoWeight(C))
6    cargoLoc(C): D --> S
```

Terms: The operator expressions use terms that can be constructed recursively from **constants** (e.g. docked), **variables** (e.g. D,C,S,X), **state variables** (e.g. shipState(S)), and **domain rule instances** (e.g. (X - cargoWeight(C))[2]).

Atomic conditions: For two terms within an *enumerative* class we use comparison relations $=$ and $\neq$ to build an atomic condition. In case of two terms within a *numeric* class we use the combinations of $=, <, >$. Standard operators of FOL $(\land, \lor, \neg, \forall, \exists)$ are used to construct more complex *conditional expressions*.

In the code above there are three *conditional expressions*. Two of them are asserting equality of *enumerative terms* (lines 2, 3) and one is asserting inequality of two *numeric terms* (line 4). There are two *transitional expressions* (lines 5, 6).

The action ordering can be enforced by introducing a *domain rule* fsaCheck, which references a finite state automaton (FSA) that describes all valid action sequences [2]. The FSA will have the following states: undocked, navigating, waiting, loading, unloading, and refueling. We replace the line 3 with:

```
3a    exists B: fsaCheck(A,B) = true
3b    shipState(S): A --> B
```

The conditional expression at line 3a constrains the variables A and B according to the FSA and the transitional expression at line 3b changes the state of the ship S from A to B. The value of the variable A is defined at line 3b.

---

[2] Arithmetic operators are binary functions.

We have showcased usage of *state variables, domain rules,* and *operators.* The resulting domain model allows user-defined extensions called domain rules to restrict action ordering. Arbitrary arithmetic functions can be defined. While NDDL permits a user to define functions as well, PDDL is less flexible which can be limiting for some real world applications [4].

7 Conclusion

The interface presented in this paper gives a different view on knowledge modeling which depends on multi-valued state variables. The value ranges of these variables are defined either by enumerative or numeric classes instead of the predicates as in the NDDL. The resulting domain model uses operators simmilar to actions from the PDDL. The closest formalism currently in existence is Action Notation Modeling Language (ANML) [5].

From the perspective of a domain modeler the interface provides an easy way to integrate various kinds of domain-specific knowledge using the domain rules.

On the planner side a problem instance is described by a finite set of state variables and domain rules, while the domain specific information is represented only in the operators.

References

1. Barreiro, J., Boyce, M., Do, M., Frank, J., Iatauro, M., Kichkaylo, T., Morris, P., Ong, J., Remolina, E., Smith, T., Smith, D.: EUROPA: A Platform for AI Planning, Scheduling, Constraint Programming, and Optimization. In: Proc. of ICKEPS (2012)
2. Barták, R., Zhou, N.-F.: On Modeling Planning Problems: Experience from the Petrobras Challenge. In: Castro, F., Gelbukh, A., González, M. (eds.) MICAI 2013, Part II. LNCS, vol. 8266, pp. 466–477. Springer, Heidelberg (2013)
3. Gerevini, A., Long, D.: BNF Description of PDDL 3.0, http://www.cs.yale.edu/homes/dvm/papers/pddl-bnf.pdf
4. Parkinson, S., Longstaff, A.P.: Increasing The numeric Expressiveness of the Planning Domain Definition Language. In: Workshop of the UK PlanSIG (2012)
5. Smith, D., Frank, J., Cushing, W.: The ANML Language. In: International Conference on Automated Planning and Schedulling - Poster Session (2008)
6. Vaquero, T.S., Costa, G., Tonidandel, F., Igreja, H., Silva, J.R., Beck, J.C.: Planning and Scheduling Ship Operations on Petroleum Ports and Platforms. In: Proceedings of the Schedulling and Planning Aplications Workshop (2012)

Finetuning Randomized Heuristic Search for 2D Path Planning: Finding the Best Input Parameters for R* Algorithm through Series of Experiments

Konstantin Yakovlev, Egor Baskin, and Ivan Hramoin

Institute for Systems Analysis of Russian Academy of Sciences, Moscow, Russia
{yakovlev,baskin,hramoin}@isa.ru

Abstract. Path planning is typically considered in Artificial Intelligence as a graph searching problem and R* is state-of-the-art algorithm tailored to solve it. The algorithm decomposes given path finding task into the series of subtasks each of which can be easily (in computational sense) solved by well-known methods (such as A*). Parameterized random choice is used to perform the decomposition and as a result R* performance largely depends on the choice of its input parameters. In our work we formulate a range of assumptions concerning possible upper and lower bounds of R* parameters, their interdependency and their influence on R* performance. Then we evaluate these assumptions by running a large number of experiments. As a result we formulate a set of heuristic rules which can be used to initialize the values of R* parameters in a way that leads to algorithm's best performance.

Keywords: path planning, grid, 2D, A*, R*, heuristic search.

1   Introduction

Ability to plan a path is one of the key features for an intelligent agent. In our work, we examine the case when an agent operates in a rectangle-bounded region of static 2D environment composed of traversable and non-traversable areas (free space and obstacles). We use 8-connected grid as a formal model of agent's environment [1, 2, 3]. Within this model the task is to find a sequence of unoccupied adjacent grid cells connecting given start and goal cells.

Heuristic search algorithm A* [4] is widely used in AI community for finding paths on grids. A* using admissible heuristics guarantees finding a shortest path [5]. Plenty of such heuristics for grid-worlds exist: Manhattan distance, octile (diagonal) distance etc. These heuristics being natural metrics in grid-worlds are the "best-available", but nevertheless A*-search exercising them explores too much of the state-space in case the goal is located beyond the obstacles. The reason of the over-exploration is that A* guided locally by one of the abovementioned heuristics necessarily (in presence of obstacles) falls into a local minimum – "the portion of state-space from which there is no way to states with smaller heuristics without passing through states with higher heuristics" [6]. There exist a number of approaches to

G. Agre et al. (Eds.): AIMSA 2014, LNAI 8722, pp. 278–285, 2014.

reduce the A* search space (and thus increase the computational effectiveness of path planning algorithm). One approach is to modify A* in some way. Using weighted heuristics [7, 8, 9], implementing iterative deepening techniques [10, 11], imposing limits on the size of the set of candidate cells for exploration [12, 13] are examples of such approach. Another approach exploits the idea of decomposition. Methods implementing this approach split the given task to the series of subtasks (local tasks) each of which is solved independently (by local planners) and final solution is constructed by the composition of local solutions. Decomposition can be performed using predefined criteria [14] or in random fashion [15, 16].

One of the well-known state-of-the art algorithms suitable for path finding in grid-worlds based on parameterized random decomposition is R* introduced in [6]. It exploits the same ideas lying behind RRT planners [16] and uses WA* (A* with weighted heuristic) as local planner. To perform a search R* needs to be provided with the values of its 3 input parameters (as well as the weight of the heuristic used for the local WA* search). Preliminary experiments show that the algorithm performance largely depends on these values. At the same time to the best of authors knowledge there is no reported research results on how exactly the choice of the parameters values influence the performance of R* and which values should be used to solve practical path planning tasks with R*. This works aims at filling this gap.

In our work we theoretically analyze the possible influence of each parameter on R* performance along with evaluating its lower and upper bounds. We show that the bounds for 2 parameters are either constants or can be expressed as functions of the 3rd parameter. At the same time, we propose that the value of the latter is the function of start and goal positions. Then we perform comprehensive experimental analysis of R* solving more than 5000 of 2D path planning tasks to estimate the coefficients of the parameters' bindings proposed before. Thus we end up with a set of rules for R* parameterization applying which leads to algorithm's best performance.

2   R* Algorithm for Grid Path Planning

2.1   Path Planning Problem

Consider a 8-connected grid which is a finite set of cells $A=(a, b, c, ...)$ that can be represented as a matrix $A_{M \times N}=\{a_{ij}\}$, where: i, j – are cell position indexes (also denoted as $i(a), j(a)$) and M, N – are grid dimensions. Each cell is labeled either traversable or un-traversable and agent is allowed to move from a traversable cell to one of its traversable neighbors.

A metric function $dist$ (also known to AI community as diagonal heuristic) is used to measure the distance between any two cells:

$$dist(a_{ij}, a_{kl}) = c_d \cdot \min(\Delta_i, \Delta_j) + c_{hv} \cdot (\Delta_i + \Delta_j - 2 \cdot \min(\Delta_i, \Delta_j)),$$

where $\Delta_i=|i-k|$, $\Delta_j=|j-l|$, $c_d=k \cdot c_{hv}$ ($1<k<2$), $c_{hv}=const \in \boldsymbol{R}^+$. c_{hv} is a distance between a cell and any of its horizontal or vertical neighbors, c_d is a distance between a cell and any of its diagonal neighbors. In our work we use integer constants $c_{hv}=10$ and $c_d=14$ for corresponding distances.

Path planning task is considered to be set if two distinct traversable cells – start and goal – s, $g \in A$ are set. The solution of the problem is a path $\pi(s, g)$, e.g. a sequence of traversable adjacent cells starting with s and ending with g. The length of the path $L(\pi)$ is the sum of the distances between all pairs of adjacent cells forming the path.

2.2   Randomized Heuristic Search Algorithm R* Overview

R* – is state-of-the art heuristic search algorithm that decomposes initial path planning task into series of subtasks, identifies ones to be solved by local planner (WA*) and tries to solve them. If the local solution is found "easily" it is stored and can be used lately to reconstruct final solution but if the local solution is "hard to find" R* postpones local search and chooses another local task. As its creators say in [6]: "R* postpones the ones [*local searches*] that do not find solutions easily and tries to construct the overall solution using only the results of searches that find solutions easily".

We encourage the reader to examine the papers [6, 17] for the detailed explanation of R* and now give only a brief overview of the algorithm.

At each step R* firstly chooses the most promising cell c from OPEN list (initially containing only start cell). Then algorithm randomly selects K traversable cells residing at the distance Δ from c and inserts them into OPEN. These cells (b_i) are called the successors of c, while c itself is called the predecessor ($pred(b_i) = c$). If $dist(g, c) \leq \Delta$ then the goal cell is also added to OPEN.

Next R* tries to find a local path $\pi(pred(c), c)$ with WA* algorithm. If the path is not found after the m steps of WA* the cell c is labeled AVOID which means it was hard to find the current local path and the local search should be postponed. Cell c is kept in OPEN list in that case. If the path is found the cell is removed from OPEN and is inserted into CLOSED list.

The process of generating successors, adding them to OPEN, choosing the best cell in OPEN and trying to find a local path is referred as the expansion of the cell c.

R* chooses the cells from OPEN using the same heuristic rule as WA*. The only difference is that R* chooses only such a cell which is not labeled AVOID (initially none of the cells has this label) and only if no such cells are left in OPEN it chooses a cell amongst AVOID ones. The stop criteria for R* is analogous to WA*.

One specific thing related to algorithm's implementation which is not addressed in original paper is the procedure of generating successors for the expanding cell. In our implementation of R* we use midpoint circle algorithm [18] to generate successors – see fig.1.

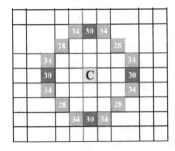

Fig. 1. Set of successors for cell C. Numbers indicate the distance from C

One can think that the circumference of radius Δ/c_{hv} (if measured in cells) with the center in the cell c is "drawn" and K traversable cells forming this circumference are randomly chosen as successors of c.

2.3   R* Parameters Influence on Algorithm Performance and Their Lower and Upper Bounds

Obviously R* performance depends on the values of its 3 parameters: K – number of successors generated for each expanded cell c, Δ – the distance between c and generated successors, m – number of steps local planner is allowed to perform before abandoning the search. Let's analyze the influence of each parameter and assess its lower and upper bounds taking into account R* working principles.

The value of m affects both execution time and memory usage. The higher the value of m is the more steps are performed by WA* while finding *each* local path. At the same time it is likely that only small fraction of all local paths compound final solution which means in case of higher m more chances are that R* wastes much time on "useless" computations. Thus high values of m should be avoided.

The influence of parameter m on path length is less evident. It is likely that the latter depends primarily on how local goal cells are chosen at each step of R* and that is independent of m.

The lower bound for m can be assessed in the following way. One can show that minimum number of steps WA* needs to find a path is $m^*=\max(|i(s)-i(g)|, |j(s)-j(g)|)$. This happens in particular when there are no obstacles in between s and g. As detailed above our implementation of R* uses midpoint circle algorithm to form the set of possible successors and it can be shown that $\max(|i(s)-i(g)|, |j(s)-j(g)|)$ is achieved when s (the cell under expansion) and g (the successor of s) lie on the same grid row (or column) and in that case $m^*=\Delta/c_{hv}$. So the value $m'=\Delta/c_{hv}(=\Delta/10)$ is the lower bound for m (if $m<\Delta/10$ WA* would simply fail to solve local path finding tasks in most cases).

Theoretically upper bound for m is the number of all traversable cells on grid but it's reasonable to limit m by some value $m''=k\cdot m'$ ($k\in N$) which means that the local planner (WA*) is allowed to perform k times more steps to solve local path planning tasks than in the most trivial cases (when there are no obstacles present).

The value of K affects both execution time and memory usage but primarily memory usage as *all* K successors for *each* expanded cell are permanently stored in memory. So, if we are interested in decreasing memory consumption high values of K should be avoided. At the same time, the higher the value of K is the more successors for each cell are generated and more chances are R* would pick up "good" candidates for further expansion (the candidates that minimize both local and overall path lengths). So setting K to high values potentially leads to better quality solutions.

Upper bound for K can be assessed in the following way. As said above, set of successors for any expanded grid cell is the subset of cells comprising the discrete circumference of radius Δ/c_{hv} (if measured in cells). The length of such circumference equals $2\cdot\pi\cdot\Delta/c_{hv}$. Thus maximum number of successors $K''\approx6\cdot\Delta/10$. Minimum possible number of successors K' apparently equals 1. But it's obvious that in that case we would likely get very awkward shaped and very long paths, so we suggest $K'=3$.

As shown above upper/lower bounds for m and K can be expressed in relation to Δ making Δ the key parameter for R* which value affects both execution time and memory usage and solution quality. Considering the influence of Δ independently one can say that the higher the value of Δ is (with $dist(s, g)$ being the maximum) the more R* relies on local path planning which makes algorithm behave more like typical A*-family algorithm (the behavior we are trying to avoid). On the other hand setting Δ to extremely small values (with $c_{hv}=10$ being the minimum) directly converges R* to A* which does not make any sense at all. So, the value of Δ should be picked from the middle of the spectrum of possible values. In other words, Δ can be represented as a positive monotone function (with known minimum and maximum) of start and goal locations, e.g. $\Delta=dist(s, g)/k$ and the "right" value for binding coefficient k should be estimated via experimental analysis but it is likely to belong to the middle range of possible spectrum (defined by minimum and maximum values of Δ).

By now we have theoretically evaluated lower and upper bounds for the R* parameters (m, K and Δ) and showed that m and K can be expressed as linear functions of Δ and Δ can be expressed as linear function of start and goal locations. The coefficients of the bindings are unknown and we are going to evaluate them experimentally.

3   Experimental Analysis

3.1   Testbed

To examine the influence of R* input parameters on the algorithm's performance and find parameters "best" values we have run 5250 of experiments on 3 types of grids:
- randomly generated grids containing rectangle shaped obstacles of different sizes (70 grids * 25 different parameters configurations = 1750 experiments);
- randomly generated grids containing tetris-shaped obstacles of different sizes (70 grids * 25 different parameters configurations = 1750 experiments);
- grids which are models of city landscape (70 grids * 25 different parameters configurations = 1750 experiments).

While generating grids containing rectangle and tetris-shaped obstacles the latter were added one by one at randomly selected positions until the total number of untraversble cells equals or slightly exceeds predefined threshold, $e.g.$ 30%. When adding each obstacle its size and orientation was chosen randomly within predefined thresholds.

Grids modeling city landscape were generated semi-automatically and the maps of the real cities were used as sources. The percentage of blocked cells on these grids equals or slightly exceeds 30% (just as on randomly generated grids).

All grids were of the size 501x501 and start and goal cells were always located on the opposite edges of grid in such a way that $dist(s, g)=5000$.

The following indicators were used to evaluate R* performance:

cells – the number of cells stored in OPEN and CLOSED (used to assess memory consumption);

time – time (in ms) used by R* to find a path;

length – length of the path found.

3.2   Results

There were conducted 3 consecutive series of experiments and a preliminary one.

Preliminary experiments were aimed at fine tuning local planer (WA*), *e.g.* estimating the best value for weight of heuristic function. For the sake of space we omit the results of experiments but they count in favor value 3 should be used as the weight. As R* is supposed to use the same heuristic function as local planner weighted by the same weight (to guarantee suboptimality) the later was also set to 3.

First, we examined the influence of parameter *m* – number of steps local planner (WA*) is allowed to commit before abandoning the search for a local path – on R* performance. Then parameter *K* – number of successors generated for each expanded cell – was evaluated with *m* being set to its best value. Finally, parameter Δ – the distance between expanded cell and its successors – was evaluated (with *m* and *K* being fixed to their best values discovered before).

The averaged results of the experiments are shown on fig. 2. These averaged results correlate well with all "individual" ones, *e.g.* results obtained on grids of specific types, and thus can be used as consistent basis for R* performance evaluation.

In the first series of experiments values of Δ and *K* were set to $dist(s, g)/10(=500)$ and $\Delta/10(=50)$ respectively and *m* was assigned a range of values: 50, 75, 100, 200, 300, 500, 750, 1000.

Obtained results (see fig. 2) support the assumption that *m* mainly affects running time and memory consumption – *time* and *cells* values differ (due to different *m*-values) ≈ 5 and 1,5 times respectively – while the influence of *m* on solution quality is less evident. Interesting case which breaks the evident tendency – the higher the value of *m* is the worse the performance of R* is – is setting *m* to 50. Worst results in that case (which are not depicted on diagrams but reported in the table) can be easily explained: when *m* is set to 50 local planner is guaranteed to almost always fail in local pathfinding (during the first attempt). So in the end more such searches are performed which in turn substantially degrades R* performance.

Based on gained results value $100(=\Delta/5)$ can be recommended to initialize *m*. This can be interpreted in the following way (see previous section): minimum number of steps local planer needs to find a path is $\Delta/10$, so to get "best" results local planner should be allowed to use 2 times more steps (than the minimum).

In the second series of experiments value of Δ remained the same, *m* was set to its best value, *e.g* $\Delta/5$, and *K* was assigned a range of values: 3, 5, 7, 10, 25, 50, 70, 100.

Obtained results support the assumption that *K* drastically affects memory consumption and running time – *cells* and *time* values differ 4 and 3 times (due to different *K*-values) respectively; they also justify that *K* has a major influence on solution quality– difference in *length* reaches 30-35%.

Results of the experiments evidently show that higher values of *K* should be avoided due to high computational costs and lower values of *K* should be avoided due to lower solution quality. On this basis the recommended value for *K* is $25(=\Delta/20)$. This can be interpreted as following: maximum number of successors for any cell $\approx 6 \cdot \Delta/10$, so to get best results R* should generate round $1/10^{th}$ -$1/12^{th}$ of that value (1 out the 10-12 possible successors should be generated).

	cells	length	time, ms
Various M			
Δ = 500 M = 50 K = 50	977	6495	747
Δ = 500 M = 75 K = 50	802	6202	160
Δ = 500 M = 100 K = 50	816	6187	192
Δ = 500 M = 200 K = 50	813	6189	281
Δ = 500 M = 300 K = 50	842	6191	355
Δ = 500 M = 500 K = 50	910	6212	468
Δ = 500 M = 750 K = 50	994	6233	636
Δ = 500 M = 1000 K = 50	1077	6241	781
Various K			
Δ = 500 M = 100 K = 3	330	9306	224
Δ = 500 M = 100 K = 5	332	7863	122
Δ = 500 M = 100 K = 7	345	7175	101
Δ = 500 M = 100 K = 10	368	6777	97
Δ = 500 M = 100 K = 25	522	6330	118
Δ = 500 M = 100 K = 50	808	6166	189
Δ = 500 M = 100 K = 70	1028	6155	247
Δ = 500 M = 100 K = 100	1353	6150	357
Various Δ			
Δ = 50 M = 10 K = 10	1576	6490	309
Δ = 100 M = 20 K = 10	959	6560	178
Δ = 200 M = 40 K = 10	462	6642	79
Δ = 300 M = 60 K = 15	452	6459	81
Δ = 500 M = 100 K = 25	514	6304	111
Δ = 750 M = 150 K = 38	624	6200	145
Δ = 1000 M = 200 K = 50	736	6178	227
Δ = 1500 M = 300 K = 75	1032	6139	987
Δ = 2500 M = 500 K = 125	1492	6165	1375

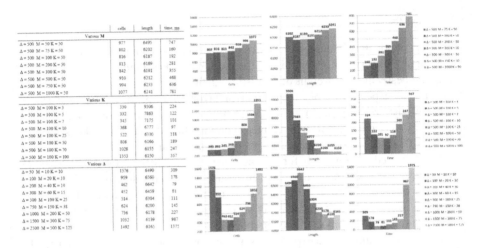

Fig. 2. Experimental results

In the third series of experiments *m* and *K* were assigned their best values, *e.g* Δ/5 and Δ/20, and Δ was consequently initialized as: 50, 100, 200, 300, 500, 750, 1000, 1500, 2500.

Obtained results (see fig. 2) verify the assumption that Δ (just like *K*) has more influence on R* performance rather than *m*: *cells* and *time* differ 4 and 10 times respectively (due to the different values of Δ) and the difference in *length* reaches 10%.

As one can see, setting Δ to higher or lower values significantly reduces algorithm's computational efficiency (the fact we have predicted earlier) so these values should be rejected and the values from the middle of the spectrum should be used. Based on the obtained results we recommend setting Δ to 500(=*dist*(s, g)/10).

Summarizing the results of experimental analysis two main conclusions can be made. First, R* performance depends *largely* on the values of its parameters and assigning them in a wrong way can lead to a dramatic fall in computational efficiency. Second, there exist a set of rules which can be used to automatically initialize R* input parameters in such a way that leads to best performance. These rules can be formalized as the set of bindings:

$$\Delta = dist(s, g)/10;$$
$$K = \max(10, \Delta/20);$$
$$m = \Delta/5.$$

Presented bindings are dependent only on start and goal locations which are known a priori and thus can be viewed as a universal (heuristic) rule of parameterizing R* when solving path planning tasks on 8-connected grids.

4   Conclusions

R* is state-of-the-art randomized heuristic search algorithm and a powerful tool to solve 2D path planning tasks at low computational costs. But to benefit from using R* in actual practice one needs to initialize 3 algorithm's parameters in a "right" way as

R* performance heavily depends on them. In presented work we analyze (both theoretically and experimentally) the nature of that dependencies and end up with the set of heuristic rules that can be used to automatically parameterize R* in order to get the best results (low computational cost and high solution quality). Presented rules are easily applicable to any path planning task in any grid-world as they do not require any additional knowledge except the positions of start and goal cells.

References

1. Elfes, A.: Using occupancy grids for mobile robot perception and navigation. Computer 22(6), 46–57 (1989)
2. Yap, P.: Grid-based path-finding. In: Cohen, R., Spencer, B. (eds.) Canadian AI 2002. LNCS (LNBI), vol. 2338, pp. 44–55. Springer, Heidelberg (2002)
3. Tozour, P.: Search space representations. In: Rabin, S. (ed.) AI Game Programming Wisdom, vol. 2, pp. 85–102. Charles River Media (2004)
4. Hart, P.E., Nilsson, N.J., Raphael, B.: A formal basis for the heuristic determination of minimum cost paths. IEEE Transactions on Systems Science and Cybernetics 4(2), 100–107 (1968)
5. Pearl, J.: Heuristics: intelligent search strategies for computer problem solving. Addison-Wesley (1984)
6. Likhachev, M., Stentz, A.: R* Search. In: Proceedings of the Twenty-Third AAAI Conference on Artificial Intelligence. AAAI Press, Menlo Park (2008)
7. Pohl, I.: First results on the effect of error in heuristic search. In: Bernard, M., Michie, D. (eds.) Machine Intelligence, vol. 5, pp. 219–236. Edinburgh University Press, Edinburg (1970)
8. Gallab, M., Dennis, A.: Aε – an efficient near admissible heuristic search algorithm. In: Proceedings of the Eighth International Joint Conference on Artificial Intelligence (IJCAI 1983), pp. 789–791 (1983)
9. Likhachev, M., Gordon, G., Thrun, S.: ARA*: Anytime A* with Provable Bounds on Sub-Optimality, Advances in Neural Information Processing Systems 16 (NIPS). MIT Press, Cambridge (2004)
10. Korf, R.E.: Depth-first iterative-deepening: An optimal admissible tree search. Artificial Intelligence 27(1), 97–109 (1985)
11. Reinefeld, A., Marsland, T.A.: Enhanced iterative-deepening search. IEEE Transactions on Pattern Analysis and Machine Intelligence 16(7), 701–710 (1994)
12. Bisiani, R.: Beam search. In: Shapiro, S. (ed.) Encyclopedia of Artificial Intelligence, pp. 56–58. John Wiley and Sons (1987)
13. Zhang, W.: Complete anytime beam search. In: Proceedings of the Fifteenth National Conference on Artificial Intelligence (AAAI 1998), pp. 425–430 (1998)
14. Botea, A., Muller, M., Schaeffer, J.: Near optimal hierarchical path finding. Journal of Game Development 1(1), 7–28 (2004)
15. Kavraki, L.E., Svestka, P., Latombe, J.C., Overmars, M.H.: Probabilistic roadmaps for path planning in high-dimensional configuration spaces. IEEE Transactions on Robotics and Automation 12(4), 566–580 (1996)
16. LaValle, S.M.: Rapidly-exploring random trees: A new tool for path planning, Technical Report, 98-11, Computer Science Dept., Iowa State University (1998)
17. Likhachev, M., Stentz, A.: R* search: The proofs. Technical Report, University of Pennsylvania, Philadelphia, PA (2008b)
18. Pitteway, M.L.V.: Algorithms of conic generation. In: Fundamental Algorithms for Computer Graphics, pp. 219–237. Springer, Heidelberg (1985)

Analysis of Strategies in American Football
Using Nash Equilibrium

Arturo Yee, Reinaldo Rodríguez, and Matías Alvarado

Computer Sciences Department, Center of Research and Advances Studies, México D.F.
{ayee,rrodriguez}@computacion.cs.cinvestav.mx,
matias@cs.cinvestav.mx

Abstract. In this paper, the analysis of American football strategies is by applying Nash equilibrium. Up to the offensive or defensive team-role, each player usually practices the relevant plays for his role; each play is qualified regarding the benefit that could add to the team success. The team's strategies, that join the individual's plays, are identified by means of the strategy profiles of a normal game formal setting of American football, and valued by the each player's payoff function. Hence, the Nash equilibrium strategy profiles can be identified and used for the actions decision making in a match gaming.

Keywords: American football, Nash equilibrium, team's strategies.

1   Introduction

Recently, the formal modeling and strategic analysis for support the matches gaming of multi-player sports like American football (AF) or baseball [1-3], have led investigations in areas of sport science [4-6], computer science, game theory [7], operation research [1, 3], and simulation models [2, 8], among others. In American football gaming, the team members are encouraged to do the best individual actions, but they must cooperate for the best team's benefit. The strategies are indicated by the team manager regarding on each player's profile as well as the specific match circumstances to obtain the most benefit [9]. A planned-strategy should include both the individual and the team motivation. The selection of strategies is an essential aspect to be considered for a whole AF automation. In [10] a formal model for automated simulation of AF gaming, using a context-free (grammar) language and finite state machine, allows for precise simulation runs on this ever strategic multi-player game.

1.1   American Football Description

American football (AF) is one of the top strategic games, played by two teams on a rectangular shaped field, 120 yards long by 53.3 yards wide, with goalposts in the end of the field. Each team has 11 players and a match lasts 1 hour divided in four quarters. The offensive team goal is advance an oval ball, by running or passing toward the adversary's end field [11-13]. The ways to obtain points are by advancing the ball,

G. Agre et al. (Eds.): AIMSA 2014, LNAI 8722, pp. 286–294, 2014.

ten yards at least, until reach to the end zone for touchdown scoring, or kicking the ball such that it passes in the middle of the adversary's goalposts for a field goal, or by the defensive tackling the ball carrier in the offensive end zone for a safety. The offensive team should advance the ball at least ten yards in at most four downs (opportunities) to get four additional downs; otherwise the defensive team that is avoiding the ten yards advance, changes to the offensive role. The current offensive team's advance starts from the last ball stop position. If the defensive catches the ball before a down is completed, it starts the offensive role at this position.

1.2   Selection of Strategies

The Nash equilibrium (NE) [14] is a widely used mathematical concept in game theory, especially in non-cooperative games. The Nash equilibrium formal account for multi-player games follows. Let $P = \{1, ..., n\}$ be the set of players, $i \in P$, $a_x^i \in \Sigma^i$ be an element of the set of simple plays, and $s_x^i = a_1^i ... a_n^i$ be a strategy of player i is a sequence of actions , $s_x^i \in S_i$, S_i the set of strategies for the i_{th} player. Let $(s_1, ..., s_n) \in S_1 \times ... \times S_n$ a strategy profile, one strategy per player, and let $\{u_1, ..., u_n\}$ be the set of every player payoff functions, such that $u_i (s_1, ..., s_n) = r \in \mathbb{R}$. Let $G = (S_1, ..., S_n; u_1, ..., u_n)$ be the game in normal form [14].

To identify the strategy profiles that satisfy the condition of Nash equilibrium, every strategy profile is evaluated with the payoff functions of the players, and the chosen profiles are those which, for every player, is the options that produces less loss for him regarding the other players' strategies; so, is the best option, for each player, but individually, in non-cooperative way. A NE strategy profile $(s_1^*, ..., s_i^*, ..., s_n^*)$ maximizes the payoff function in equation (1):

$$u_i(s_1^*, ..., s_i^*, ..., s_n^*) \geq u_i(s_1^*, ..., s_i, ..., s_n^*) \quad \forall i \in P, s_i \in S_i \qquad (1)$$

We use the normal game formal account of American football in order to analyze the strategies of a team in both, offensive and defensive roles. Each strategy profile involves all the players' actions in certain moment of a match, and NE provides a formal way to measure and identify the strategy profile that satisfy the expectative of the players, both, given certain action and by regarding the other players' actions.

Next, in Section 2 concerns an overview of American football, the team-roles, as well as the each role's most relevant plays. Section 3 describes the each role's utility function for worth the strategy profiles. Section 4 presents the strategy profiles that satisfy the Nash equilibrium condition; they are selected and used during actions decision making in a match. Some remarks feed the Discussion and Conclusion.

2   Strategies by Team-Role and the Average Occurrence of Plays

In this section, we present some AF plays being divided according to the team-role, offensive or defensive (Table 1).

Table 1. Offensive and defensive plays

Off. plays	Description	Def. plays	Description
kb	Kick the ball	tl	Tackling
cb	Catch the ball by product of a pass	sf	Safety
rb	Run with the ball	sb	Stop the ball
pb	Pass the ball	in	Interception
fd	Scoring yards	qs	Tackling the quarterback
td	Touchdown	yb	Roll back the contraries
p	Extra point (1 point by product of a kick)	fb	Fumble the ball
re	Conversion (2 points)	fr	Turnover the ball
fg	Field goal	tb	Touchback

Using real statistical from NFL (National Football League) see http://gametheory. cs.cinvestav.mx/NFL_statistics.pdf, the probability of occurrence of each play above is calculated and listed in descending order in Table 2. These values come from performing statistical averages of values in tables showing data by player-role and not by specific player, but for a specific player his individual statistics can be used.

Table 2. Probability of ocurrence of AF plays

Play	Average	$p(play)$	Play	Average	$p(play)$
sb	1050.5	0.232884067	p	39.4375	0.008742851
yb	845	0.187327022	fg	26.96875	0.005978669
tl	775.6875	0.171961218	td	25.125	0.005569931
pb	566.75	0.125642118	in	15.6875	0.003477743
rb	348.484375	0.077255077	fb	15.09375	0.003346115
cb	346.9375	0.076912152	fr	9.625	0.002133755
fd	319.125	0.070746433	tb	5.875	0.001302422
kb	78.40625	0.017381786	re	1.03125	0.000228617
qs	40.46875	0.008971468	sf	0.625	0.000138555

2.1 Offensive Team Plays

- Offensive linemen players OL have two major tasks: 1) block the defensive team members which try to tackle to the quarterback (QB), and 2) open ways in order to runners can pass. The OL players are, the center, left guard, right guard, left tackle and right tackle. We defined these players as OL and the plays to consider are $OL_{plays} = \{tl, yb\}$.
- The quarterback (QB) is the offensive leader, whose plays follows, $QB_{plays} = \{rb, pb, fd, td, re, tb\}$.
- The backfield players BF are: the halfback, tailback the fullback. The BF plays follow, $BF_{plays} = \{rb, fd, td, re, tb, tl\}$.

- Receiver's role RC is to catch the ball passed by the QB; RC players are the tight end and wide. The RC plays follow, $RC_{plays} = \{cb, rb, fd, td, re\}$.

2.2   Defensive Team Plays

- The defensive linemen players DL are: the defensive end, defensive tackle and nose tackle, their main task is to stop running plays on the inside and outside, respectively, to pressure the QB on passing plays. The DL plays follow, $DL_{plays} = \{tl, sf, sb, qs, yb, fb, fr\}$.
- The linebacker players LB have several tasks: defend passes in shortest paths, stop races that have passed the defensive line or on the same line and attack the QB plays penetration; they can be three or four. The LB plays follow, $LB_{plays} = \{tl, sf, sb, qs, fb, fr\}$.
- The defensive backfield players DS are: the cornerbacks and safeties, which major task is to cover the receivers. The DS plays follow, $DS_{plays} = \{tl, in, fb, fr\}$.

2.3   Special Team Plays

- Kicker player K kicks off the ball and do field goals and extra points. The kicker's plays follow, $K_{plays} = \{kb, p, fg\}$.
- The kickoff returner R is the player on the receiving team who catches the ball. The plays are $R_{plays} = \{rb, td, tb\}$.

3   Setting-Up of Payoff Functions

The each role's payoff function to value the strategy profiles, selects the own convenience value by regarding:

- For QB is important to make a pass, his characteristic move, even with a touchdown scoring can generate a greater personal gain.
- The basic action of RC is to increase the score, but to make it happens he must catch the ball and run to the touchdown line.
- The OL main function is tackling the adversary to allow QB send pass; as well, open space for RC ball runs, or, in some cases, push back the opposing team.
- The BF preferred score is touchdown or conversion, and should run to get there. Other option is to get a first down, or tackling a player of the opposing team.
- The DL should be tackling the opposing QB, roll back yards to the opposing team or get a safety; in descent order of importance the following is to stop the ball, tackling and cause fumbles and try to recover it by the opponent.
- The main function of LB is to recover a lost ball and then could be to generate a safety.
- For DS intercepting a pass would be best, but it is also important to get the other team loses control of the ball.

- For K, the most important is to make a field goal, followed by an extra point and typically perform the corresponding kicks.
- For R, the best choice is to score a touchdown with the return of the kick, but usually just run until stopped, or perform touchback for time.

We propose that the player-roles' skills are qualified on the base of the player-roles' performance on certain plays, and the statistics resumes these qualifications. Let $u_i(x_1, \ldots, x_i, \ldots, x_n) = V_1(x_1) * p(x_1) + \cdots + V_i(x_i) * p(x_i) + \cdots + V_n(x_n) * p(x_n)$ be the payoff function of the player-role i, $(x_1, \ldots, x_i, \ldots, x_n)$ is a strategy profile such that x_i is one play of player-role i, The factors in the payoff function are: $V_i(x_i)$ represents the player-role i's preference on the play x_i, and $p(x_i)$ is the average statistics of the player-role on play x_i, by regarding the NFL statistics [15]; as well, the other elements in the formula are the contributions of the other player-roles whom directly share the play.

3.1   Offensive Team

Let define the strategy profile for offensive team as (w, x, y, z), with $w \in QB_{\text{plays}}$, $x \in RC_{\text{plays}}$, $y \in OL_{\text{plays}}$, $z \in BF_{\text{plays}}$.

- For QB, we should consider the QB plays as well as the OL plays, the payoff function (2) follows.

$$u_{QB}(w, x, y, z) = V_{QB}(w) * p(w) + V_{OL}(y) * p(y) \tag{2}$$

- For RC, we should consider the RC plays, the QB as well as the OL plays, the payoff function (3) follows.

$$u_{RC}(w, x, y, z) = V_{RC}(x) * p(x) + V_{QB}(w) * p(w) + V_{OL}(y) * p(y) \tag{3}$$

- For BF, we should consider the BF plays, the QB plays as well as the OL plays, the payoff function (4)   follows.

$$u_{BF}(w, x, y, z) = V_{BF}(z) * p(z) + V_{QB}(w) * p(w) + V_{OL}(y) * p(y) \tag{4}$$

- For OL, we should only consider the OL plays, the payoff function (5) follows.

$$u_{OL}(w, x, y, z) = V_{OL}(y) * p(y) \tag{5}$$

3.2   Defensive Team

Let define the strategy profile for defensive team as (x, y, z) where $x \in DL_{\text{plays}}$, $y \in LB_{\text{plays}}$, $z \in DS_{\text{plays}}$.

- For DL and LB, we should consider DL plays as well as LB plays, the payoff function (6) follows.

$$u_{DL|LB}(x, y, z) = V_{DL}(x) * p(x) + V_{LB}(y) * p(y) \tag{6}$$

- For DS, we should only consider the DS plays, the payoff function (7) follows.

$$u_{DS}(x, y, z) = V_{DS}(z) * p(z) \tag{7}$$

3.3   Special Team

- For K, the payoff function (8) follows.

$$u_K(x) = V_K(x) * p(x) \text{ where } x \in K_{\text{plays}} \tag{8}$$

- For R, the payoff function (9) follows.

$$u_R(x) = V_R(x) * p(x) \text{ where } x \in R_{\text{plays}} \tag{9}$$

4   Experiments

We use the set of values in Tables 3 – 4 that are assigned according to each player's preference values on each of the own plays, and are used to calculate the payoff functions, that in turns are used to find out the strategy profiles that fit the Nash equilibrium condition.

Table 3. Values of offensive plays by player

QB	RC	OL	BF
$V_{QB}(rb) = 0.5$	$V_{RC}(cb) = 0.7$	$V_{OL}(tl) = 0.5$	$V_{BF}(rb) = 0.7$
$V_{QB}(pb) = 0.7$	$V_{RC}(rb) = 0.7$	$V_{OL}(yb) = 0.6$	$V_{BF}(fd) = 0.4$
$V_{QB}(fd) = 0.6$	$V_{RC}(fd) = 0.6$		$V_{BF}(td) = 0.9$
$V_{QB}(td) = 0.8$	$V_{RC}(td) = 0.9$		$V_{BF}(re) = 0.8$
$V_{QB}(re) = 0.9$	$V_{RC}(re) = 0.8$		$V_{BF}(tl) = 0.5$
$V_{QB}(tb) = 0.5$			

Table 4. Values of defensive plays by player

DL	LB	DS
$V_{DL}(tl) = 0.5$	$V_{LB}(tl) = 0.5$	$V_{DS}(tl) = 0.5$
$V_{DL}(sf) = 0.8$	$V_{LB}(sf) = 0.8$	$V_{DS}(in) = 0.9$
$V_{DL}(sb) = 0.6$	$V_{LB}(sb) = 0.6$	$V_{DS}(fb) = 0.7$
$V_{DL}(qs) = 0.9$	$V_{LB}(qs) = 0.9$	$V_{DS}(fr) = 0.8$
$V_{DL}(yb) = 0.8$	$V_{LB}(fb) = 0.7$	
$V_{DL}(fb) = 0.5$	$V_{LB}(fr) = 0.8$	
$V_{DL}(fr) = 0.4$		

Now, we define the set of strategy profiles. The set of strategy profiles for offensive team, (w, x, y, z) where $w \in QB_{\text{plays}}, x \in RC_{\text{plays}}, y \in OL_{\text{plays}}, z \in BF_{\text{plays}}$ is $OffensiveT_{\text{profiles}} = \{(rb, rb, tl, rb), (rb, rb, tl, fd), ..., (tb, re, yb, tl)\}$. The set of strategy profiles for defensive team, (x, y, z) where $x \in DL_{\text{plays}}$, $y \in LB_{\text{plays}}$, $z \in DS_{\text{plays}}$ is $DefensiveT_{\text{profiles}} = \{(tl, tl, tl), (tl, tl, in), \dots, (fr, fr, fr)\}$.

Using the payoff functions defined in Section 3, each strategy profile is valued by respective player's payoff function. Some illustrative examples follow.

Offensive Team

- For QB, $u_{QB}(rb, rb, tl, rb) = V_{QB}(rb) * p(rb) + V_{OL}(tl) * p(tl)$ is the payoff function that only embrace QB and OL plays. Reason is that RC and BF plays do not relevant impact the QB plays, so not the QB payoff function valuations.
  $u_{QB}(rb, rb, tl, rb) = 0.5 * 0.077255077 + 0.5 * 0.171961218$
  $u_{QB}(rb, rb, tl, rb) = 0.124608$.
- For RC, $u_{RC}(rb, rb, tl, rb) = V_{RC}(rb) * p(rb) + V_{QB}(rb) * p(rb) + V_{OL}(tl) * p(tl)$ is the payoff function, that only embrace RC, QB and OL plays. Reason is that the BF plays do not relevant impact the RC plays, so not the RC payoff function valuations.
  $u_{RC}(rb, rb, tl, rb) = 0.7 * 0.077255077 + 0.5 * 0.077255077 + 0.5 * 0.171961218$. $u_{RC}(rb, rb, tl, rb) = 0.1786867014$.
- For BF, $u_{BF}(rb, rb, tl, rb) = V_{BF}(rb) * p(rb) + V_{QB}(rb) * p(rb) + V_{OL}(tl) * p(tl)$ is the payoff function, that only embrace BF, QB and OL plays. Reason is that the RC plays do not relevant impact the BF plays, so not the BF payoff function valuations.
  $u_{BF}(rb, rb, tl, rb) = 0.7 * 0.077255077 + 0.5 * 0.077255077 + 0.5 * 0.171961218$. $u_{RC}(rb, rb, tl, rb) = 0.1786867014$.
- For OL, $u_{OL}(rb, rb, tl, rb) = V_{OL}(rb) * p(rb)$ is the payoff function, that only embrace the OL plays. $u_{OL}(rb, rb, tl, rb) = 0.5 * 0.171961218$.
  $u_{OL}(rb, rb, tl, rb) = 0.085980609$.

Defensive Team

- For DL and LB, $u_{DL|LB}(tl, tl, tl) = V_{DL}(tl) * p(tl) + V_{LB}(tl) * p(tl)$ is the payoff function, that only embrace the DL and LB plays. Reason is that the DS plays do not relevant impact the DL and LB plays, so not the DL and LB payoff function valuations. $u_{DL|LB}(tl, tl, tl) = 0.5 * 0.171961218 + 0.5 * 0.171961218$.
  $u_{DL|LB}(tl, tl, tl) = 0.171961218$.
- For DS, $u_{DS}(tl, tl, tl) = V_{DS}(tl) * p(tl)$ is the payoff function, that only embrace the DS plays. $u_{DS}(tl, tl, tl) = 0.5 * 0.171961218$. $u_{DS}(tl, tl, tl) = 0.085980609$.

We should calculate all payoff values on strategy profiles using the players' payoff functions. For the offensive team the strategy profile (pb, rb, yb, tl) satisfies the Nash equilibrium condition, and for the defensive team, the NE profile is (yb, sb, tl). So, the best combination of offensive plays is by combining a pass from QB, the RC running with the ball and the OL and BF opening the way and stopping their opponents. Notice that the Nash equilibrium strategy profile depends on the player's prefe-